S0-BAQ-984

Instruments and Experimentation in the History of Chemistry

DIBNER
INSTITUTE
FOR THE HISTORY
OF SCIENCE AND
TECHNOLOGY

Dibner Institute Studies in the History of Science and Technology
Jed Z. Buchwald, general editor, Evelyn Simha, governor

Jed Z. Buchwald and I. Bernard Cohen, editors, *Isaac Newton's Natural Philosophy*

Anthony Grafton and Nancy Siraisi, editors, *Natural Particulars: Nature and the Disciplines in Renaissance Europe*

Frederic L. Holmes and Trevor H. Levere, editors, *Instruments and Experimentation in the History of Chemistry*

Agatha C. Hughes and Thomas P. Hughes, editors, *Systems, Experts, and Computers: The Systems Approach in Management and Engineering, World War II and After*

N. M. Swerdlow, editor, *Ancient Astronomy and Celestial Divination*

INSTRUMENTS AND EXPERIMENTATION
IN THE HISTORY OF CHEMISTRY

———

edited by Frederic L. Holmes and Trevor H. Levere

The MIT Press
Cambridge, Massachusetts
London, England

This book was set in Bembo by Asco Typesetters, Hong Kong and printed and bound in the United States of America

Library of Congress Cataloging-in-Publication Data

Instruments and experimentation in the history of chemistry / edited by Frederic L. Holmes and Trevor H. Levere.
    p.   cm. — (Dibner Institute studies in the history of science and technology)
  Includes bibliographical references and index.
  ISBN 0-262-08282-9 (hc. : alk. paper)
  1. Chemical apparatus—History. 2. Chemistry—Experiments—History.
I. Holmes, Frederic Lawrence. II. Levere, Trevor Harvey. III. Series.
QD53.I57   1999
540'.28'4—dc21   99-21064
                                                                      CIP

Contents

INTRODUCTION: A PRACTICAL SCIENCE   vii
Trevor H. Levere and Frederic L. Holmes

CONTRIBUTORS   xix

I   THE PRACTICE OF ALCHEMY   1

1   THE ARCHAEOLOGY OF CHEMISTRY   5
Robert G. W. Anderson

2   ALCHEMY, ASSAYING, AND EXPERIMENT   35
William R. Newman

3   APPARATUS AND REPRODUCIBILITY IN ALCHEMY   55
Lawrence M. Principe

II   FROM HALES TO THE CHEMICAL REVOLUTION   75

4   "SLIPPERY SUBSTANCES": SOME PRACTICAL AND
CONCEPTUAL PROBLEMS IN THE UNDERSTANDING OF
GASES IN THE PRE-LAVOISIER ERA   79
Maurice Crosland

5   MEASURING GASES AND MEASURING GOODNESS   105
Trevor H. Levere

6   THE EVOLUTION OF LAVOISIER'S CHEMICAL
APPARATUS   137
Frederic L. Holmes

7   "THE CHEMIST'S BALANCE FOR FLUIDS": HYDROMETERS
AND THEIR MULTIPLE IDENTITIES, 1770–1810   153
Bernadette Bensaude-Vincent

8   "FIT INSTRUMENTS": THERMOMETERS IN EIGHTEENTH-
CENTURY CHEMISTRY   **185**
Jan Golinski

9   PLATINUM AND GROUND GLASS: SOME INNOVATIONS IN
CHEMICAL APPARATUS BY GUYTON DE MORVEAU AND
OTHERS   **211**
William A. Smeaton

III   THE NINETEENTH AND EARLY TWENTIETH
CENTURIES   **239**

10   MULTIPLE COMBINING PROPORTIONS: THE EXPERIMENTAL
EVIDENCE   **243**
Melvyn C. Usselman

11   ORGANIC ANALYSIS IN COMPARATIVE PERSPECTIVE:
LIEBIG, DUMAS, AND BERZELIUS, 1811–1837   **273**
Alan J. Rocke

12   CHEMICAL TECHNIQUES IN A PREELECTRONIC AGE: THE
REMARKABLE APPARATUS OF EDWARD FRANKLAND   **311**
Colin A. Russell

13   BRIDGING CHEMISTRY AND PHYSICS IN THE EXPERIMENTAL
STUDY OF GUNPOWDER   **335**
Seymour H. Mauskopf

14   LABORATORY PRACTICE AND THE PHYSICAL CHEMISTRY OF
MICHAEL POLANYI   **367**
Mary Jo Nye

INDEX   401

Introduction: A Practical Science
Trevor H. Levere and Frederic L. Holmes

Thematic collective volumes enable field specialists to focus attention on promising areas for future research as well as to display in one place the early results of new explorations in existing scholarship: that is, they can explore subjects or issues that appear topical to a discipline but are not yet ready for full synthetic treatment. Typically, the organizers of such a volume invite contributions from scholars whose research deals with topics relevant to the chosen theme. The contributors are encouraged to converge on common questions and approaches to the subject, often (as was the case for this volume) by coming together in a conference at which preliminary versions of the essays are presented and discussed. Sometimes these discussions induce individual authors to add or amplify comments in their papers that resonate with the topics presented in other papers. The result of this process is a set of essays clustered around the central theme.

Although such collective volumes inevitably make compromises between individual and group interests, that outcome may be not a drawback but an advantage. The creative tension between the unity sought and the diversity of individual projects can significantly enrich the resulting publication. There are many fruitful ways to pursue and to illustrate through specific cases the broad themes in the history of science. To represent some of these in a volume such as this one is to suggest directions without prescribing them, to attract participation without constricting it to fit particular historiographical or interpretative preferences.

The theme of this volume—the role of instruments and experiments in the history of chemistry—allows ample scope for individual scholarly perspectives. Its boundaries are nearly as large as chemistry itself. This science has been, from its very beginnings, defined by instruments and apparatus comprising a repertoire of laboratory operations that its practitioners have employed to examine experimentally natural materials or the fabrications of human culture. Historians of chemistry have always written, therefore, about activities that rely on instruments and experiments.

Often, however, the role of the instruments and experiments has remained in the background, while historians directed their primary attention to such questions as the origins, structures, and transformations of theory or to the careers of chemists and the institutional organizations in which those careers were built. Indeed, until very recently, although with striking exceptions,[1] the history of chemistry has been overwhelmingly a history of chemical theory, with practice little considered and with the apparatus that rendered that practice possible almost entirely ignored. The imbalance is particularly striking in view of the serious attention given to practice in other areas of natural science, especially physics. Most of the few studies of chemical instruments have provided inventories, a necessary starting point, rather than analytical or synthetic accounts.[2] Historical neglect of apparatus has existed partly because historians of science have tended to concentrate on ideas over things, although much good literature covers, for example, astronomical instruments. We therefore need to look at chemical practice and apparatus. The aim of this volume is simply to move the instruments and experiments into the foreground of our concern.

One reason for the historical neglect of chemical apparatus has been the dearth of material evidence. Chemical apparatus, for the most part, has been made of glass, earthenware, and stoneware. It is seldom signed and seldom is especially beautiful; therefore, it is of much less appeal to collectors than are astrolabes or microscopes. More than most laboratory equipment, chemical apparatus is disposable. Platinum crucibles are melted down, to be turned into new instruments or components. Glass breaks and, after it has been used several times in stressful conditions, it becomes even more fragile and then can't survive long. As Pierre-Joseph Macquer noted in the mid-eighteenth century, in his *Elements of Theoretical Chemistry*, "Vessels intended for chemical operations should, to be perfect, be able to bear without breaking the sudden application of great heat and cold, be impenetrable to every substance and inalterable to any solvent, be unvitrifiable and capable of enduring the most violent fire without fusing. But up to the present no vessels are known which combine all these qualities."[3] Apart from accidental breakage and disposal because of chemical contamination, an additional problem is that much chemical apparatus has been modular, especially since the 1780s, when brass couplings became increasingly common.[4]

Although solid ground-glass stoppers were used in the eighteenth century, they were individually ground for each vessel; interchangeable ground-glass joints had to wait until this century. William Smeaton (chapter 9) discusses the problem of such joints, among others. Brass couplings, however, did render possible the assembly of units into a complex arrangement for one experiment, then disassembly (where possible) and reuse in different combinations in another experiment.

Faraday advocated such practices and the use of "small, temporary, and generally useful apparatus," as opposed to apparatus built for a single purpose.[5] What constitutes the original apparatus then becomes a metaphysical as much as a historiographical problem. Nonetheless, a great deal is found in museums, especially in Europe but also in the United States. The Smithsonian Institution, for example, has Joseph Priestley's laboratory equipment. The kind of detective work that historians are used to performing with archives can and should be done in museums, too.

What survives in museums or in the basements of laboratories is neither representative nor comprehensive: With some extremely valuable exceptions, we have a much higher proportion of brassware and ironware and a much lower proportion of glassware than that contained in eighteenth- or early nineteenth-century laboratories. Lavoisier's laboratory had some 6,000 pieces of glassware, of which less than 1% survives, but the Conservatoire des Arts et Métiers in Paris has most of his dramatic pieces, with lots of metal in them, including Lavoisier's gasometer, his ice calorimeter (in two versions), and his precision balances. In constructing a natural history of chemical apparatus, we need to be guided by all the available evidence—manuscript, printed, engraved, and material.

Wherever possible, examining the apparatus for most of the eighteenth century is desirable, as it is for the history of chemistry and alchemy (as discussed by Robert Anderson in chapter 1) because, until the last quarter of the eighteenth century, chemists were not in the habit of describing their apparatus in great detail in writing and even less so in engraved plates, which were expensive. (Of course, exceptions occur, such as Agricola's descriptions of furnaces and the wonderful plates in the *Encyclopédie*.) More typical, in different ways, are Macquer, who says that words suffice to describe instruments, so that plates are unnecessary; Tiberius Cavallo, who tells us a lot about instruments but not everything because, as he said, you will discover what you really need to know about instruments when you use them;

and J. G. Children, who says in his translation of Berzelius's book on
the use of the blowpipe, that he had eliminated much of Berzelius's
detailed description of apparatus, which he found "so very minute, that
though such may be desirable in Sweden, in Britain I am sure it is not
wanted, abounding as this country does in skillful artists, from whom
every species of philosophical apparatus may be had, of the best work-
manship and construction."[6] He then refers the reader to his favorite
supplier, Newman. Worth noting is that illustrated catalogs of chemical
apparatus are scarce before the mid–nineteenth century.

That chemistry is above all a science of practice is stressed in numerous eighteenth-century definitions. Gabriel François Venel, in the
article about chemistry in the *Encyclopédie* of Diderot and d'Alembert,
explained that "Chemistry is a science which occupies itself with the
separations and union of the constituent principles of bodies, whether
these are effected by nature, or the result of the operations of art, with a
view to discovering the properties of bodies, or to render them suitable
for a variety of uses."[7] Georg Ernst Stahl, generally associated with
Johann Joachim Becher as cofounder of the phlogiston theory, said that
chemistry was "the art of resolving mixt, compound or aggregate
Bodies into their Principles and of [re]composing such bodies from
those Principles." Guillaume François Rouelle also emphasized practice: "Chemistry is a physical art which, by means of certain operations
and instruments, teaches us to separate the various substances which
enter into the composition of bodies, and to recombine these again,
either to reproduce the former bodies, or to form new ones from
them."[8] Operations and instruments are essential parts of the definition
of eighteenth-century chemistry.

Lavoisier was clear about their role, not only in research and
demonstration (two categories that require careful demarcation) but in
the process of persuasion that was essential if his reforms or revolution
were to succeed. Frederic L. Holmes (chapter 6) redresses a common
imbalance by focusing on Lavoisier's research apparatus rather than on
his apparatus for demonstration and persuasion. Perhaps, Lavoisier
pondered, "[i]f, in the various memoirs that I gave to the Academy, I
had said more about the detail of manipulations, I would have made
myself more readily understood, and the science [of chemistry] would
have made more rapid progress."[9] When he wrote his manifesto and
textbook of the chemical revolution, his *Traité élémentaire de chimie*
(1789), he ensured that one-third of the volume was devoted to the

instruments and operations of chemistry—an unprecedentedly large part of such a work.

The chemical laboratory through the first two-thirds of the eighteenth century has been described as a stable environment,[10] for the most part experiencing incremental change rather than significant innovation. Most of the apparatus illustrated in the plates of the *Encyclopédie* was similar to the apparatus described by Libavius in his *Alchemia* of 1597, which in its turn incorporated a great amount that would have been familiar to an alchemist of a century previously. Lavoisier's laboratory, or Berzelius's or even Kekulé's, well into the nineteenth century, still contained much that would have caused chemists of previous generations to feel at home. The earlier history of alchemy and chemistry reveals, through its apparatus, a striking degree of stability.

Robert Anderson (chapter 1) shows how many instruments remained for centuries essentially unchanged in shape or function. The art of chemistry, chemical manipulation, continued to involve various heat treatments and so continued to require a variety of furnaces and water and sand baths and apparatus for distillation, condensation, fermentation, calcination, crystallization, solvent extraction, and many more operations. Improvements were made in several of these: through the identification of sources of glass least likely to contaminate reactions, an increasing recognition of the importance of pure reagents and, consequently, a greater emphasis on the procedures needed to obtain them; and through improved lutes (compounds or mixtures used for sealing joints between different vessels in the days before uniform extruded glass and rubber tubing).

Stability is one of the central themes in this volume; not surprisingly, the same instruments are discussed in successive chapters, bridging alchemy to the eighteenth and even the nineteenth century. The balance, for example, emerges in William Newman's essay on alchemy and assaying (chapter 2) as a key instrument, just as it was for later analytical chemists.

The theme of stability stands, however, in tension and contrast with another and equally central theme: rapid change, as in the eighteenth-century development of gasometers, eudiometers, and hydrometers (Trevor H. Levere, chapter 5; Bernadette Bensaude-Vincent, chapter 7). Change could take the form of the development of existing apparatus (e.g., the dramatic improvements made to the precision balance in the last half of the eighteenth century), or it could involve truly novel apparatus, as was the case in pneumatic chemistry, which entailed significant conceptual change (Frederic L. Holmes, chapter 6).

Innovation can also be more technical than conceptual (i.e., it can lead to an improvement in technique without necessitating any conceptual adjustment). Such was the case with Guyton de Morveau's use of ground glass and platinum (as discussed by William A. Smeaton, chapter 9) and, more ambiguously, with Justus von Liebig's *Kaliapparat* (Alan J. Rocke, chapter 11) and Edward Frankland's remarkable range of new apparatus (Colin A. Russell, chapter 12).

As several of the chapters show, instruments and skills generally traveled swiftly once innovations were introduced. Dissemination of instruments and the concomitant dissemination of practices and theories were important aspects of chemical activity and an important product of the instrument makers' trade. Stability and innovation are as inevitably central to a consideration of chemical practice as being and becoming were to Greek philosophers, and they are individually or together constant themes in this volume.

Precision is another important theme that runs through the essays. Almost every chemist mentioned in this volume was a highly skilled practitioner, and many of the techniques described placed heavy demands on experimental skill. Lawrence Principe (chapter 3) makes this clear in his account of apparatus and reproducibility in alchemy, a discipline where rigor and reproducibility are often questioned. Precision and experimental skill are also apparent in the development of organic analysis (Alan J. Rocke, chapter 11) and in the case of Edward Frankland (Colin A. Russell, chapter 12). The reliability and precision of laboratory analyses carried out by John Dalton, Thomas Thomson, William Hyde Wollaston, and Jacques Bérard are central to Melvyn Usselman's assessment of experimental skill in the history of multiple combining proportions (chapter 10).

Exact measurement of reactants and products was an appropriate goal for chemists. It is not clear, however, that precision was a goal in and of itself, except perhaps in the case of the precision balance. This instrument emerged alongside the new and excellent dividing engines that underlay the work of the finest mathematical practitioners, such as Jesse Ramsden in England. An interesting note is that, although precision balances soon were regarded as essential instruments for the research chemist, the best balances were more accurate than the skill of their operators and the purity of chemical samples would demand through the nineteenth century. The balance fits very well with the enthusiasm for precision that characterized the Enlightenment and that the volume edited by M. Norton Wise nicely delineates.[11] However, that volume does not adequately acknowledge the importance—indeed, we believe, the driving role—exercised by the instrument makers of the day. The adoption of new techniques in chemistry, often based on new or improved instruments, generally arose from the improved precision that those techniques and instruments rendered possible. The measurement of the goodness of respiration gases (Levere, chapter 5) was at first most important for health and medicine, but the newly invented apparatus was quickly transferred to chemical assays of gases. As it was elaborated, modified, and incorporated into chemical discourse, it was also incorporated into the drive for exact measurement. The story of eudiometry is a story of a transition from qualitative to precise quantitative science.

Another general theme that runs through much of this volume is the way in which novel apparatus, designed in response to a particular theoretical or experimental concept, spread until it became a standard laboratory device or failed to gain such acceptance. The thermometer (Jan Golinski, chapter 8) changed its function significantly as theories of heat in relation to chemistry evolved and changed; from an instrument whose function was matter for debate and controversy, it became, if not "black-boxed," at least taken for granted, so that it could be used in uncontested ways. The hydrometer (chapter 7), in contrast, never achieved this status and ended by becoming marginal to precise chemical quantification. The eudiometer (chapter 5) was invented in response to questions arising from a new conception of gases. Frederic Holmes (chapter 6) demonstrates how some of Lavoisier's instruments were innovative precisely "because they were designed to confront problems that had not arisen in prior chemical investigations." Similarly, Frankland's innovative apparatus (chapter 12) was designed to solve specific problems. Liebig's *Kaliapparat* (chapter 11) was an immediate success, transforming the practice of organic analysis and accelerating its pace within the first few years of its introduction. Innovations in instrument design were dictated by laboratory problems, by new concepts, or by a combination of both.

Several chapters examine problems of communication, transfer, and adaptation to different purposes and practices when apparatus was imported into chemistry from other fields, such as physics, or when chemical apparatus was exported into other fields. The spread of the

Kaliapparat (chapter 11) and of Van Marum's gasometers for demon-strating Lavoisier's principal pneumatic experiments (chapter 5) are two such instances. The bridging of chemistry and physics, through instruments and concepts alike, is most elaborately explored in Seymour Mauskopf's chapter on the experimental study of munitions (chapter 13) and in Mary Jo Nye's chapter on Michael Polanyi's chemistry (chapter 14). These twin themes—dissemination and bridging—are also explored in many other chapters here. It becomes abundantly clear that instruments and techniques shaped chemistry as they spread and that they frequently served as a bridge between chemistry and other disciplines, thereby contributing to the very definition of the boundaries of chemical science.

The principal themes running through these chapters, then, are change and stability; precision; the construction and transformation of apparatus; the dissemination of instruments; and the bridging of disciplines through instruments. Other themes, important to the history of chemistry, emerge as significant as one considers these chapters.

The virtues of simplicity in apparatus, weighed against the advantages of refinement and consequent elaboration, are noted by several contributors and were debated by many of the chemists they discuss. From these discussions, what becomes clear is the impossibility of separating the design of an apparatus from the skills needed to use it effectively. Cavendish's pneumatic apparatus was among the simplest of his day; yet he obtained with it some of the most exact and consistent results. In contrast, precision in his apparatus for determining the gravitational constant was obtained with relatively complex apparatus. The most elaborate eudiometers were frequently the least reliable. Van Marum's gasometers were much simpler than were Lavoisier's but were just as reliable. Liebig's *Kaliapparat* was a device of dramatic design simplicity that helped to transform organic analysis.

The relative merits of volumetric and gravimetric techniques are explored by several contributors. Precision and reproducibility are the two factors most relevant in deciding between them, but they leave also the question of what counts properly as a chemical (as opposed to a physical) laboratory technique, yet another question that rests on problems of the definition of disciplines. Mauskopf (chapter 13) classifies Lavoisier's technique of volume determinations, which was used by Proust in his chemical analysis, as a chemical method, whereas Bensaude-Vincent's approach (chapter 7) would suggest that this can be seen as a physical method. Here, as often elsewhere, practical, material, conceptual, and institutional factors are brought into interaction.

Yet another recurrent theme is the inextricability of theoretical research and practical applications, such as in Frankland's analyses (chapter 12) and in the experimental study of munitions by Robert Bunsen, Frederic Abel, and others (chapter 13). Chemistry is very much a mixed science.

Such unifying themes are developed throughout the volume, partly in response to a set of questions that defined the framework for the workshop at the Dibner Institute (from which these chapters developed) and partly because chemistry, for all its protean nature, has a coherent foundation in practice and in its apparatus. However, to consider only thematic unity would be an error. Of at least equal significance is the suggestive richness of the diversity of these chapters. To the initial set of questions, the authors often added or substituted their own questions, adopting different approaches and supplying different answers. This diversity should stimulate us to ask further questions about the relation between the cases chosen and the generalities they illustrate.

Suppose some of our authors were to exchange their cases without changing their individual perspectives. For example, can the history of the eudiometer, which Levere has used to follow the changing demands for precision, have been used as well by Golinski to illustrate the trajectory that he has described for the thermometer? The project appears promising because of an equivalent question about what the apparatus and the operation were measuring. When Joseph Priestley invented the nitrous air test, it measured quantitatively a qualitative property, the respirability of air. After the acceptance of Lavoisier's matured conception of the atmosphere, the same test measured the proportion of oxygen in a sample of air. What did this test measure during the transition between the conceptual framework within which Priestley worked and that in which Lavoisier and his followers eventually worked?

Reciprocally, we might ask whether the history of the thermometer in chemistry could be used also to illustrate the increasing requirements for precision and the evolution of the instrument to meet these demands. During the period that Golinski treats, this was not a central question, because chemists had little need even for the degree of accuracy attainable by existing thermometers. If the story were prolonged into the nineteenth century, however, that issue might emerge. By then, the use of specific boiling and melting points to identify compounds, and other new procedures, might well have introduced new demands for precision in measurements of temperature that had not confronted eighteenth-century chemists. In that case, Levere's approach to the successive designs of eudiometrical methods intended to improve the precision of gasometric analysis might be fruitfully transferred to the history of thermometry in chemistry.

One can easily imagine other exchanges of approach among the various chapters included in this volume. Precisely this open-endedness, this incompleteness, invites further inquiry and justifies the genre of the collected thematic volume. Seldom definitive, such collections can be immensely provocative. That is our intention, and our hope, for these chapters on the role of apparatus and experimentation in the history of chemistry.

NOTES

1. Exceptions include Maurice Daumas, *Lavoisier, Théoricien et Expérimentateur* (Paris: Presses Universitaires de France, 1955); Jon Eklund, *The Incompleat Chymist* (Washington, DC: Smithsonian Institution Press, 1975); and F. L. Holmes, *Eighteenth-Century Chemistry as an Investigative Enterprise* (Berkeley: Office for the History and Philosophy of Science and Technology, University of California, 1989).

2. R. G. W. Anderson, *The Playfair Collection and the Teaching of Chemistry at the University of Edinburgh 1713–1858* (Edinburgh: The Royal Scottish Museum, 1978), is a model of combining catalog with narrative history. A different but useful balance is struck by Ferenc Szabadváry, *History of Analytical Chemistry* (Oxford: Pergamon, 1966; reprinted in 1992 by Yverdon and Langhorne, eds., PA. Gordon and Breach).

3. Translated in Eklund, *The Incompleat Chymist*, p. 7, from Pierre-Joseph Macquer, *Elémens de Chymie Théorique* (Paris, 1751), 275.

4. The nineteenth-century development of extruded glass tubes and rubber tubes rendered connecting and disconnecting vessels much simpler.

5. Michael Faraday, *Chemical Manipulation* (London: John Murray, 1829), vii.

6. Pierre-Joseph Macquer, *Dictionnaire de Chymie*, 2nd ed., (Paris, 1778), has detailed descriptions but no plates; Tiberius Cavallo, *A Treatise on the Nature and Properties of Air, and Other Permanently Elastic Fluids* (London, 1781); and J. G. Children, in J. J. Berzelius, *The Use of the Blowpipe in Chemical Analysis, and in the Examination of Minerals ...* (London, 1822), v.

7. G. G. Venel, in Diderot and d'Alembert, eds., *Encyclopédie, ou Dictionnaire Raisonné des Sciences, des Arts et des Métiers*, vol. 3 (Paris, 1753), 417.

8. Eklund, *The Incompleat Chymist*, 2.

9. A.-L. Lavoisier, *Oeuvres* (Paris, 1862), vol. 1, 14.

10. Holmes, *Eighteenth-Century Chemistry as an Investigative Enterprise*.

11. M. Norton Wise, ed., *The Values of Precision* (Princeton: Princeton University Press, 1994).

CONTRIBUTORS

Robert Anderson has been Director of the British Museum from 1992. Having studied chemistry as an undergraduate and graduate at Oxford in the 1960s he became a curator in the field of history of science and has continued his interest in the material culture of science ever since. From 1982 to 1997 he served as President of the Scientific Instrument Commission of the International Union of the History and Philosophy of Science and from 1988 to 1990 as President of the British Society for the History of science. He has been awarded the Dexter Prize of the American Chemical society and holds honorary doctorates from the Universities of Edinburgh and Durham. He has written on the history of museums as well as on eighteenth-century chemistry and scientific instrumentation.

Bernadette Bensaude-Vincent is Professor of History and Philosophy of Science at the University of Paris X. Her publications on the history of chemistry include *Lavoisier, mémoires d'une révolution* (Paris, Flammarion, 1993); *A History of Chemistry*, (I. Stengers coll.), Harvard University Press, 1996; *Lavoisier in European context (ed. F. Abbri. coll.); Eloge du mixte* (Paris: Hachette littératures, 1998).

Maurice Crosland has taught at the University of Leeds, England and at Berkeley, Cornell and Pennsylvania. From 1974–1994 he was Director of the History of Science Unit at the University of Kent at Canterbury. He is the author of *Historical Studies in the Language of Chemistry* (2nd edn., 1978, Spanish translation, 1988); *Gay-Lussac, Scientist and Bourgeois* (1978, French translation, 1992); *The Science of Matter*, Reprint, 1992; and *In the Shadow of Lavoisier: The "Annales de Chimie"* (1994). He is now at the School of History, Rutherford College, University of Kent.

Jan Golinski is Associate Professor of History and Humanities at the University of New Hampshire. He is the author of *Science as Public Culture: Chemistry and Enlightenment in Britain, 1760–1820* (Cambridge University Press, 1992); and *Making Natural Knowledge: Constructivism* *and the History of Science* (Cambridge University Press, 1998); and co-editor (with William Clark and Simon Schaffer) of *The Sciences in Enlightened Europe* (University of Chicago Press, 1999).

Frederic L. Holmes is chair of the Section of the History of Medicine at the Yale University School of Medicine. Among his publications on topics related to his contribution to this volume are: *Eighteenth Century Chemistry as an Investigative Enterprise* (Berkeley, 1989); *Lavoisier and the Chemistry of Life* (University of Wisconsin, 1985); and *Antoine Lavoisier — The Next Crucial Year* (Princeton, 1998).

Trevor Levere is Professor of History of Science at the University of Toronto, and editor of *Annals of Science*. His previous books in the history of chemistry are *Affinity and Matter* (Clarendon Press, 1971) and *Chemists and Chemistry in Nature and Society 1770–1878* (Variorum, 1994). Other books include *Poetry Realized in Nature: Samuel Taylor Coleridge and Nineteenth-Century Science* (Cambridge University Press, 1981) and *Science and the Canadian Arctic: A Century of Exploration: 1818–1918* (Cambridge University Press, 1993). His current research is on chemical instrumentation in the eighteenth and early nineteenth centuries, and on the Chapter Coffee House Society, a philosophical society that met in London through the 1780s and was also known as the Chemical Society.

Seymour H. Mauskopf is Professor of History at Duke University. He was the first Edelstein International Fellow in the History of Chemical Sciences and Technology (1988–1989). His relevant research interests include the history of atomic and molecule theories (*Crystals and Compounds*, 1976) and his current project, the scientific study of explosives and munitions, 1775–1900. He edited *Chemical Sciences in the Modern World* (1993). He is also interested in the history of unconventional science, co-authoring with Michael R. McVaugh *The Elusive Science: Origins of Experimental Psychical Research* (1980) and editing *The Reception of Unconventional Science* (1979).

William R. Newman is a Professor in the Department of the History and Philosophy of Science at Indiana University, Bloomington. He has written *Gehennical Fire: The Lives of George Starkey, An American Alchemist in the Scientific Revolution* (Cambridge, Mass: Harvard University Press, 1994), and *The Summa Perfectionis of Pseudo-Geber: A Critical Edition, Translation and Study* (Leiden: Brill, 1991). Presently, Newman and Lawrence Principe are editing and translating Starkey's laboratory notebooks and writing a study of his collaboration with Robert Boyle.

Mary Jo Nye is Horning Professor of the Humanities and Professor of History at Oregon State University in Corvallis where she teaches the history of science. She has written on the history of the physical sciences in intellectual, political and social context, most recently in the book *Before Big Science: The Pursuit of Modern Chemistry and Physics, 1800–1940* (Harvard University Press, 1999 paperback) and in several articles on P. M. S. Blackett, Linus Pauling, and Michael Polanyi.

Lawrence M. Principe is an Associate Professor in the Department of the History of Science, Medicine, and Technology and the Department of Chemistry at The Johns Hopkins University. His research interests focus on early modern "chymistry," and he has recently published *The Aspiring Adept: Robert Boyle and His Alchemical Quest* (Princeton, 1998).

Alan J. Rocke is H. E. Bourne Professor of History at Case Western Reserve University in Cleveland, Ohio. He is the author of *Chemical Atomism in the Nineteenth Century* (Ohio State University Press, 1984); *The Quiet Revolution* (University of California Press, 1993); and *The "French Science": Adolphe Wurtz and Chemistry in the Nineteenth Century* (MIT Press, 2000).

Colin A. Russell is Emeritus and Visiting Research Professor in History of Science, the Open University, UK, and Past President of the British Society for the History of science. His most recent book is *Edward Frankland: Chemistry, Controversy and Conspiracy in Victorian England* (Cambridge University Press, 1996). Forthcoming books include *Chemistry, Society and Environment: A New History of the British Chemical Industry* (Royal Society of Chemistry, Cambridge [ed. and senior author]; due 1999), and a sequel to *Recent Developments in the History of Chemistry* [co-ed. with G. K. Roberts]; Royal Society of Chemistry, Cambridge; due 2000).

William A. Smeaton lectured in chemistry at the Northern Polytechnic, London before joining the department of history of science at University College London. On his retirement he became Emeritus reader in the History and Philosophy of Science in the University of London. He now lives in Ely, Cambridgeshire, and is an affiliated research scholar in the department of history and philosophy of science, University of Cambridge.

Mel Usselman teaches chemistry and the history of chemistry at the University of Western Ontario and (with co-workers) has recently completed a series of reconstruction experiments on the combining proportion investigations discussed in his chapter.

I

The Practice of Alchemy

Chemistry as a recognizably modern discipline emerged from a complex matrix of practices and theories in which alchemy was for centuries a dominant focus. It depended necessarily on its practitioners' use of apparatus and techniques, and these have proved difficult to find and difficult to recapture. We have few detailed descriptions of apparatus, and accounts of experiment and of the theories on which experiment rests have often seemed fanciful, impenetrable, and even deliberately obscure. If we are to obtain any detailed understanding of alchemical experimentation or of other early modern precursors to chemistry, such as assaying and herbal distillation, we need (wherever possible) to supplement the written and iconographic record with direct knowledge of the apparatus.

Part I of this volume consists of three chapters that consider material and iconographic evidence, the written record, and the issue of reproducibility. What becomes very clear is that, whatever the theoretical distance between alchemy and chemistry, the material evidence and much of the practice in these two related disciplines have much in common.

Robert Anderson explores the range of material evidence for chemical experiments and practices in the ancient and medieval world. His account is literally an archaeological one, remarkable both for what it uncovers and for the astonishing absences in artifacts that it records. He shows how the pictorial record can be misleading and thus confirms the importance of studying the artifacts wherever possible. He underscores the ambiguity that has always surrounded chemistry as a mixed science with a wide range of motivations, uses, and goals. He stresses chemistry's technical origins in metallurgy, medicine, agriculture, and myriad other practical activities. Further, he demonstrates by constructing a kind of genealogy of objects that many instruments of chemistry and alchemy remained essentially unchanged over centuries. The early record is imperfect, and the surviving objects are thus all the more important, albeit infrequently unproblematic.

Anderson's archaeological account, however, is strongly supportive of an interpretation emphasizing continuity and stability in the history of alchemy and cognate areas of inquiry and practice. His assessment of the evidence available from the material remains of early chemical objects is distinct from the themes elucidated in the other chapters in this volume, but his project is a prerequisite to future efforts to ask questions for the early historical period similar to the questions asked in this volume for the modern era.

William Newman moves wholly into the complex world of alchemy. He uses the written and pictorial record to argue for continuities of practice, as Anderson did with the archaeological evidence. In the process, he also addresses other central issues in the history and historiography of alchemy, including the precise locus of "modernization" and the role of the reproducibility of experiments, language, and public discourse. He argues not only that two important instruments of early chemistry were used by alchemists but that they were used for experimental investigations of nature. Flame tests, which achieved striking prominence in late eighteenth-century chemical and mineralogical analysis, were in wide use by the alchemists and at least "provided an environment that could contribute to the development of the blowpipe in the context of analysis." Similarly, the balance, which has too exclusively been claimed for chemistry in the decades leading up to its revolution, was in regular use by alchemists, even if they did not accord it the privileged position that it later acquired.

Lawrence Principe extends our understanding of the common ground between alchemy and chemistry yet further by taking some dramatic and seemingly allegorical imagery and arguing that "the admittedly culturally influenced metaphorical clothing ... may (in more than a few cases) cover a solid body of repeated and repeatable observations of laboratory results." His reproduction of one such experimental sequence is presented with the dramatic modern imagery of a photograph of "the Tree of the Philosopher." He goes beyond the specific and remarkable achievement of that reproduction to argue more generally that alchemists were "involved in daily practical experimentation that they expected to be reproducible." These experiments are such fundamental attributes of the chemistry that later emerged and distanced itself from its roots in alchemy that they can remain unspoken premises for discussions of the role of instruments and experiments. For Principe and Newman, however, they are central issues.

Their investigations of alchemy have led both Principe and Newman to stress continuity over discontinuity in the history of chemistry and to advocate a more modulated and less abrupt interpretation than is commonly advanced. They take us beyond metaphors and rhetoric and present alchemy as not only the pursuit of the elixir or the philosopher's stone—keys to health and to the transmutation of metals—but as an activity concerned, as is later chemistry, with testing, experimenting, classifying, systematizing, and explaining.

1

The Archaeology of Chemistry
R. G. W. Anderson

Inadequate attention has been paid by historians to the material culture of science and to the skills of scientific practices. Comparing the various scientific disciplines, chemistry has done even worse than average. The balance may have been redressed a little in recent years; still, differences remain in levels of interest between, say, astronomy and chemistry. Books and papers have explored telescopes, planetaria, and astrolabes,[1] but little or nothing has appeared concerning furnaces, burning lenses, and alembics.

The case of alembics suggests why this might be. The alembic, used for distillation, was certainly known to Greek chemists working in Alexandria and was used by Arabs from the rise of Islam.[2] It was extensively employed in the West until relatively recently and may continue to be used in out-of-the-way pharmacies even today. It furnishes a striking example of continuity and stability in the history of chemical instruments.

Evidence for the history of the alembic can be found in texts, both in manuscript and in printed form. However, it must be treated with caution. Few alembics are known, or rather known about, and most are fragmentary.[3] They are utilitarian and not objects of high status. Glass examples, being fragile and relatively cheap to manufacture, are likely to have been discarded, and metal alembics rarely survive. They contain relatively little information. It is exceedingly difficult—in fact, nearly impossible—to determine who made them and where they were made. They incorporate little or no contemporary chemical theory. If texts and objects are scrutinized, one finds that the texts are often obscure and illustrations are diagrammatic or inexact. For all these reasons, alembics are scarcely studied by science historians at all. Their presence is unacknowledged. In fact, they may exist in some specialized museum collections but, as they reveal little (or seem to), they are largely ignored.

A worthwhile endeavor might be to reassess this situation and to consider what early chemical apparatus does survive alongside literary description. This would be a major research project and, in this chapter, it can be attempted only in a cursory manner. Only apparatus and texts earlier than the arbitrary date of 1750 are considered, and the argument concentrates on distillation apparatus. Distillation is a process used primarily in earlier times for concentrating alcohol from mixtures of water and alcohol. The product was often used, sometimes with other substances, for medicinal purposes, though distinguishing these from social uses of alcoholic beverages can be difficult. Perfumes were also extracted using distilling techniques.

Nearly all treatises on chemistry make reference to distillation. Distillation apparatus includes some of the most distinctively shaped chemical vessels. Caution is necessary when surveying previous work on this subject because, from time to time, apparatus that was never intended for distillation is identified as such, this being true particularly of spouted cupping glasses and breast relievers, which are sometimes called *alembics*.[4]

Most examples of texts and apparatus considered are from western Europe, though references are made to the Near East. A small fraction of the surviving archaeological evidence is assessed here: At least 70 single items or groups from a single location are to be found in the literature and, of these, approximately one-fourth are considered very briefly. Only a small proportion of the evidence comes from sources that can be dated even approximately. One might think that a clear evolution of styles of apparatus would emerge, though even if this were the case (which it probably is not), too few examples have a secure chronological base on which to make comparisons. Texts are of some help, but they must be treated with caution, as printed sources became available only some time after the development of chemical practices in western Europe, and manuscripts have so far not proved to be very useful.

Before dealing directly with the material evidence, we consider a little of the literature of early chemical apparatus. The first problems were identified by R. J. Forbes in 1948: "We must remark ... that illustrations of apparatus in medieval manuscripts are neither very numerous nor clear ... [i]t is always well to mistrust their [book] illustrations as they usually refer to the time when the printed edition was issued, which may be several centuries later than the original manuscript."[5]

This accounts for the derivative nature of many illustrations and of the descriptions of them. For example, the double pelican (used for long reflux distillations) appears in similar form in many works. An

Ꝑe muſt haue alſo croked glaſſes na·
med retozte/ and alſo glaſſes with two
armes named pellycane / faſcyoned as
this figure ſheweth

Figure 1.1
Pelican ("Glasses with two armes named pellycane") and a retort, from Brunschwig
(1527). See note 28.

early description and an illustration appear in Hieronymus Brunschwig's
Liber de arte Distallandi de Compostis of 1512, and this may act as a source
for later authors. The pelican is a good example of the disparity that in
all likelihood exists between the frequency of description in texts and
the actual production and usage of certain items of chemical apparatus.
Nearly every sixteenth- and seventeenth-century text includes a description and a (usually poor) woodcut of the vessel that reveals little of
the mode of construction (figures 1.1 and 1.2). Yet very few complete or
even fragmentary double pelicans survive.[6] Some illustrations are fanciful, such as those published by G. B. della Porta in 1609,[7] wherein it is
difficult to believe that the tortoiselike retort was not engraved to show a
closer resemblance to the animal than to its actual form. Other anthro-
pomorphic devices in this work include a tower of alembics intended
for fractional distillation and constructed in the form of the multi-
headed hydra and an alembic and cucurbit resembling a rampant bear.

Written descriptions vary from the deliberately obscure to the
intensely practical. A number of chemical texts incorporated the so-
called *Tabula Smaragdina*, where we find the following description:[8]

> Ascend with the greatest sagacity from the earth to heaven, and then
> again descend to the earth, and unite together the powers of things

The form of a Pellican.

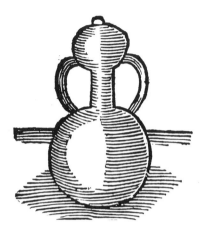

Figure 1.2
Pelican, from French (1651). See note 22.

superior and things inferior. Thus you will obtain the glory of the whole world, and obscurity will fly far away from you.

This alchemical precept refers to distillation or sublimation and provides a revealing comparison with the following clear instructions in a seventeenth-century English translation of Geber:[9]

> The distance from the fire, the magnitude of the vent holes of the furnace, and the structure of the furnace are important in achieving and maintaining desired temperatures. The apparatus is placed in an earthen pan full of ashes for some distillations, or wrapped in hay or wool and placed into water, for others. If upon ashes, the apparatus rests upon a layer one finger thick, and is covered with ashes almost as high as the neck ...

Undoubtedly, detail was deliberately omitted on occasion, a practice that was sometimes even admitted. Thomas Norton's *Ordinal of Alchemy* was written in verse form circa 1477. In chapter 6, he lists the uses of a furnace of his own design, but he declines to describe it:[10]

> An other fornace for this operacion
> By me was fownde bi ymaginacion;
> Nobilly servynge for separacion
> Of dyuydentis, and for altificacion;

It will for some thyngis serve desiccacion,
It servith full well for preparacion;
So for vj thingis it servith well,
And yet for all at oons as I can tell.
This is a newe thynge which shall not be
Sett owte in picture for all men to see.

What historians can see, or at least can seek, is the physical evidence. This might well precede any written evidence—*might* because some doubt must surround the two earliest groups claimed: bowl-shaped vessels from Mesopotamia and retortlike apparatus from the Indian subcontinent.

The Mesopotamian evidence is found in material excavated from Tepe Gawra on the River Tigris.[11] Fragments of four ceramic bowls were excavated from levels dating to 3500 BC. The common feature of all is the double rim, or gutter, around the top. Martin Levey postulated that lids would fit around the circumference, though these have not been found. A vapor could condense on the inner surface of the lid and dribble down into the gutter, whence it could be removed in batches. Two of the vessels, now in Baghdad, have holes in the circumference. Levey further speculated that the gutters might have been filled with comminuted plants, the liquid being distilled leaching out products, presumably of pharmaceutical use.

Distinctive vessels, quite probably for distillation, have been found at widely separated sites in India and Pakistan. The first group to be reported had been excavated in 1930 at Sirkap, site of Taxila in the Indus Valley.[12] The city originated on this site circa the second century BC, and it remained in occupation for three centuries. The still itself was found in levels of the time of the Sáka invasions (i.e., 90 BC to AD 25). The complete apparatus was composed of four elements, all made of ceramic material, and the way in which they fit together provides the argument that they form a distillation unit. The parts are a globular pot; a shallow vessel with a spout close to the rim, which fits on the pot; a tapered tube, which fits into the spout; and a globular vessel with an opening to receive the tube. The archaeologist Sir John Marshall speculatively added a further conjectural component: a water bath in which the globular vessel, the receiver, is placed to increase efficiency of condensation. The still head could not be described as an alembic, as it had no internal gutter. Rather, it is a proto-retort. Large numbers of this type of receiver bottle were discovered in the Vale of Peshawar in 1963 and 1964, from levels datable from 150 BC to AD 350.[13] Some of the receivers had an impressed "tanga" mark that led the excavator to suggest that this was a royal stamp or license and that such marks indicated ownership, or at least a license, for the ownership of the distilled product that they may have contained. This theory must remain highly speculative, however.

The earliest literary references to chemical practices relate to workshops in Hellenistic Alexandria of the third and fourth centuries. They have been carefully studied and published by Robert Halleux.[14] As regards physical evidence for chemical practices, the best known are those texts that include diagrams of distillation processes. These manuscripts are, however, late medieval copies of earlier texts and afford no certainty that the diagrams illustrate ancient apparatus. They may simply have been brought into the text by the medieval scribe. This possible discrepancy must be examined carefully. The diagrams show alembics, some with two or three separate spouts, examples of which do not survive from any physical evidence that remains (none whatsoever having been reported from Alexandria).[15]

Generally agreed is that the culture of Hellenistic Egypt and Byzantium was transmitted to the West via Arabic cultures of the Near East. This seems to have involved an intermediate stage when Greek texts, including alchemical ones, were first translated from Greek into Syriac in the city of Edessa, just to the east of the upper Euphrates.[16] The absence of apparatus from Greek sites might lead to the conjecture that somewhat later sites in the Near East might provide clues.

This material is also disappointing, however, because although a number of probable Arabic and Iranian pieces are known in museums, their origin is unclear. Sketchy reports cite discovery of glass distillation apparatus at Fustat in Egypt in eighth- to eleventh-century levels[17] and at Tell Hesban in Jordan of ninth- to tenth-century date.[18] Perhaps most frustrating of all are four remarkable pieces of apparatus in the Science Museum, London, of devitrified glass.[19] They include an alembic (figure 1.3) and associated cucurbit for distillation and what may be a pair of vessels for sublimation. They were purchased in 1978 from a dealer who was not able to offer their source other than to suggest that they were from Lebanon. Glass historians have suggested dates ranging from the eighth to the twelfth century, though an earlier period has been suggested by one scholar who wondered whether they might be Sasanian, which would date them from the third to the seventh centuries.[20]

At least six other unprovenanced Near Eastern alembics in museum collections have been noted, and doubtless many more await

Figure 1.3
Alembic, devitrified glass, from Near East. Courtesy Trustees of the Science Museum, London.

discovery. One, however, has excellent provenance (figure 1.4). It was collected in 1975, along with a cucurbit from the alchemist Azad Manesh, who was operating at the time in Isfahan, Iran. When it was acquired, the cucurbit was wrapped in clay, but this fell off when the vessel was en route to London.[21] The technique of surrounding vessels of inferior-quality glass with clay to prevent them cracking during heating is mentioned in many texts. No fundamental difference distinguishes the design of this Isfahan apparatus from that which could date from a millennium earlier. Stability over time in the design of instruments can scarcely go further.

Turning from Eastern to Western practices and concentrating particularly on the later medieval period, we find manuscripts and early printed books produced at this time in some numbers. Four are examined here before the archaeological evidence is considered.[22]

Portions of the first have been quoted already: Thomas Norton's *Ordinal of Alchemy*, which was written in 1477. Norton (?1433–1513 or 1514) was a significant figure from Bristol, a member of King Henry VII's household. The second has also been mentioned: Hieronymus Brunschwig's *Liber de arte Distallandi de Compositis* of 1512, sometimes known as the *Grosses Distillierbuch*, which went through many editions, three of them in English. The first English edition of 1527 has been

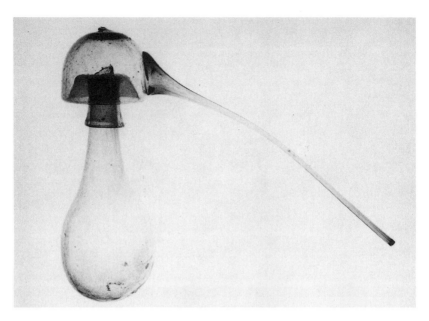

Figure 1.4
Alembic and cucurbit, glass, from Isfahan, twentieth century. Courtesy Trustees of
the Science Museum, London.

used and is readily available in reprint. Brunschwig (ca. 1430–1512 or
1513), an exact contemporary of Norton, came from Strasbourg; his
name is spelled in a variety of ways. The third is John French's *The Art
of Distillation* of 1651. French (1616–1657) was an English military
physician. Last is Johan Joachim Becher's *Tripus Hermeticus Fatidicus* of
1689, which concentrates particularly on chemical furnaces. Becher
(1635–1682), a well-traveled German physician from Speyer, came to
England in 1679 and probably died in London.[23]

The Norton manuscript offers some very practical details. A section is devoted to the shape and material of chemical vessels, which
relates to their specific purpose:[24]

> The thridd concorde to many us full derke,
> To ordeyne Instrument is according to ye werke.
> As every Chapter hath dyuesse ententis,
> So hath thei dyuesse Instrumentis;
> Both in matere, and allso in shappe,
> In concorde that no thynge myshappe.

Norton goes on to explain that for division and separation, small vessels are used, whereas for soaking and circulation, they must be broad. For precipitation, they should be long, whereas for sublimation, the vessels might be long or short. For purification from adulterants ("correccion"), narrow vessels 4 inches tall are used. Of the materials from which ceramic vessels might be manufactured, Norton refers to "dedd claye," or highly fired ceramic. Many clays will split ("leepe in fyre") when heated. For heating at high temperatures, stoneware is recommended, though it has to be imported:[25]

> Othir vessels be made of stone,
> For fyre sufficient, but few or noon
> Amonge werkmen as yet is fownde
> In any contray of english grownde
> Which of watire no thynge drynke shall,
> And yet abide drye fyre with all.

In fact, stonewares (low-fired protostoneware fabric) were imported into England from the Rhineland in the thirteenth century and in increasing quantities as a refined body over the period from the fourteenth to the seventeenth centuries. Not until the 1670s was the first English stoneware industry established, in London.[26]

Norton also discusses glass in a little detail: "All other vessels be made of glasse, That spirituall maters shuld not owte passe." Norton recommends that to make glass of the better kind, the components should be annealed overnight; harder glass contains cullet (waste glass melted down with fresh ingredients: "The hardir stuffe is callide freton, off crippynge of other glassis it com."

Then furnaces are dealt with. Norton says that in the past, individual chemists would devise a furnace to their own design, but these were often unsuccessful, having the wrong dimensions. He considers how the heat can be regulated and suggests that stoppers can be used to adjust air flow. Experience is vitally important:[27]

> The more is the stoppell, The lasse is the hete;
> Bi manyfolde stoppellis degrees ye may gete.
> Who knowith the power, ye worchyng and kynde
> Of every fornace he may well ye trouth fynde;
> And he which thereof dwellith in ignorance,
> All his werkis fallith uppon chaunce.

Bruschwig's *Grosser Distillierbuch* is also very practical and makes the point that though alchemists are secretive, he will be open:[28]

> Now be it that the lerned and experte maysters of the scyence of Alkemye hereof have a knowledge/yet it is not upon to all maner of people wherefore I shall make here as thus, the fyrste rehersall.

The first part of Brunschwig's book deals with the apparatus of distillation in a systematic and detailed way; for example:

> [y]e must have capellys of whyte claye/suche as the goldesmythes crowl bales [crucibles] is made of/some leded and some not leded/ comonly half a yerde wyde and depe/or more or lesse as beyoweth acordynge to the proporcion of the fornayse.

Brunschwig lists apparatus for different kinds of distillation. For that at low temperatures, he specifies the bain-marie, a glass cucurbit bound to a lead plate with four rings. The plate weighs the cucurbit down so that when it is lowered into a copper kettle filled with water, it sinks instead of floating.

The type of glass is specified: Cucurbits "must be made of venys [Venetian] glasse bycause they shoulde the better withstande the hete of the fyre." Elsewhere, he recommends "bohemy [Bohemian] glas." For performing the distillation itself, Brunschwig mentions helms made of clay, copper, tin, or lead and retorts and alembics. For the process of circulation (reflux distillation), he lists pelicans, "blind helms" (one flask inverted into the neck of a larger one), and "circulatories" (shaped like two gourds joined at their necks, with a tube projecting from the lower vessel at an angle).

A good deal of discussion concentrates on furnaces and the rate at which distillation should proceed ("as ye telle one two thre by the clock/so softly must your droppys fall"). The heat produced by a furnace can be controlled by adjusting the ventilation ("And to every smoke hole ye shal make a plogge or tappe to governe your fyre with/ great or smal as it is nedefull"). The construction of the furnaces themselves is addressed, both the masonry and the grills ("above the holowe place ye shall laye rounde or square yron barrys every barre a great ynche of thyckenesse and they must be layde an ynche from eche other").

Many of the basic forms of vessels had not changed when John French described them in his *Art of Distillation* approximately 150 years

mynge downe to the foure rynges that
be fast on the leden rynge and bynde the
fast eche to the other as here is fygured.

℣Than set the glas with the lede in the
water and standed upryght/and is sure
from fallyng on the one syde or the other
b.iii.

Figure 1.5
Bain-marie, from Brunschwig (1527). See note 28.

after Brunschwig's description. The helm (which French calls the "common cold Still"), the bain-marie with its lead plate (figures 1.5 and 1.6), and the pelican are very similar. The change that took place was in the means of condensation, which became more efficient over time.

Though the serpentine coil of copper tubing passing through a barrel of water is described in the thirteenth century, Brunschwig does not illustrate it, nor does he illustrate the Moor's head still, in which the still-head is surrounded by a basin into which cooling water is introduced (figure 1.7). This improvement seems to have been introduced at the end of the fifteenth century.[29]

Furnaces had become more sophisticated by French's time, and there were good descriptions:[30]

The forms also of Furnaces are various. The fittest form for Distillation is round; for so heat of the fire being carried up equally diffuseth in every way, which happens not in a Furnace of another figure, as four square or triangular, for the corners disperse and separate the force of the fire.

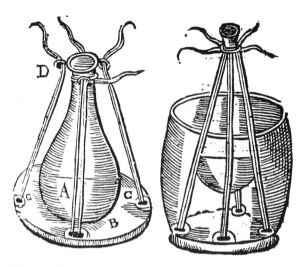

Figure 1.6
Bain-marie, from French (1651). See note 22.

The Form of an Alembick

Figure 1.7
Moor's head still, from French (1651). See note 22.

Revealingly, French includes a list of 22 hints or wise words ("Rules to be considered in Distillation") for the practical chemist. Some refer to apparatus ("When thou takest any earthen, or glasse vessell from the fire, expose it not to the cold aire too suddenly for feare it should break"), whereas others are more general ("Try not at first experiments of great cost, or great difficulty, for it will be a great discouragement to thee and thou wilt be very apt to mistake").[31]

Becher's book on the portable furnace[32] has been considered because of the widely recognized problem of inflexibility of the traditional masonry type of furnace, some being incorporated into the fabric of the laboratory. Engravings of chemistry laboratories often show this (e.g., that depicting the 1682 laboratory of the University of Altdorf, near Nuremburg).[33] French already says, "[I]n defect of a Furnace, or fit matter to make one, we may use a Kettle, or a Pot set upon a Trefoot . . . The truth is, a good Artist will make any shift, yea and in half a dayes time make a Furnace or something equivalent to it for any operations."[34] For operations in which lower levels of heat are adequate, he illustrated portable "lamp furnaces" that use candles; he claims, "[T]he truth is that if your candles be big . . . you may have as strong a heat this way as by ashes in an ordinary furnace."[35]

Becher designed a furnace that was both portable and versatile. Each chemical process—fusion, cupellation, calcination, reverberation, cementation, distillation, digestion (and the bain-marie), sublimation, fierce distillation, and distillation per ascendum—needed a particular kind of furnace. Becher's simple solution was to provide an eight-piece kit of parts that could be put together in six ways. The furnace was made of earthenware. According to Joseph Black's commentary published as late as 1803 (though his notes date from 1766 onward):[36]

> "This [Becher's] furnace is constructed with much ingenuity and judgment, and does honour to this early chemist. When accurately made, so that the joinings of its different parts can be made tight with a little chalk, I believe that it may be easily managed in all its functions. But such earthen ware is very subject to crack . . . A pottery, also, which is sufficiently strong for the furnace, will transmit the heat so much as to become extremely disagreeable, and it becomes very troublesome to keep receivers cool enough for condensing the vapours."

Becher's ideas were pirated by Peter Shaw, translator of Boerhaave, who in 1731 published *An Essay for Introducing a Portable Labo-*

ratory with instrument maker Francis Hauksbee. "A Principle Obstacle to the general exercise of Chemistry in England, being the difficulty of procuring proper Furnaces, Vessels, Utensils and Materials for the purpose; a Portable Laboratory, ready for Business, is here recommended to the Publick."[37] An advertisement appears just before the plates: "THE PORTABLE LABORATORY, ready fitted for Business, may be had of Mr *Hauksbee*, in *Crane-Court, Fleet-street, London*." This is an early example of a named supplier of chemical apparatus, though one Glisson Maydwell (surely the name under which he traded rather than his true name) of The Strand, London, supplied copious quantities of chemical glassware from, at the latest, 1726.[38] As with ceramicware, glassware likely was imported to England from the Continent at an earlier date. One record cites 300 "stilling glasses" being imported into England from Germany in 1587 to 1588.[39]

Returning now to the archaeological evidence from Europe, the majority of reports of chemical ceramic-, and glassware derive from English sites (figure 1.8).[40] However, two particularly interesting groups, from Paris and near Vienna, have been discovered and reported in the last few years.

Glassware and ceramicware can conveniently be assessed separately; few but the largest groups discovered include vessels of both pottery and glass. This may be a reflection on the small number of pieces with chemical associations identified in any excavations. Any extant glassware occurs mainly in three types of medieval site: glasshouses, castles, and monasteries. Ordinary urban locations are unusual, though that may indicate the relative paucity of excavations carried out in towns of medieval origin.

Considering glasshouses, the most substantial group was discovered at Knightons, Alford, Surrey, where four fragments of alembics were found, identified by the rims with their collecting channels or gutters.[41] Additionally, an alembic spout was discovered, six rims of cucurbits, and pieces of tubing. This site, at which four glass kilns were discovered, is described as a Wealden glasshouse, situated in the afforested area between the North and South Downs (in the counties of Surrey and Sussex), where nearly all English medieval glass was made.[42] The date appears to be the 1550s. If this is confirmed to be earlier than records of German imports, English chemical ware may not have been technically satisfactory, may have been unable to keep up with demand, or may not have been economically competitive.

The castle category includes the largest number of sites where distillation apparatus has been discovered. The most substantial group

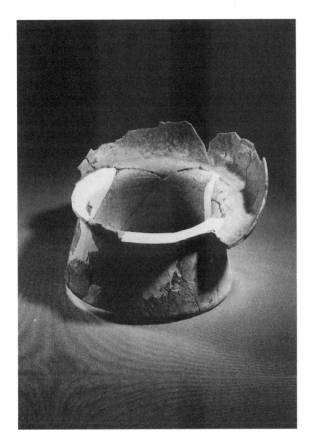

Figure 1.8
Alembic fragment, devitrified glass, from Manor of the More, Rickmansworth,
sixteenth or seventeenth century, showing characteristic double curvature around
rim. Courtesy Trustees of the Science Museum, London.

has been reported from Sandal Castle, near Wakefield in Yorkshire,
and has been dated to the period 1400 to 1450.[43] Sandal Castle was
originally built in the early years of the twelfth century. From then
until the fourteenth century, its importance steadily grew. The chemical ware dates from a period of stagnation, if not decline. More than
200 green glass fragments, representing a minimum of 40 vessels, were
found in the barbican ditch. Two of these fragments are alembic spouts,
and six of the vessels are cucurbits. Of the ceramic material found at
Sandal Castle, two complete cucurbits have been reconstructed.

Another spectacular group of chemical apparatus has been recovered from excavations near the Louvre in Paris during preparations for

the Grand Louvre project which marked the bicentenary of the French Revolution in 1989.[44] As visitors who descend beneath I. M. Pei's glass pyramid can discover, the original royal palace was a medieval castle. The chemical ware was found a few yards away, in gardens belonging to the College of St. Nicholas. It has been argued that siting a laboratory away from other buildings would have been sensible because of the noxious nature of the products of chemical reactions. Another explanation for the siting is that the gardens may have provided ingredients for the processes. Large numbers of fragments of glass flasks, cucurbits, still-heads, and alembics were discovered, along with ceramics that included spouted cucurbits, distillation bases, aludels for sublimation, a remarkable pelican (discussed later), crucibles, strainers, and possibly the remains of a furnace.

The chemical ware was found in a circular masonry structure that was used for dumping rubbish over a long period. The apparatus was found together in a single stratum, and hence it was dumped at one given time. The date can be ascertained by the evidence of associated coins: It appears to be shortly after 1350. The same layer contained tiles and demolition rubbish, and one is tempted to think that the distillation laboratory and apparatus were abandoned together—and suddenly.[45] The purpose and products of the laboratory are unclear. The possibilities have been discussed in terms of professional activity, such as military science, goldsmithing, metallurgy, medicine production, and alchemy, but no clear conclusion emerges. With such a large amount of evidence on such a prominent and well-known site, this must be considered disappointing.

Another remarkable discovery associated with a castle was made in 1980 at Schloss Oberstockstall at Kirchberg am Wagram, Lower Austria.[46] Under the floor of the castle in a waste dump, fragments of 800 objects were found. The glass and ceramic objects were pieces of alembics, cucurbits, aludels, crucibles, flasks, a muffle furnace, and much else besides. Contrasted with the uncertainty of the function of the laboratory of St. Nicholas's College, the purpose of the Oberstockstall laboratory is much clearer: It was primarily intended for metallurgical processes. Some 280 crucibles have been recovered, all having been used in smelting processes. Included are also cupels for assaying. Many residues have been recovered, which should eventually clarify which distillation, sublimation, and melting processes were being conducted. The date of these operations has not been as easy to define as those that took place near the Louvre. The date of 1550 has been suggested, and research is being conducted into the activities of Christoph von Trenbach, Vicar of Kirchberg and Canon of Passau, the likely patron.[47]

Monasteries are the final category for glassware. Three English establishments have produced chemical vessels: Selbourne Priory in Hampshire,[48] Pontefract Priory in Yorkshire,[49] and St. Leonard's Priory in Stamford, Lincolnshire.[50] Monastic foundations are surely likely sites in which to find chemical ware because of the medical role they played; many had botanical gardens to provide the ingredients of pharmaceutical products. Rather discouraging is that the very intense work conducted in recent years at a medieval hospital site at Soutra, near Edinburgh, has produced no chemical vessels.[51]

Ceramic evidence is more widespread than that of glass. Moorhouse classified pottery distillation apparatus into two basic groups: that which parallels glass apparatus (comprising cucurbit, alembic, and receiver) and a variation of this, in which the dome of the alembic form is drawn into a cone and the cucurbit is reduced to a bowllike form with a grooved flange around the rim, adapted to support the alembic and forming a seal (possibly without the need for luting). Paralleling the case for glassware, pottery apparatus is found on sites of kilns, castles, and religious houses, though later examples are also found at urban domestic sites. True pottery alembic forms are uncommon; far fewer have been recorded as compared with those of glass. One of the closest parallels is the nearly complete example found during excavations near the Radcliffe Camera in Oxford (figure 1.9).[52] This example has a very large volume and a wide neck; the cucurbit must have been huge and would have been used for large-scale distillation, perhaps reflecting the requirements of Oxford's medieval scholars (figure 1.10). The material from which it is made, pink and sandy, is typical of later medieval Oxfordshire wares. In contrast, an example found during excavations at Surrey Street in London was almost certainly an import from the Raeren workshops south of Aachen, which dates it as pre-1630.[53] A remarkable find was made in 1976 at a colonial site, Martin's Hundred, near Jamestown, Virginia.[54] This discovery was of a similar alembic with a particularly pronounced finial. Associated evidence indicates that it was made locally, the most likely candidate being one Thomas Ward, who was active at Martin's Hundred between 1623 and 1635. What is surprising is that such a sophisticated vessel, surely not in great demand in the New World, could have been made by the local potter.

Figure 1.9
Alembic, ceramic, from Radcliffe Camera site, Oxford. Courtesy Visitors of the
Ashmolean Museum, Oxford.

A Retort and its receiver before they be set on worke.

Figure 1.10
Retort and receiver with operator, perhaps indicating scale of operations, from
French (1651). See note 22.

This is the form of a common cold Still.

Figure 1.11
Alembic and distillation base, from French (1651). See note 22.

A number of ceramic cucurbits, matching glass examples in shape, are known. Some pottery cucurbits have pronounced shoulders, however, and may have been made in this shape to increase the volume-to-height ratio, thereby allowing greater batches of liquid to be distilled. A fair number of cone-shaped ceramic alembics are also known. Here, the characteristic double curvature of the rim seen on glass alembics is unnecessary, as the gutter rests on the flange of a distillation base, a squat cylindrical vessel. Distillation bases abound (figures 1.11 and 1.12). Their general shape can be seen in an illustration in French's *Art of Distillation*.[55] However, a certain ambiguity surrounds the function of this type of vessel. Other uses can be postulated, and some pointed out that in certain regions of Britain in the fifteenth century, jars, cooking pots, and storage vessels have a flanged rim as the seating for the lid. However, in other instances, alembic and base have been found in the same archaeological context (e.g., at a kiln site at Chilvers Coton, Nuneaton, Warwickshire).[56] An example of a distillation base that is datable is one that was used to contain a hoard of more than 1000 silver coins from Hartford near Huntingdon.[57] The most recent coins date from 1503 (11 in all), and this is not likely to be long before the date of burial.

Figure 1.12
Alembic and distillation base with receivers, set on stove (note smoke holes), from
Brunschwig (1527). See note 28.

Pelicans, according to the texts, were used for a process called
circulation, a long-term reflux distillation. John French recommends a
pelican "to make the Magistery of Wine, which will be one of the
greatest Cordialls, and most odoriferous Liquor in the World." The
pelican has two side tubes that recycle the condensed vapor down to
the base of the vessel for redistillation. The illustrations of pelicans
make them appear somewhat unlikely as practical vessels. They would
be very difficult to make and very difficult to fill and seal. As far as I am
aware, no surviving datable medieval example of traditionally illustrated
shape exists. The Louvre pelican is a much better, stronger design, the
return tube being incorporated into the fabric of the base of an alembic,[58] though very effective luting would be needed if the vapor were
not to escape over, for example, the course of 1 month (the period
needed for the preparation of French's Magistery of Wine). Sensible
though this design is, it does not appear to be illustrated in texts. Here,
perhaps, is an example of an artifact adding to our previous understanding of distillation techniques from literary sources alone.

No mention has yet been made of surviving examples constructed from metal; yet, all the texts make reference to vessels made of iron, tin, copper, and lead.[59] The reason is that none appears to have come from an archaeological context. No example is known of the lead plates with rings that were used to weigh down cucurbits being gently heated in boiling kettles, though they are frequently illustrated in treatises on distillation. A fairly usual phenomenon is for metal to be absent from scrapped material. Though ceramic waste could not be recycled and glass may not have been worth the effort, metal was valuable and could be melted down. A wise saying refers to the pot stills used in Highland malt-whisky distilleries today: They are all as originally installed, and yet not a scrap of them is original. Worn-out parts are simply sent back to the coppersmith for melting down and reworking.

However, a number of metal vessels may be early, yet have no extended provenance. A Rosenhut (a variety of alembic) of pewter, complete with finial, has an early appearance[60] (figure 1.13), and a large copper boiler with still-head and receiver likewise might be old; it has certainly been well used, as can be seen by the way it has been patched.[61] These are located in the Science Museum in London. The Museum of the History of Science at Oxford possesses two Moor's head stills, one in copper (figure 1.14) and the other in copper and pewter.[62] Though they have no certain provenance, they are said to come from the laboratory of Antoine Baumé (1728–1804). Both are better made and more sophisticated than are earlier examples.

Little has been said about surviving evidence of furnaces. This is because very little does appear to survive. Wonderful examples seen in museums inevitably turn out to be "reconstructions," wood and plaster conjectures based on illustrations. Laboratories whose basic external fabric survives (e.g., that of the Ashmolean Museum, Oxford, constructed 1683, or that of the College of Surgeons of Edinburgh, constructed 1696) now contain no evidence of the furnaces that were once installed. A few furnace parts exist from the Louvre and Schloss Oberstockstall establishments but not enough to provide any real enlightenment. Strangely, none of the portable furnaces of the Becher-Shaw variety is known. Even the improved portable furnace of Joseph Black, devised in the 1750s and produced probably in the thousands (a supplier's catalog of 1912 still advertising it) has vanished without trace.[63] Here, we are forced to accept the evidence of the texts alone.

The exception is two curious furnaces with alchemical association. The first is that made for the Landgraf Moritz von Hessen-Kassel

Figure 1.13
"Rosenhut" alembic, pewter, of unknown origin (cf. figure 11). See note 60.

in 1610. It is a glamorous gilded *tour de force* and is engraved with symbols pertaining to alchemy.[64] Its purpose is gentle heating, and it was probably used as a bain-marie for low-temperature distillations. The second is more mysterious, a curious ceramic stove, probably French and of the seventeenth century.[65] Internally, it has various channels and connections, and it is difficult to interpret.

Finally, two fairly early chemical furnaces are located in the Science Museum, London. The first is a brass, barrel-shaped vessel mounted on a tripod with curving legs. It is likely to be Dutch, possibly of the eighteenth century, and it may have been intended for use in an apothecary's establishment.[66] The other is a simple iron muffle furnace, possibly of the early eighteenth century and associated with Sir Isaac Newton.[67] As it comes from the Royal Mint, this is not impossible, Newton being Master of the Mint at that time.

This mass of disparate evidence can now be cautiously summed up. References to texts on chemical operations may not give an accu-

Figure 1.14
Moor's head still, copper, perhaps eighteenth century (cf. figure 1.7). Courtesy
Museum of the History of Science, Oxford.

rate or balanced picture of what took place in chemical laboratories. In some cases, information is withheld, though some texts seem free with detailed information. The latter must be considered critically. The illustrations may be diagrammatic or symbolic. Pelicans, if used in large quantities, might be expected to have survived in the archaeological record. Their working life likely would not be extensive, as they surely were difficult to use, and broken examples should be common. Perhaps they were not, in fact, commonly used, though possibly they and other specialized vessels have not been recognized by archaeologists.

A further reason we should not take texts at face value is the importance that many authors place on defining a taxonomy of chemical operations. Brunschwig delineates two methods of distillation: the first without cost (by the use of fermenting dung as the heat source) and the second with the cost of fuel for a fire. The first method is subdivided into four techniques, the second into five. Yet, we have no way of knowing relatively how much each was adopted by the chemist, and maybe it was systematic description that was important to the author rather than an assessment of the practical situation.

The current archaeological record is patchy and is not sufficiently large to have statistical significance. The preponderance of records involve English sites, reflecting simply that archaeology has been an active pursuit in England or perhaps that publication has more frequently followed excavation than elsewhere. Most sites explored are the more obvious and traditional: castles and monasteries. Urban medieval sites are often inaccessible because of the buildings that stand on them today. Having declared that, distilling apparatus sometimes is discovered at the known centers of power—castles and monasteries—in medieval Europe. By a rough statistical count, approximately twice as many sites can be classified as castles or fortified houses than as any other kind of site, well defined or not. Glasshouses or kilns, which are special cases, are not included in the count.

To conclude, both the archaeological and the written record indicate relative stability and suggest that this change generally embodies a significant element of continuity. More alembics survive than might originally be thought. There is good potential to discover considerably more about early chemical practice from the surviving archaeological evidence. As further material comes to light, a better understanding of early chemistry will become possible.

NOTES

1. The earliest comparative paper on the history of the astrolabe is that of William H. Morley, *Description of a Planispheric Astrolabe* (London: Williams and Norgate, 1856). The major study of preserved astrolabes, both in the Eastern and Western traditions, remains R. T. Gunther, *Astrolabes of the World* (London: Oxford University Press, 1932). See also S. L. Gibbs, J. A. Henderson, and D. J. de Solla Price, *A Computerized Checklist of Astrolabes* (New Haven, 1973).

2. R. J. Forbes, *A Short History of the Art of Distillation* (1948; reprinted in 1970, Leiden: E. J. Brill). This work remains the most extensive survey of distillation literature. A particularly interesting group of distillation vessels from eleventh-century Moorish Toledo has been discovered; see Carmen Bosch Ferro and Marina Chinchilla Gómez, "Formas Cerámicas Auxiliares: Anafes Arcaduces y Otras," in *Arqueologia Medieval Española. II Congreso, Madrid 19–27 Enero 1987. Tomo II: Comunicaciones* (Madrid, 1987), 491–500.

3. The most complete survey of the archaeological evidence of distillation, used extensively throughout this study, is Stephen Moorhouse, "Medieval Distilling-Apparatus of Glass and Pottery," *Medieval Archaeology* 16 (1972): 79–121. This paper excludes all non-Western examples and those not discovered in archaeological contexts.

4. Emilie Savage-Smith, "A Medical, Pharmaceutical or Perfumery Utensil," in Francis Maddison and Emilie Savage-Smith, *Science, Tools & Magic* Part One (London: The Nour Foundation, 1997), 42–47. Some chemical apparatus is indistinguishable from domestic ware; indeed some *is* domestic ware. Joseph Black, the eighteenth-century Scottish chemist, recorded that he used tea cups, beer glasses, and wine glasses on various occasions [R. G. W. Anderson, *The Playfair Collection and the University of Edinburgh (1713–1858)* (Edinburgh: Royal Scottish Museum, 1978), 146, note 33]. The famous engraving of Joseph Priestley's laboratory at Bowood in Wiltshire appears positively domestic, which indeed it was. [Joseph Priestley, *Experiments and Observations on Different Kinds of Airs*, vol. 1 (London: J. Johnson, 1774), plate 2]. Here, a fireplace acts as the source of heat for the production of oxygen, though this level of domesticity was probably unusual: In most experiments, the heat supply was a furnace built into the laboratory or specially designed portable furnaces, though we know little about them.

5. Forbes, *A Short History of the Art of Distillation*, 74, 76. Genre paintings of laboratories have been considered as a source of evidence, see C. R. Hill, "The Iconography of the Laboratory," *Ambix* 22 (1975): 102–110; the author urges caution.

6. For two rare double pelicans (probably of relatively recent—?eighteenth- or nineteenth-century—construction) that survive from Spanish pharmacies, see Consejo Social de la Universidad Complutense de Madrid, *El Museo de la Farmacia Hispana* (Madrid, 1993), 86, plate 67; and Ministerio de Defensa [Madrid], *La Casa de la Quimica: Cienca, Artilleria e Ilustración* (Madrid, 1992), unnumbered plate. The Spanish term for the pelican is *Mujer en jarra*, or "woman with arms akimbo."

7. G. B. Della Porta, *De Destillatione Libre IX* (Rome: Lazari Zetzneri, 1609).

8. John Read, *Prelude to Chemistry* (London: Bell, 1936), 54.

9. E. J. Holmyard, *The Works of Geber, Englished by Richard Russell, 1678* (London: J. M. Dent & Sons, 1928), 82.

10. John Reidy, ed., *Thomas Norton's Ordinal of Alchemy* (London: Oxford University Press, 1975), lines 2865–2876, 89.

11. Martin Levey, *Chemistry and Chemical Technology in Ancient Mesopotamia* (Amsterdam: Elsevier, 1959), 31–41.

12. John Marshall, *Excavations at Taxila* (Cambridge, UK: Cambridge University Press, 1951), 420, 421. See also A. Ghosh, "Taxila (Sirkap), 1944–45," *Ancient India* 4(1948): 41–84.

13. F. R. Allchin, "Evidence of Early Distillation at Shaikhan Dheri," in Maurizio Taddei, ed., *South Asian Archaeology 1977,* vol. 1 (Naples: Naples Istituto Universitario Orientale, 1979). A total of 108 receivers, one retort spout, and a connecting tube were discovered.

14. Robert Halleux, *Les Alchimistes Grecs,* vol. 1, *Papyrus de Leyde Papyrus de Recettes Stockholm Fragments de Recettes* (Paris: Belles Lettres, 1981); and Robert Halleux, *Indices Chemicorum Graecorum I Papyrus Leidensis Papyrus Holmiensis* (Rome: Edizioni dell'Ateneo, 1983).

15. F. Sherwood Taylor, "A Survey of Greek Alchemy," *Journal of Hellenic Studies* 50(1930): 109–139.

16. Joseph Needham, *Science and Civilisation in China,* vol. 5, part 4 (Cambridge, UK: Cambridge University Press, 1980), 410.

17. R. H. Pindar-Wilson and George T. Scanlon, "Glass Finds from Fustat," *Journal of Glass Studies* 15(1973): 17.

18. Sidney M. Goldstein, "Glass Fragments from Tell Hesban," *Andrews University Seminary Studies* 14(1976): 127–132.

19. R. G. W. Anderson, "Early Islamic Chemical Glass," *Chemistry in Britain* (1983): 1822.

20. Private communication, Sidney M. Goldstein, Corning Museum of Glass, 30 March 1983.

21. Private communication, A. J. Turner, 11 September, 1981. See S. H. Nasr, *Islamic Science: An Illustrated Study* (London: Festival of Islam, 1976), 205, plate 114.

22. A comprehensive survey of chemical treatises has not been attempted, this not being the purpose of this chapter. Their number is daunting, in any case, and French himself sets out the problem in his preface "To the Reader":

There is a glut of Chymicall books, but a scarcity of chymicall truthes: Nature & Art affords variety of Spagyrical preparations, but they are as yet partly undiscovered, partly dispersed in many bookes and those of diverse languages, and partly reserved in private mens hands. When therefore I considered what need there is of, and how acceptable a generall treatise of Distillations might bee, especially to our English nation (and the rather since Baker upon Distillations is by reason of the description of a few furnaces, and vessels therein, besides which there is small variety either of preparations, or curiosities sold at such a high rate) I thought I could do no better service than to present them with such a treatise on that subject ...

[John French, *The Art of Distillation or a Treatise of the Choicest Spagyricall Preparations performed by way of Distillation with the Description of the Chiefest Furnaces and Vessels used by Ancient and Moderne Chymists* (London: T. Williams, 1651).] The "Baker" mentioned in this quotation is Johan Joachim Becher (see note 32).

23. Details of the 31 surviving manuscripts of Norton's *Ordinal* are found in the introduction to Reidy, ed., *Thomas Norton's Ordinal of Alchemy,* ix–xxi. Bibliographical details of Brunschwig's and Becher's publications are to be found in J. R. Partington, *A History of Chemistry,* vol. 2 (London: Macmillan, 1961), 83, 84, and 640–642, respectively. Bibliographical details of French's work are found in Alan Pritchard, *Alchemy: A Bibliography* (London: Routledge & Kegan Paul, 1980), 45, item 149. An extremely valuable translation from Latin into German of Libavius's *Die Sceuastica Artis* of 1606 has recently appeared. This treatise deals specifically with the subject of chemical instruments. See Bettina Meitzner, *Die Gerätschaft der chymischen Kunst. Der Traktat "Die Sceuastica Artis" des Andreas Libavius von 1606* (Stuttgart: Franz Steiner Verlag, 1995).

24. Reidy, ed., *Thomas Norton's Ordinal of Alchemy,* 86, lines 2773–2778.

25. Ibid., 87, lines 2795–2880.

26. David Gaimster, "The Supply of Rhenish Stoneware to London 1350–1600," *London Archaeologist* 5(1987): 339–347; and Dennis Haselgrove, "Imported Pottery in the 'Book of Rates.' English Customs Categories in the 16th and 17th Centuries," in David Gaimster and Mark Redknap, eds., *Everyday and Exotic Pottery from Europe c.650–1900* (Oxford: Oxbow Books, 1992), 324–335.

27. Reidy, ed., *Thomas Norton's Ordinal of Alchemy,* 89, lines 2891–2896.

28. Hieronymus von Braunschweig, *The Vertuose Boke of Distyllacyon of the Waters of all Maner of Herbes* (London: Laurens Andrewe, 1527). I have used the facsimile issued as *The English Experience,* Number 552 (Amsterdam: Theatrum Orbis Terrarum, 1973). Pages are unnumbered.

29. Hieronymus Brunschwig, *Book of Distillation. With a New Introduction by Harold J. Abrahams* (New York: Johnson Reprint, 1971), xiv. The Moor's head still appears in some late editions of Brunschwig but not in the 1527 English translation.

30. French, *The Art of Distillation,* 2.

31. Ibid., 12–16.

32. Johann Joachim Becher, *Tripus Hermeticus Fatidicus, seu I Laboratorium Portabile* (Frankfurt: Johann Georg Schiese, 1689). See also Pamela H. Smith, *The Business of Alchemy. Science and Culture in the Holy Roman Empire* (Princeton: Princeton University Press, 1994), 277.

33. A. V. Simcock, *The Ashmolean Museum and Oxford Science 1683–1983* (Oxford: Museum of the History of Science, 1984), 8.

34. French, *The Art of Distillation,* 3.

35. Ibid., p. 153.

36. Joseph Black, *Lectures on the Elements of Chemistry, delivered in the University of Edinburgh by the late Joseph Black, M.D.,* vol. 1 (Edinburgh: William Creech, 1803), 332.

37. Peter Shaw and Francis Hauksbee, *An Essay for Introducing a Portable Laboratory* (London: Osborn and Longman, 1731), 13.

38. Anderson, *The Playfair Collection,* 143, 145; note 14.

39. Eleanor S. Godfrey, *The Development of English Glassmaking 1560–1640* (Oxford: Clarendon Press, 1975), 247.

40. This may occur simply because the most comprehensive compilation of data on the subject (though published a generation ago) was written by an English archaeologist [Moorhouse, "Medieval Distilling-Apparatus of Glass and Pottery,"], or it may be an indication of the level of archaeological activity in England rather than anything else.

41. Eric S. Wood, "A 16th Century Glasshouse at Knightons, Alford, Surrey," *Surrey Archaeological Collections* 73(1982): 32, 33.

42. G. H. Kenyon, *The Glass Industry of the Weald* (Leicester: Leicester University Press, 1967).

43. P. Mayes and L. A. S. Butler, *Sandal Castle Excavations 1964–1973* (Wakefield: Wakefield Historical Publications, 1983). See particularly S. A. Moorhouse, "Vessel Glass," 225–230.

44. Isabelle Rouaze, *Un Atelier de Distillation du Moyen Age* (Paris: Editions du Comité des Travaux historique et scientifiques, 1989). This is a reprint of *Bulletin Archéologique du C.T.H.S.,* nouv. sér., fasc. 22, *Antiquités nationales* (1989): 159–271.

45. An interesting speculation is that this happened because of an edict of Charles V forbidding the use of furnaces and the practice of alchemy. Unfortunately, the date of this is 1380.

46. R. Werner Soukoup and Sigrid von Osten, "Das Alchemistenlaboratorium von Oberstockstall. Ein Vorbericht Zum Stand des Forschungsprojekts," *Mitteilungen der Fachgruppe 'Geschichte der Chemie'* 7(1992): 11–19; and R. W. Soukoup,

S. von Osten, and H. Mayer "Alembics, Cucurbits, Phials, Crucibles: A 16th Century Docimastic Laboratory Excavated in Austria," *Ambix* 40(1993): 39.

47. Also of interest is a castle that is better described as a fortified house. This is the Manor of the More in Rickmansworth, just outside London [M. Biddle, L. Barfield, and A. Millard, "The Excavation of the Manor of the More, Rickmansworth, Hertfordshire," *Archaeological Journal* 116(1959): 136–199]. In the 1520s, Cardinal Wolsey owned the property and organized its enlargement, including the digging of a trench for the garderobe. In the silt of this pit, part of an alembic was found in a high devitrified state. It consists of a wide neck and part of the gutter. The piece may have been discarded at any time between 1520 and the destruction of the house in 1650. This example led Frank Greenaway of the Science Museum to propose that a glass fragment that had a double curvature is likely to be from the rim of an alembic. [See figure 1.8, and Frank Greenaway, "Introduction," in Moorhouse, "Medieval Distilling-Apparatus of Glass and Pottery," 82.] Possibly chemical vessels have not been identified as such in archaeological reports because of the specialized knowledge that is needed.

48. Moorhouse, "Medieval Distilling-Apparatus of Glass and Pottery," 99–101.

49. Ibid., 91–95.

50. Unpublished, but see Mayes and Butler, *Sandal Castle Excavations 1964–1973*, 225.

51. Soutra Hospital Archeothnopharmacological Project, *Sharp Practice*, vols. 1–5 (1986–1995), a series of volumes subtitled: *The First (Second . . . Fifth) Report on Researches into the Medieval Hospital at Soutra, Lothian Region.*

52. Unpublished. The alembic is in the Antiquities Department of the Ashmolean Museum, Oxford, registration number 1910.311.

53. Unpublished, but see Moorhouse, "Medieval Distilling-Apparatus of Glass and Pottery," 108, no. 5, and 116. The alembic is in the collections of the Museum of London and bears the number A4784.

54. Ivor Noel Hume, *Martin's Hundred* (New York: Alfred A. Knopf, 1982), 101–105, 194.

55. French, *The Art of Distillation,* 17 (captioned "This is the form of the common cold Still").

56. Medieval Archaeology 12(1968): 208–210.

57. P. G. M. Dickinson, "The Treasure Trove from Hartford, Huntingdon," *Proceedings of the Cambridge Antiquarian Society* 58(1965): 138–140. The distillation base is on deposit in the department of Medieval and Later Antiquities at the British Museum.

58. Rouaze, *Un Atelier de Distillation du Moyen Age,* plate 16.

59. For example, Brunschwig, *Book of Distillation*: "After that ye must have helmys made of whyte claye such as before is specyfyed & they must be leded within & without or ellys they must be copper tynne or lede of the fascyn after folowyng" (aii, v).

60. Science Museum, London, Rosenhut still, pewter, inventory number 1954–85.

61. Science Museum, London, boiler, still-head and spouted receiver, copper, inventory number 1976–126.

62. Museum of the History of Science, Oxford, two Moor's head stills, copper and pewter, and copper, registration numbers 56–47 and 56–89, respectively. C. R. Hill, *Catalogue 1. Chemical Apparatus* (Oxford: Museum of the History of Science, 1971).

63. R. G. W. Anderson, "Joseph Black and His Chemical Furnace," in R. G. W. Anderson, J. A. Bennett, and W. F. Ryan, eds., *Making Instruments Count: Essays on Historical Scientific Instruments* (Aldershot: Variorum, 1993), 118–126.

64. Ludolf von Mackensen, *Die Naturwissenschaftlichtechnische Sammlung* (Kassel: Georg Wendeeroth Verlag, 1991), 124, plate 98. (This reference provides only a very brief description.)

65. Science Museum, London (Wellcome Museum of the History of Medicine), inventory number A634411.

66. Science Museum, London, inventory number 1954–105.

67. Science Museum, London, inventory number 1876–3.

2

ALCHEMY, ASSAYING, AND EXPERIMENT
William R. Newman

Often, those who write about the history of science assume that alchemy had little to do with the development of chemistry as an exact science. The image of alchemists as insouciant empirics who cobbled their mixtures together with little regard to purity or quantity is an ingrained feature of modern historiography.[1] Hence, technical processes concerned with the refining of metals from ores and their subsequent testing for purity are often consigned by modern historians to the realm of mineralogical and metallurgical literature, as distinct from alchemy.[2] Although it is certainly true that the sixteenth century witnessed the birth of an autonomous literature about mining and metallurgy, as evinced by the works of Vannoccio Biringuccio, Georg Agricola, Lazarus Ercker, and others, it does not follow that alchemists were unconcerned with the purification, testing, and exact measurement of their own materials. Indeed, I show here that two of the most important analytical tools of the early chemist—the blowpipe and the precision balance—were associated with alchemy long before the early modern development of the mining and metallurgy genre. I then pass more generally into a discussion of the relationship between alchemical practice and assaying techniques and still more generally into a consideration of the role of testing and experiment in alchemy. As we shall see, the techniques of mineral testing and analysis were not employed by medieval alchemists merely as empirical means for attaining precious metals. By the late Middle Ages, these techniques had already evolved into tools for the experimental investigation of nature. I begin briefly with the blowpipe.

Most writers about the history of blowpipe analysis are aware of the fact that the blowpipe has an ancient artisanal history. Without doubt, the forced-air tube was used by jewelry makers and metal workers, even in ancient Egypt, for fusing and soldering metals.[3] It was long employed in the manufacture of fine glass as well, as Ferenc Szabadváry has pointed out.[4] However, after a statement made by

Hermann Kopp in his *Geschichte der Chemie* (1844), the commonly accepted opinion was that the blowpipe did not appear in chemical literature until the seventeenth century, when Johann Kunckel suggested that it be used to reduce metallic calces on a hollowed piece of charcoal.[5] Soon after that, the popularity of the blowpipe mushroomed under the promotional umbrella of Berzelius, becoming an extremely important analytical tool among chemists. Kopp therefore viewed the eighteenth-century use of the blowpipe as being effectively a "miniaturization" of earlier analytical techniques by "the dry way," such as cupellation.[6]

Reliable evidence challenges the claim that the blowpipe was absent from chemical literature before the eighteenth century. The blowpipe definitely was used as an aid to analysis early on and not merely as a means of facilitating the manufacture of jewelry. The tenth-century Greek encyclopedia, the *Suda*, referred to its use in oxidizing lead and removing excess litharge during cupellation, as Robert Halleux has pointed out.[7] A fuller description of cupellation with the blowpipe is given by the alchemical *Theorica et Practica* of Paul of Taranto,[8] probably written around the end of the thirteenth century:

> Let very well sieved cinder be taken, and mixed with water of salt: let a vessel be made from it, in which silver or whatever metal that you seek to test in the cupel be put on a very violent fire. With the metal fused, let a sixth part of lead be thrown on; this is especially done in the case of silver. Let a pipe of iron or reed be had, through which one can blow on the surface of the fused metal. The lead fused on the metal will be seen smoking due to this—that it has volatile flight as well as the loss of its substance from its badly fixed principles. Thence it is that, passing into smoke, it will draw with it all which in the metal to be purged is imperfect. The purged metal with the lead added to it will be recognized not to be vaporized, but it will seem to be boiling, and to eject froth—as it were flying forth; then let no more lead be added.

We see then that despite the common view, the blowpipe did find some representation in the alchemical literature of the Middle Ages and that this was in the context of analysis. Nonetheless, Paul of Taranto clearly did not have in mind the use of the blowpipe as an analytical tool in itself; he merely used it to facilitate the production and removal of litharge from the assaying cupel. This information, in a sense, confirms Kopp's greater point that blowpipe analysis was the microscale descendent of much older analyses by "the dry way," such as cupellation and cementation. From this vantage point, the medieval examples of the blowpipe add little that is new.

Other evidence, however, throws a fairly different light on the subject. Kopp seems to have been unaware of the extensive medieval literature on salts and alums, which forms one of the main genres of alchemy. The origin of this material is clearly Arabic; indeed, the founder of the genre seems to have been Muḥammad ibn Zakariyya al-Rāzī (c. 854–925), a well-known physician and philosopher who also wrote on alchemy. Rāzī's genuine *Kitāb al-asrār* or *Book of Secrets* was translated into Latin as the *Liber Secretorum*. Although this became an influential text in Latin, Rāzī acquired his greatest fame among Western alchemists by means of a pseudonymous work, the *Liber de Aluminibus et Salibus.*[9]

Rāzī and his successors were keenly interested in the classification of salts, alums, and atraments or vitriols. The *De Aluminibus et Salibus* describes rock salt (*sal gemma*), table salt (*sal panis*), a "bitter salt" (*sal amarus*), a "Nabatean salt" (*sal Nabataeus*), alkali (*sal alkali*), ammoniac salt (*sal armoniacus*), and others, giving detailed instructions for their purification.[10] Within this literature on salts, one sometimes finds a type of test unmentioned by Kopp, though it has some bearing on the early history of the blowpipe.

The *Ars Alchemie* of Michael Scotus, a thirteenth-century text, is attributed to the court philosopher of Frederick II von Hohenstaufen, the great Holy Roman emperor whose wide-ranging interests earned him the title of *stupor mundi*. The *Ars Alchemie*, despite its rather general title, is largely a treatise on salts and alums, in the tradition of Rāzī. In the first folios of the text, the author launches into a description of the salts that are indispensable to alchemy. The author gives, for example, three different types of *sal nitrum*: niter *de puncta*, "leaved" niter, and "depilated" niter. Niter *de puncta* comes from India, from the sea near "Marioth," and from the area around Alexandria. Leaved niter, on the other hand, comes from the area near Narbonne, and from Aleppo. Depilated niter comes from Hungary, in the area of "Strigo" (Esztergom?), and from Barbary. Each of the three niters has peculiar characteristics, such as the facts that leaved niter tastes like vinegar and depilated niter has the saltiness of dried pork. These obvious characteristics are not of present concern, however; rather, we focus on the author's insistence that one test niter by placing it on a piece of glowing charcoal:[11]

> If you wish to know it perfectly, keep it in a brass [vessel] for a while, and then put it on a coal by means of a little spike. If it jumps off of the coal and emits water above, and boils, put it in the brass [vessel] again.... If it rests on the coal for a while and does not jump off of it nor crackle, it will no doubt be good.

Niter *de puncta*, then, either jumps about and boils when placed on a burning coal or remains there inert. Compare this with the author's description of leaved niter: "[i]t is not very salty, but has as it were the taste of vinegar when you touch it with your tongue, and it gives a flame on the fire."[12] It is clear that the *Ars Alchemie* is here describing a test for differentiating two chemicals that were both known to the medievals as *nitrum* or *sal nitrum*—soda (sodium carbonate)—and salt-peter (potassium nitrate).[13] Sodium carbonate, called *natron* by Greek writers and *nitrum* by the Latins, was used for washing even in remote antiquity.[14] By the thirteenth century, however, the terms *nitrum* and *sal nitrum* were commonly used for saltpeter as well.[15] That the *Ars Alchemie* is differentiating the two is confirmed not only by the fact that the first salt is found to be nonflammable on the coal while the second burns but by the following identification of "niter from Alexandria" in a closely related text also linked to Michael Scot:[16]

> One species of sal nitrum is that which is brought from Alexandria, and this is truly sal nitrum. Indeed, the Saracen women wash their linen clothes in it and make them white as snow. They also wash their faces and their bodies [with it] in baths.

Surely this can only be a description of Egyptian natron, which the *Ars Alchemie* also decribes as coming from Alexandria. The *Ars Alchemie*'s flammable niter, on the other hand, is probably our saltpe-ter.[17] The *Ars Alchemie* then proceeds to give a number of additional examples of the test using a burning coal. In a number of these, the author was concerned with the ash or cinder left by a salt, such as *sal napiticum*, probably identical to the *sal napthe* of Arabic sources: "If you wish to know it, put it on a coal and it will burn up at once. It will leave a white ash that will seem to be an earth, and will not crackle."[18] We can see, then, that the characteristic test with hot coals requires the alchemist to go through a mental checklist for smoke or vapor, saltation, crackling, and a combustion product. One additional bit of information emerges: In a rather confused fashion, the *Ars Alchemie* repeats the test for leaved niter, saying, "If it is put upon coals and you blow on them, it will be inflamed and will jump off of the coals. . . ."[19] Here the author explicitly noted that one must blow on the coal for the test to succeed. Was he using a blowpipe, a bellows, or simply his mouth? One cannot know. What is important, however, is that the *Ars Alchemie* refers to the test with glowing coals a total of some 10 times, as a means of distinguishing different salts. In most of these cases, the salt is placed on the hot coal, and the observer is told to examine it for smoke or vapor, hopping about, crackling, and ash when the salt burns. True, no mention is made of the reduction of a metal oxide or even of the purification of a metal, but the use of the charcoal is otherwise suspiciously similar to Kunckel's suggestions in the *Ars Vitraria*. Examination, moreover, of William Jensen's breakdown of the tests for which the blowpipe was actually used in the eighteenth and nineteenth centuries will reveal that reduction of calces was only part of the picture. One finds, in addition to reduction, color of flame, thermal decomposition and oxidation product, and a host of other indicators.[20] These correspond rather closely, in some instances, to the revelations of the glowing coal.

A reasonable supposition is that the *Ars Alchemie* and related texts provided an environment that could contribute to the development of the blowpipe in the context of analysis. At the least, the glowing coal provided an early form of testing that has been overlooked by historians of chemistry and deserves further examination.[21]

THE BALANCE AND ASSAYING IN ALCHEMY

A widespread belief holds that alchemy was fundamentally nonquantitative and that its practitioners did not make much use of the balance. This viewpoint finds succinct expression in a recent article by Anders Lundgren:[22]

> The chemical (as opposed to the hydrostatic) balance does not appear in illustrations of laboratories in the 17th and 18th centuries. Chemists did not use it in their daily work. Only in commercial mining, which typically involved amounts of material far larger than anything of interest to the chemist or the assayer, was the balance at home.

This statement contains three claims: first, that chemical illustrations of the seventeenth and eighteenth centuries are devoid of the chemical balance; second, that chemists of the period made little use of it; and third, that the only place where the balance was widely used was in the mining business. In fact, however, all three claims are subject to substantial correction.

First, the *Theatrum Chemicum Britannicum* of Elias Ashmole, F.R.S., printed in 1652, contains a detailed picture of an alchemical laboratory with an alembic, a pelican, and an analytical balance (figure 2.1). The printed illustration is a more-or-less verbatim copy taken from the manuscripts of a fifteenth-century alchemist, Thomas Norton (?1433–1513 or 1514). At least one fifteenth-century codex of this work survives with the balance illustration: British Library MS Additional 10302, an important presentation copy prepared in the 1480s or 1490s (figure 2.2 and color plate). Because MS Additional 10302 was copied during Norton's lifetime (and probably under his direct care), the manuscript illuminations command more authority than is common with alchemical illustrations.[23] Norton's balance illumination is sometimes said to be the oldest illustration of a balance in a case, and we have seen no evidence to contradict the same.[24] Beneath the Norton balance, one sees a trunk containing what appear to be concave conic sections. Here the virtues of expensive color photography become rapidly apparent, for what may appear rather like the calibrated weights that accompany the balance in black and white copies of this illustration are clearly chemical vessels. The blue and red contents may indicate that they are crucibles rather than cupels, for one would expect only the yellow of litharge or the color of the metal being tested if the vessels were the latter. The curious upper object in the middle of the trunk is very likely the "monk" or plunger that was to be driven into clay to make a crucible (figure 2.3). On the table, the master alchemist has before him a piece of silver (represented by the conventional crescent moon), a ball (perhaps of gold), and a gold-colored vessel. We suggest that the latter is a cupel that appears to be emitting the spume of litharge mentioned already by Paul of Taranto.

The appearance of an enclosed analytical balance in the Norton illustration raises an interesting question about the accuracy of the instrument. Appropriate here is a consideration of an *Ordonnance* of Philip VI of France, represented as having been issued in 1343:[25]

> The general or particular assayer must have good, light balances, faithful and exact, which do not decline to either side. . . . When one carries out the assay, it must be done in a place where there is neither wind nor cold, and one must see that his breath does not affect the balance.

Figure 2.1
Printed illustration of Thomas Norton's balance from Elias Ashmole, *Theatrum Chemicum Britannicum* (London: Nathaniel Brooke, 1652), 51.

Figure 2.2
Illumination from Thomas Norton's *Ordinal of Alchemy*, British Library MS. Add.
10302, f. 37v (late fifteenth century).

Plate

Illumination from Thomas Norton's *Ordinal of Alchemy*, British Library MS. Add. 10302, f. 37v (late fifteenth century).

Figure 2.3
M and *L* are the "monk" and "nun," the two parts of a crucible mold. From Lazarus Ercker, *Treatise on Ores and Assaying*, Anneliese G. Sisco and Cyril S. Smith, transl. (Chicago: University of Chicago Press, 1951), figure 4.

From this, one can see, even in the first half of the fourteenth century, the recognition that a precise balance must be relatively light (i.e., have a beam of low mass).[26] Although we cannot give an absolute measure either to the weight of the balance or its precision, the fact that its accuracy could be affected by one's breath suggests a fair degree of precision, as Szabadváry has noted.[27] One can assume that Norton's encased alchemical balance was of an accuracy similar to that in the cited *Ordonnance*.

One may now object that an appeal to assaying literature for descriptions of precise balances has no bearing on alchemy as such. This opinion is reflected in Lundgren's statement that chemists before the eighteenth century did not employ the balance in their daily work. Although one could draw forth countless alchemical recipes that call for absolute measurements of reagents, a more relevant consideration for the moment is the role of assaying in alchemy. As mentioned, even contemporary historians tend to state as a bald fact that assaying was part of a distinct mining and minting tradition that had nothing to do with alchemy.[28] Some sources, however, are usually overlooked.

One of the most influential alchemical works available to the Latins of the High Middle Ages was the *De Anima in Arte Alkimiae*, attributed falsely to the Persian philosopher Avicenna (d. 1037).[29] The *De Anima*, although actually a forgery by an anonymous Arabic author, was translated at an early date into Latin, where it achieved immediate success, becoming, for example, the main alchemical source of Roger Bacon.[30] Pseudo-Avicenna's *De Anima* is devoted largely to the decomposition of animal, vegetable, and mineral substances by means of fractional distillation. In a way that is fairly similar to the iatrochemistry of the sixteenth and seventeenth centuries, the *De Anima* provides extensive recipes for separating the four elements from such diverse products as apples, milk, and cheese. In the case of milk, for example, the author says that when it is distilled, a clear, tearlike substance passes over first: This is the element of water. When the distillation continues, a yellow water follows, and a black, burnt substance is left in the flask. The yellow water corresponds to air, the black residue to earth, and the smokey vapors that appear during the process to fire. Hence, all four elements are said to have been separated from the initial milk.[31]

Given the analytical orientation of the *De Anima*, perhaps one should not be surprised that it relates a full complement of assaying tests for the precious metals. Seven of these are given for gold: They include dissolution in "salts," which is a sign of artificial gold; use of the touchstone; weight (gold that is heavier or lighter in specie than normal gold is a fake); loss of its color when fired; ability to sublime; glowing or boiling on fusion; and taste.[32] In all of these, the goal is to discern natural gold from its artificial imitations, so as to measure the success of the alchemist.

Although several of the *De Anima*'s assaying tests are rather unorthodox, it is not at all unusual to find the more mainstream processes of cupellation and cementation given prominent descriptions in alchemical texts. We have already encountered a plenary rendition of cupellation in the *Theorica et Practica* of Paul of Taranto. As Paul points out in another passage, this process was capable of separating the noble metals from the base but could not be used to isolate gold from silver.[33] Before the advent of the mineral acids, it was necessary to employ another dry process—cementation—to remove the silver from gold. This was done by taking leaves of the silver-gold alloy and placing them in layers within a crucible, separated by strata of so-called cement, often a mixture of brick dust and salt. The crucible would then be sealed shut and heated to a high temperature (but beneath that of the melting point of the alloy). The chloride that would be liberated would attack the silver, while leaving the gold unscathed. One important work of the early thirteenth century, the *De Perfecto Magisterio* (misattributed to either Aristotle or Rāzī) gives a clear description of this process:[34]

> Separation of gold. Make leaves of it as thick as your fingernail, and cement them with this powder. Take two parts of common salt, one part of old brick found on the banks of rivers, or on the seashore, which is better, grind well and sieve through a thick cloth. Then leave the leaves thus cemented on a tripod in the middle of an athanor for a day and a night, filling the athanor, that is, the furnace, with live coals. When they are diminished, add more. When the fire is out, open the cool crucible and you will find the gold very well separated.

As in the case of the *De Anima*, the goal of the *De Perfecto Magisterio*'s cementations and cupellations may well have included the testing of alchemically produced gold and silver. But it is important to note that other alchemical authors had another use for such assaying tests. The *Summa Perfectionis* of pseudo-Geber, written around the end of the thirteenth century (probably by Paul of Taranto himself), was arguably the most influential alchemical text of the Middle Ages. As Szabadváry has noted, the *Summa* is keenly concerned with the specific weights of the different metals, although giving them only in relative terms.[35] At the end of the text, the *Summa* describes a battery of assaying tests that include cupellation and cementation along with a number of others. The author distinguishes between two types of examination. First, he refers to "manifest tests ... which are known to all." These include "the practices of determining weight, color, and extension by the hammer."[36]

Because such tests are commonplace, the *Summa* does not describe them as such, though they underlie much of the practice behind the text. Geber then launches into a discussion of assaying tests that employ methods more properly alchemical. These include cupellation, and cementation but also firing or "ignition" to see whether a metal incandesces before fusion; inspection of the fused metal after cooling to see whether it has blackened; exposure to acid vapors with subsequent inspection of any efflorescence; extinction of the hot metal in salty or aluminous water or sulfur to produce color changes; burning with sulfur again to induce color changes or alterations of weight; the repetition of calcination and reduction for changes in color, weight, or volume; and the easy or difficult amalgamation with quicksilver.[37]

Now, the reader may well question why the *Summa* spends so much time on these multifarious tests when cupellation and cementation would have sufficed to determine the genuineness of alchemically produced silver and gold. The answer is, in part, that the *Summa* uses these tests not only for the goal of assaying but rather to determine the fundamental constituents of the metals. Unlike the *Ars Alchemie* or the *De Perfecto Magisterio*, the *Summa*'s practice does not consist merely of a collection of indicators; instead, the author has molded these tests into experimental tools for revealing the nature of matter.[38]

TESTING AND EXPERIMENT IN THE *SUMMA PERFECTIONIS*

The *Summa*'s experimental use of assaying tests to determine the nature of the metals is evident already in the definitions that Geber gives to the metals. Tin, for example, is "a metallic body, white but not purely, slightly bluish, little participating in earthiness, sounding a small creak, soft, possessing in its root a rapid liquefaction without firing, not waiting through cupellation and cementation."[39]

Here we see that failure to withstand cupellation and cementation is viewed as a defining characteristic of tin, along with such manifest properties as its impure whiteness or the creak that it makes when bent. In addition, Geber refers to the test of firing, pointing out that tin cannot be made to incandesce before fusion. In another passage, the *Summa* expands considerably on these observations. Geber claims that tin is composed of a fixed (i.e., nonvolatile), white sulfur, an unfixed (i.e., volatile), white sulfur, a fixed quicksilver, and an unfixed quicksilver. He has already demonstrated that sulfur can be fixed by calcination (intense heating, which drives off the volatile components) and mercury by repeated sublimation.[40] He wants to prove, then, that tin is composed of these four ingredients:[41]

> You will find the evidence of these if you calcine tin, since you will smell the stench of sulfur going forth from it, which is a sign of unfixed sulfur.... One sulfur is made known to reason by the first test [*experientiam*]. The other is proven by the persistence of it in its calx which it has upon the fire, since a more fixed sulfureity does not stink. But a two-fold substance of quicksilver is also proven to be in tin, one of which is not fixed. This is because it creaks before its calcination, but after its double calcination, it does not creak, which is due to the fact that the substance of the fugitive quicksilver causing the creak has escaped. But that the substance of fugitive quicksilver adduces the creak is proven by the washing of lead with quicksilver. Since if you melt lead with quicksilver after its washing with the same, and the fire does not exceed that of its fusion, part of the quicksilver will remain with it, which will adduce a creak from the lead, and convert it to tin.

This elaborate series of experiments consists first of repeated calcinations of tin. The author says that the metal emits a smell of sulfur only in the initial calcination, revealing the presence of unfixed sulfur. The tin also loses its well-known creak after being calcined twice, showing that a substance has escaped, which the author will demonstrate to have been unfixed mercury. He knows this because lead, which he has independently determined to contain less quicksilver than tin if it is mixed with quicksilver, gains the very creak that the calcined tin has lost. This is quite an interesting example of Geber's experimental procedure. The initial loss of the tin's creak upon calcination told him that it must have been due to the escape of a volatile component. How does he know that this was unfixed quicksilver rather than unfixed sulfur? Because he can induce the very phenomenon of the creak by adding normal, volatile quicksilver to a metal that he believes to lack it: lead.

As we can see, Geber's determination of the constituents of tin is based on repeated calcination of that metal, again one of the assaying tests prescribed at the end of the *Summa*. However, the author is not satisfied with having demonstrated the mere presence of fixed and unfixed principles in tin; he wants to determine the relative quantity of mercury and sulfur. For this, he must employ yet another of his assaying tests, the ability of the metal to amalgamate with quicksilver:[42]

> [I]n it [tin] is equality of fixation of the two components quicksilver and sulfur, but not equality of quantity, since quicksilver predominates in their mixture, the sign of which is the easy penetration of quicksilver in its own nature into that. Therefore, if the quicksilver in that were not of greater quantity, it would not—having been taken up in its own nature—have adhered to that easily. For this reason quicksilver does not adhere to mars [i.e. iron] or venus [i.e. copper] except by means of the subtlest craft, due to the paucity of quicksilver in them in their intermixture.

The ready amalgamation of tin with quicksilver allows Geber to conclude that it, like gold, is composed primarily of that principle. Copper and iron, on the other hand, because they amalgamate with difficulty, contain less quicksilver than does tin. One could go on to relate other tests of this sort, for Geber determines the principles of all the metals by employing the same assaying techniques. He even goes so far as to compile comparative lists of the metals' ability to amalgamate and of their components' ability to sublime.[43] In short, the very core of the *Summa Perfectionis* lies in its attempt to arrive at the nature of the metals by exposing them to a variety of tests that are extensions of an age-old tradition of assaying.

From the experimental determination of the metals' components given by the *Summa Perfectionis*, we should be able to see the difficulties in dissociating alchemy from the technology of assaying. Even in the early modern period, alchemy was closely conjoined to the technology of purifying and testing metals. The *Bergwerck* and *Probierbuechlein* tradition of early sixteenth-century Germany illustrates this handily, for there one finds handbooks of mineral technology with such titles as *Rechter Gebrauch d'Alchimei* (1531), *Bergwerck und Probirbuechlin. Fuer die Bergk und Feuerwercker/ Goldschmid/ Alchimisten und Kuenstner* (1533), and *Alchimi und Bergwerck. Wie alle Farben/ Wasser/ Olea/ salia und alumina/ damit mann alle corpora/ spiritus und calces preparirt/ sublimirt/ und fixiert/ gemacht sollen werden* (1534).[44] All these works predate the far more famous *Pirotechnia* of Vannoccio Biringuccio, which was published in Venice in 1540. Any attempt to dissociate alchemy from this tradition would obviously prove fruitless, for even the earliest works in this genre, such as the *Nutzlich Bergbuchley*(n) of Ruelein von Kalbe (1505) openly acknowledged their debt to alchemical sources.[45]

CONCLUSION

I may be accused of having belabored the point in bringing the full weight of the *Summa Perfectionis* to a discussion of Thomas Norton's innocuous inclusion of an analytical balance in his alchemical laboratory. However, given the overriding presuppositions of those who have seen everything from a protofeminist Edenic *Weltanschauung* to simple lunacy in alchemy, what is imperative to stress is that the discipline was above all an attempt to understand and manipulate matter. That this attempt exploited existing tests for the purity of metals and encouraged the development of others for the sake of more general analysis should not really be an object of surprise. Norton's balance was not an anomaly but reflected the fact that determination of weight was one of many means used by alchemists to identify a variety of mineral substances and their components. From an alchemical perspective, it belongs in the same category as Michael Scot's testing for salts by means of a glowing coal.

If one wanted to ascertain the way in which the alchemical use of the balance and the primitive flame test differed from more modern counterparts, perhaps the simplest response would be to say that the alchemists chose not to privilege one over the other. As shown by the definitions of metals given in the *Summa Perfectionis* and many other texts, both weight and combustibility were factors, along with color, malleability, ductility, and point of fusion, that differentiated one metal from another. The difference between the alchemical perspective and that of the eighteenth-century chemical revolution does not lie in the putative fact that the alchemists neglected weight but rather in the genuine observation that they did not give weight the primacy that it acquired in the works of Lavoisier.

NOTES

1. A good example of the pervasive (and misleading) character of this view can be seen in the reasoning that led both Richard Westfall and Betty Jo Teeter Dobbs to misattribute the authorship of the *Clavis* or *Key* by George Starkey to Isaac Newton, who had merely transcribed it. Westfall argued that the *Key*, a detailed recipe for a "philosophical mercury" (see chapter 3 in this book) could not have been

written by an alchemist, as Starkey was, because of the precision of its directions. Westfall's evidence was that the *Key*, "unlike virtually all other alchemical literature ... described laboratory procedures in detailed operative terms that could be repeated today." Speaking of Newton's precise chemical practice in general, Westfall claimed that "alchemy had never known anything like this before. It was indeed more than alchemy could survive."

Dobbs pronounced much the same idea, saying that the *Key* betrayed "the fine-grained quality [of] the instructions for achieving the best results" that one associates with Newton's experimentalism, as opposed to the supposedly rough-and-ready character of alchemical practice [Richard Westfall and Betty Jo Teeter Dobbs, as quoted from William Newman, "Newton's *Clavis* as Starkey's *Key*" *Isis* 78(1987): 564–574; cf. 565]. For a corrective to the view that alchemy and chemistry can be clearly distinguished before the eighteenth century, see William R. Newman and Lawrence Principe, "Alchemy vs. Chemistry: The Etymological Origins of a Historiographic Mistake," *Early Science and Medicine* 3(1998): 32–65.

2. A good example may be found in Marco Beretta, *The Enlightenment of Matter* (Canton, MA: Science History Publications, 1993), 76–78, 134–136, 330–367. The same approach leads Dietlinde Goltz to the absurd conclusion that the medieval Geber, author of perhaps the most influential alchemical treatise of the Latin Middle Ages, was not an alchemist. Goltz argues that "Alchemie ist unabdingbar mit irgendeiner Art von Weltanschauung verknuepft und stellt eine Naturphilosophie dar." When she fails to find the *Weltanschauung* of the alchemists in the highly mineralogical work of Geber, Goltz concludes that he writes in a fashion that is "nicht alchemistisch." See Dietlinde Goltz, "Versuch einer Grenzziehung zwischen 'Chemie' und 'Alchemie'," in *Sudhoffs Archiv* 52(1968): 30–47; cf. 34, 39–40.

3. William B. Jensen, "The Development of Blowpipe Analysis," in John T. Stock and Mary Virginia Orna, eds., *The History and Preservation of Chemical Instrumentation* (Dordrecht: Reidel, 1986), 123–149; cf. 126. See also Charles Singer et al., *A History of Technology*, vol. 1 (Oxford: Clarendon Press, 1954), 578.

4. Ferenc Szabadváry, *History of Analytical Chemistry* (Oxford: Pergamon Press, 1966), 51. See also Singer, *A History of Technology*, vol. 2, 329–332.

5. Szabadváry, *History of Analytical Chemistry*, 51, and Jensen, 127. Compare Hermann Kopp, *Geschichte der Chemie*, vol. 2 (Braunschweig: Friedrich Vieweg and Son, 1844), 44, in which Kopp suggests rather tentatively that Kunckel's *Ars Vitraria* (1679) contains the first "Anwendung in der Chemie" of the blowpipe. In a recent article, Ulrich Burchard argued that Robert Hooke and Erasmus Bartholin used blowpipes for general mineralogical and crystallographical purposes, but he too adduced Kunckel as the first to advise its employment among chemists. See Ulrich Burchard, "Geschichte und Instrumentarium der Loethrohrkunde," in *Deutsches Museum, Wissenschaftliches Jahrbuch*, 1992/93, 7–72 (cf. 9–11).

6. Kopp, *Geschichte*, vol. 2, 42.

7. Robert Halleux, "Méthodes d'essai et d'affinage des alliages aurifères dans l'Antiquité et au Moyen Âge," in *Cahiers Ernest Babelon*, 2(1985): 39–77. I quote

his translation from p. 50: "Manqua le souffleur, manqua le plomb. Le souffleur c'est l'appareil dont on se sert pour souffler et l'homme qui manie l'appareil: on fond du plomb avec l'argent pour que, à la fusion, il monte à la surface et attire à lui l'impurité." The term that Halleux translates as *souffleur* is in Greek *physētēr*. One cannot exclude the possibility, however, that *physētēr* meant "bellows" rather than "blowpipe."

8. William R. Newman, *The Summa Perfectionis and Late Medieval Alchemy*, vol. 4 (doctoral dissertation, Harvard University, Cambridge, MA, 1986), 165; vol. 3, 222–224:

Sumatur cinis optime cribratus et cum salis aqua commixta fiat vas in quo recipi possit ad ignem impiisimum argentum sive quodcunque metallum quod in cineritio ponere et examinare quesieris. Et fuso metallo, eiiciatur ibi plumbi pars sexta; et hoc maxime fit in argento. Et habeatur cannolum vel de ferro vel canna, per quem sufflari possit super faciem fusi metalli. Plumbum super metallum fusum videbitur fumans ex eo—quod a principiis suis male fixis habeat fugam, volatilitatem, et sue deperditionem substantie. Inde est quod in fumum resiliens, omne quod in metallo purgando imperfectum extiterit secum trahet. Tunc enim bene purgatum metallum noscetur cum, addito plumbo in eo, non fumari videbitur, sed ebulliri et quasi spumams evolantes eiicere: et tunc plumbum ulterius non addatur.

9. Julius Ruska, *Das Buch der Alaune und Salze: Ein Grundwerk der spaetlateinischen Alchemie* (Berlin: Verlag Chemie, 1935).

10. Ruska, *Das Buch*, 80–83.

11. S. Harrison Thomson, "The Text of Michael Scot's *Ars Alchemie*," *Osiris* 5(1938): 523–559; cf. 535:

Si vis ipsum perfecte cognoscere retine ipsum in ere aliquantulum et cum aliquantulo spita mitte supra prunam, et si de pruna saliet et aquam superius dimiserit et bullierit, eum in ere iterum mitte … si supra prunam diu steterit et de pruna non salierit nec stridorem fecerit erit bonum sine dubio.

See also Charles H. Haskins, "The 'Alchemy' Ascribed to Michael Scot," *Isis* 10(1928): 350–359.

12. Thomson, "Text," 535: "Sal nitrum foliatum est aliquantulum longum et grossum, non nimis salsum sed quasi saporem aceti trahit quando ipsum cum lingua tetigeris, et flammam super ignem facit."

13. See J. R. Partington, *A History of Greek Fire and Gunpowder* (Cambridge, UK: W. Heiffer & Sons, 1960), 87–89, in which Partington discussed the *Ars Alchemie*. Seemingly, Partington rather missed the point when he said that the flame test was employed primarily to distinguish pure saltpeter from its adulteration with common salt. The main goal of the test was clearly to distinguish soda from saltpeter. See also Robert Multhauf, *The Origins of Chemistry* (London: Oldbourne, 1966), 33; and Marcelin Berthelot, *La Chimie au Moyen Âge*, vol. 1 (Paris, 1893; reprinted in 1967, Osnabrueck: Otto Zeller), 98.

14. Joseph Needham, *Science and Civilization in China*, vol. 5 (Cambridge, UK: Cambridge University Press, 1980), 179–194; Multhauf, *Origins*, 17; and Partington, *History of Greek Fire*, 298–309.

15. For the history of saltpeter, see Partington, *History of Greek Fire*, 298–339; and Multhauf, *Origins*, 328–332.

16. J. Wood Brown, *The Life and Legend of Michael Scot* (Edinburgh: David Douglas, 1897), 247: "Una species est salis nitri que apportatur de Alexandria et ille est vere sal nitrum cum illo lavant mulieres sarracenorum pannos lineos et faciunt eos albissimos ut nix, lavant etiam facies earum et corpora sua in balneis."

17. Potassium nitrate, sodium nitrate, and calcium nitrate are formed naturally by the action of bacteria found in manure. Some believe that these produce nitric acid as an intermediate product, which then acts on whatever bases are available (e.g., calcium oxide or carbonate, sodium carbonate, or potassium carbonate), thus yielding the three common types of "niter." The niter tends to effloresce out of soil or basement walls in areas where animals have deposited large quantities of manure. In the late Middle Ages, it was purified either by recrystallization or by solution with potash. The latter method would result in the production of potassium nitrate even when the initial ingredient was the nitrate of calcium, as is common. See George P. Merrill, *The Non-Metallic Minerals* (New York: John Wiley and Sons, 1904), 306–312; Ernest R. Lilley, *Economic Geology of Mineral Deposits* (New York: Holt, 1936), 679–685; and Sydney and Margery Johnstone, *Minerals for the Chemical and Allied Industries* (New York: Wiley, 1961), 429–432. For methods of purification, see Partington, *History of Greek Fire*, 314–339.

18. Thomson, "Text," 536: "Si vis ipsum cognoscere mitte supra prunam et statim comburetur et faciet cinerem lucidum et videbitur terra et non stridebit."

19. Thomson, "Text," 536: "Si ponatur supra prunas et suffles inflammabitur et saliet de prunis. . . ."

20. Jensen, "The Development of Blowpipe Analysis," 124.

21. The test with burning coals appears also in the *Breve Breviarium* of pseudo–Roger Bacon, written circa the end of the thirteenth century. The author of the *Breve Breviarium* interestingly refers to his niter as *sal petrae*. See *Sanioris Medicinae Magistri D. Rogeris Baconi* (Frankfurt: Johann Schoenwetter, 1603), 251: "Talis autem naturae est, quod si immediate ignitos carbones tangat, statim accensum cum impetu evolat."

22. Anders Lundgren, "The Changing Role of Numbers in 18th-Century Chemistry," in Tore Frangsmyr et al., eds., *The Quantifying Spirit in the 18th Century* (Berkeley: University of California Press, 1990), 245–266; cf. 1–2.

23. John Reidy, ed., *Thomas Norton's Ordinal of Alchemy* (London: Oxford University Press, 1975), x–xiv. Reidy was able to date the manuscript with considerable precision on the basis of the clothing worn by the alchemists in Norton's illustrations.

24. John T. Stock, *Development of the Chemical Balance* (London: Her Majesty's Stationery Office, 1969), 2.

25. Jean Boizard, *Traite des Monoyes* (Paris: La Veuve de Jean Baptiste Coignard, 1692), 166–167: "*Le General Essayeur, ou l'Essayeur particulier doit avoir ses balances bonnes & legieres, loyaux & justes, qui ne jaugent d'un coste ne d'autre. . . . Quand on poise les essays, il doit estre en lieu, ou il n'y ait vent ne froidure, & garder que son halaigne ne charge la balance.*"

26. Stock, *Development*, 7.

27. Szabadváry, *History of Analytical Chemistry*, 17: "The balances employed for these tests must have been quite sensitive for breathing to have any effect on them."

28. Beretta, *Enlightenment*, 74–93. However, see the much-needed correction to this view by Robert Halleux, "L'alchimiste et L'essayeur," in Christoph Meinel, ed., *Die Alchemie in der europaeischen Kultur-und Wissenschaftsgeschichte* (Wiesbaden: Herzog August Bibliothek in Kommission bei Otto Harrasowitz, 1986), 277–292.

29. Julius Ruska, "Die Alchemie des Avicenna." *Isis* 21(1934): 14–51.

30. William Newman, "The Philosophers' Egg: Theory and Practice in the Alchemy of Roger Bacon." *Micrologus* 3(1995): 75–101.

31. ARTIS CHEMICAE PRINCIPES, AVICENNA ATQUE GEBER, HOC Volumine Continentur. *QUORUM ALTER NUNQUAM hactenus in lucem prodijt: alter vero vestutis exemplaribus collatus, atque elegantioribus & pluribus figuris quam antehac illustratus, doctrinae huius professoribus, hac nostra editione tum iucundior tum utilior evasit* Adiecto Indice rerum & verborum copioso. *Cum gratia & privilegio.* BASILEAE PER PETRUM PERNAM M. D. LXXII., 14–15: The first product is

clarius lacryma: & illud est aqua, post destillabis inde alia non tam clara, & vertitur in colore citrino, & illud est aer, postea remanebit terra combusta in fundo nigra. Modo habes aquam, aerem & terram: ignis evanuit in vaporibus & efficitur fumus, & fumus versus est in aquam. Et intellectum est, quia omnia quae sunt in mundo de vitali & herbali sunt composita de istis quatuor elementis, & ad illa se solvunt.

32. Avicenna, *De anima*, in *Artis chemicae principes*, 125–126:

Modo dicam tibi, ut cognoscas aurum cuiusmodi sit. Primum in solutione, secundum in lapide, tertium in pondere, quartum in ore, ut gustes, quintum in igne, sextum in sublimatione, septimum in fusione. Et in unoquoque eorum est magisterium temtandi. Si vis scire cuius naturae sit aurum, funde cum salibus, & si solvetur, est de magisterio: Et si vis eum tentare in lapide si est de quinto, aut de sexto, aut de octavo, aut de medietate, aut de tertio: aut de quarto, secundum quod est judicabis. Et si vis temtare ad pondus, vide si est leve aut ponderatum magis alio: & si est, est de lapide, si sustineat omnes alias tentationes. Si vis tentare ad ignem, iacta in ignem. Si permanebit in colore suo, est aurum: sin autem, est falsum aurum. Et tenta in sublimatione pulverizatum cum aliis speciebus: Si sublimatur, ita quod ante non fiat calx, nec lavetur. Et si facias eum pulverem, &

proijcias in aludel, & sublimetur: scies quod est de nostro auro. Et in fusione est magna scientia: quia quando fundis debes videre si ferveat, aut si non ferveat. Si ferveat, est de nostro lapide, & in gustu potes cognoscere salsum, & extrahi.

33. Newman, *The Summa Perfectionis and Late Medieval Alchemy*, vol. 4, 225.

34. Pseudo-Aristotle, *"De perfecto magisterio,"* in *Bibliotheca Chemica Curiosa*, J. J. Manget, ed., vol. 1 (Geneva: Chouet, de Tournes, 1702), 644:

Auri separatio. Fac de eo laminas ad modum tuae unguis, & eas cementa cum hoc pulvere: Recipe salis communis separati partes duas, lateris antiqui in ripis fluviorum, vel in littore maris reperti, quod melius est, partem unam, tere optime & cribretur per setaceum spissum: tunc laminas dictas sic cementatas fac morari in medio athnor super tripodem per diem & noctem unam, & implendo athanor, id est furnellum carbonibus vivis, & quum minuuntur addendo semper de aliis, tunc igne remoto & infrigidato aperi crucibulum, & invenies aurum optime separatum.

35. Szabadváry, *History*, 15.

36. William R. Newman, *The Summa Perfectionis of Pseudo-Geber* (Leiden: Brill, 1991), 769.

37. Newman, *The Summa Perfectionis of Pseudo-Geber*, 769–783.

38. See Halleux, "L'alchimiste et L'essayeur," 289–290.

39. Newman, *The Summa Perfectionis of Pseudo-Geber*, 675.

40. Newman, *The Summa Perfectionis of Pseudo-Geber*, 666, 728. That the "fixed sulfur" would not really be sulfur at all but rather an impure residue is revealed by the fact that Geber says that its calcination results in a loss of 97% of its substance (666). "Fixed mercury" on the other hand refers either to mercury oxide produced by repeated sublimation of quicksilver alone (714) or to mercury chloride manufactured by sublimation of mercury with salt (691–693).

41. Newman, *The Summa Perfectionis of Pseudo-Geber*, 733.

42. *Ibid.*, 734.

43. *Ibid.*, 656, 722.

44. Paul Walden, "Mass, Zahl und Gewicht in der Chemie der Vergangenheit," in *Sammlung chemischer und chemisch-technischer Vortraege Begruendet von F. B. Ahrens, Herausgegeben von Professor Dr. H. Grossmann-Berlin, Neue Folge Heft 8* (Stuttgart: Ferdinand Enke, 1931), 3. For a short overview of the *Bergwerck* and *Probierbuechlein* tradition, see Ernst Darmstaedter, "Berg-, Probir- und Kunstbuechlein," *Muenchener Beitraege zur Geschichte und Literatur der Naturwissenschaften und Medizin* 2/3(1926): 101–206.

45. Darmstaedter, "Kunstbuechlein," 118–119.

3

Apparatus and Reproducibility in Alchemy
Lawrence M. Principe

Alchemy has long proved a difficult topic for historians to understand. Since the disappearance of its last serious followers in Germany in the latter half of the eighteenth century, alchemy has been subject to numerous schools of interpretation, all of which have endeavored to describe its real meaning and significance. One element common to several of these diverse attempts to explain alchemy as a historical phenomenon is a tendency to distance alchemy from the realm of laboratory science and, in the most extreme form, to remove it altogether from any meaningful contact with chemical materials. Two such interpretations have played—and continue to play—major roles in informing views of serious historians of science in regard to alchemy. These two interpretations are the Jungian or psychological school and what might be called the *discontinuity view*.

The Jungians, following the formulations of Carl G. Jung and Herbert Silberer, claim that alchemical texts do not primarily describe actual transformations of chemical substances but instead record the projection of the unconscious of the alchemical worker: Alchemical writings are first and foremost descriptions of psychic or psychological events occurring invisibly within. The discontinuity school, on the other hand, recognizes in a rather presentist or positivist fashion a few piecemeal contributions of the alchemists to the rise of modern chemistry—the discovery of mineral acids and better assaying or distilling techniques, for example—but reject the majority of alchemical operations with such nebulous words as *mysticism*, and fragment the totality of alchemy by largely or wholly neglecting its theoretical or explanatory aspects. The dismissive attitude of such writers mimics the sentiments of Enlightenment authors who tried to emphasize a clean break from the past by vilifying alchemy and portraying it as something distinct from chemistry. This tendency has several motivations, the chief of which is to quarantine the difficult-to-comprehend or even unsavory parts of alchemy from the origins of modern chemistry. This latter position, and its related presentism, was prevalent during the early development of the professional history of science and still persists in some quarters.[1]

I argue here, however, that historical evidence does not support these interpretations and, at the end of this chapter, advance a rather unorthodox historiographical methodology that powerfully opposes the Jungian interpretation of alchemy. Readers tempted to think that I am invoking past historiographies as straw men for the target of the present study need only consult several recent publications that continue to espouse such views of alchemy. Jungianism is still alive and well despite what would seem to be fatal blows against Jung and his "cult" and is so pervasive that some authors may not even be aware of their allegiance to it.[2] The late B. J. T. Dobbs began her first book on Newton's alchemy with a fairly detailed exposition of the Jungian position, and her work contains numerous reliances on it.[3] She did not repudiate such views in her more recent *Janus Faces of Genius*, which, in light of the importance of her studies, was an excellent opportunity missed.[4] According to a census of works on alchemy that I recently carried out, more than one-third of all publications on alchemy since 1970 are explicitly or implicitly based on Jungian interpretations. Popularizing works on alchemy seldom fail to espouse a Jungian position.

The discontinuity view is at least as prevalent, being perhaps more widely acceptable to historians, and undergirds the very common and casual use of alchemy as an uncomplicated foil against which to set off modern science. One recent example is Steven Shapin and Simon Schaffer's *Leviathan and the Air-Pump*, wherein "alchemical secretism" and purported lack of verifiability are contrasted with the program of the Royal Society, overlooking the fact that more than a few of the early (and even founding) members of the Royal Society were themselves "alchemical secretists" (most prominently Robert Boyle but also Elias Ashmole and Edmund Dickinson), a topic I discuss at length elsewhere.[5]

The tendency of both these schools of interpretation to distance alchemy from laboratory science is perpetuated in a significant array of studies. The noted scholar of the occult and Renaissance studies, Brian Vickers, has suggested that alchemy was primarily a *"verbale Kunst"* and that the sphere of its action is predominantly philological and textual rather than experimental.[6] A much bolder view was propounded by Marco Beretta in his *Enlightenment of Matter*.[7] Beretta, still strongly addicted to the notion of universal revolutionism, assailed what he calls the *myth of continuity*: the presence of any continuity between alchemy and chemistry. He advocated the total separation of alchemical thought from scientific thought on what amounts to Jungian grounds and criticizes "modern historians of alchemy" for having "failed to provide a correct historical picture of the conceptual structure of alchemy as a whole."[7] Indeed, Beretta declared that "if we want to avoid regarding this obsessive insistence on the transmutation of metals into gold as a weakness, or worse, as a pathological state of mind, it would be better to stop regarding alchemy as a scientific or experimental endeavor. . . ."[8] Beretta's argument, were it more clearly laid out, seemed to be this: The transmutation of metals is the "fundamental goal" of alchemy, and transmutation is a physical impossibility (as we now know), yet the alchemists did not seem to notice that it was impossible; therefore, they were either obsessively blind or pathological or simply made no experiments, observations, or tests. Had they done so, they would have realized their error, and thus they have no commonality at all with what we call *science*.

Thus, the dismissal of alchemy from the realm of experimentalism and natural philosophy is clearly not yet a thing of the past. The key in opposing such views as those expressed here is to demonstrate that alchemists not only dealt directly on a daily basis with material substances and their transformations but did so circumspectly and in a way that involved the interplay between hand and mind, practice and theory. Though many means of approaching this issue are possible, I here explore the related topics of apparatus and reproducibility.

If the Jungians are right in attributing alchemical imagery and apparent results to psychic states rather than to chemical laws, and if such discontinuists as Beretta are right in claiming that alchemy was not an experimental endeavor, we should find that alchemical operations were not—and are not—reproducible. Further, we should also find that alchemists were unconcerned about, or at least uninterested in, the use and development of apparatus designed with specific, practical laboratory goals in mind. However, if on examination we find that alchemists actively and rationally attempted (and perhaps accomplished) the replication of their predecessors' processes and find further that these alchemical processes continue to be recreatable in a modern chemical laboratory, the psychological school must be wrong in attributing a nonexperimental origin to alchemical texts, and the position of those who would dismiss alchemy as legitimate and organized experimentalism must be severely undermined.

To forestall the criticism that such a study would pick and choose at nuggets of recognizable "chemical technology"—assaying, production of chemical products, and so forth—and would ignore the context and a "fundamental goal" of alchemy, let us be specific in restricting ourselves for this study to the subset of alchemy known as *chrysopoeia*, the making of gold, in this case specifically by means of the Philosophers' Stone. The now-recognized impossibility of the end product of this important subset of alchemy has too often urged a precipitate rejection of the entire endeavor as imaginary or deceptive. *Prima facie*, however, the quest for the Philosophers' Stone must be intimately tied to questions of reproducibility, for the alchemists all agreed that the "ancient sages" possessed the Stone, and consequently they endeavored to extract its recipe from their writings fully expecting the classical results to be replicable by them.

By the seventeenth century, several well-defined but currently little-recognized schools of alchemical thought had developed, crystallizing around different choices of starting material and propagating along lines of descent through different authoritative authors.[9] However, such general arguments, like the countless alchemical concordances that endeavor to extract alchemical truths or practice from a painstaking compilation of numerous authorities, might be dismissed as merely literary phenomena. Though such rather diffuse evidence manifests a general belief in reproducibility, we should look instead at concrete and specific instances of chrysopoetic concerns over reproducibility in action.

An appropriate first step is to define what we actually mean by *reproducibility*. The simplest answer is that a process or event is able to be recreated and made to produce identical results repeatedly in the hands of either the same or different operators. Now, a potential problem arises in terms of chrysopoetic texts: they are written with a view toward concealment. The noble art of chrysopoeia had to be kept secret for several moral and religious—not to mention economic—reasons. Does this insistence on secrecy mean that the processes described in such books are inherently irreproducible or meant to be so? It certainly does not, any more than the argument would apply to closely guarded trade secrets, the very value of which lies in their continued reproducibility.

Reproducibility seems to have two parts. The first part falls on the author, who must express an operation in a fashion sufficiently complete to allow, in the second part, the reader to interpret the text accurately enough to carry out the described operation and to produce the desired product or effect. A study of important alchemists shows clearly that they were routinely concerned with both reproducibility and instrument design, both as authors and as readers, even in the face of the need to preserve strict secrecy about their *arcana maiora*.

For example, in the fifteenth-century *Ordinall of Alchimy*, Thomas Norton devoted one of his verse chapters to "concords" necessary for the Great Work. The third of these concords is that the "Warke accordeth with Instruments" or, less poetically, that apparatus be accommodated to its purpose.[10] Norton, like a man accustomed "to ordeyne Instruments according to the Werke," recites the differing length and shapes of vessels for circulation, precipitation, sublimation, and so forth. He then details the differences in types of clay for earthenware vessels and the types of ashes and frits for making differing qualities of glass; he also describes various contrivances for furnaces, of which he provides illustrations.[11] Norton is clear about why he has treated this topic: he wants to provide readers with rules so that they may "by this Doctrine chuse or refuse" various types of glass and earthen vessels with which they might be presented.[12] Norton apparently knew how to choose good apparatus by virtue of his experience of having worked experimentally with various sorts of materials.

An example combining both right apparatus and right practice comes from the corpus published under the name of *Basil Valentine*. This corpus is extremely heterogeneous, and questions of authorship, now debated for more than three centuries, are still far from resolved.[13] Without distracting ourselves with more contentious issues, a safe and logical approach is to divide this collection into three groups.

The first, corpus A, is composed of works published from 1599 to 1604 by Johann Thoelde, a salt boiler and Paracelsian of Hessen, who is often cited as their true author. This group includes the chrysopoetic *Von dem Grossen Stein der Uhralten* (1599) and the famous *Triumph-Wagen der antimonii* (1604). Corpus B is composed of works published in the 1620s, shortly after Thoelde's death, including the *Letztes Testament*. Corpus C encompasses later works yet more distantly attributed.

A short tract belonging to corpus B is entitled *Offenbahrung der verborgenen Handgriffe* or *Revelation of the Hidden Manipulations*; here the title with the word *Handgriffe* promises to tell us something of manual alchemical laboratory work.[14] This work, published under the name of Basil Valentine, initially was considered "genuine" (i.e., it flowed from

the same pen as that of the earlier works of corpus A, which it pro-
ported to unravel). Yet, more discriminating readers eventually noticed
the very different style of the *Offenbahrung* and decided that the author
was certainly not the same as that (or those) of corpus A. This observation was made as early as 1727 by the alchemical collector and publisher Friedrich Roth-Scholtz.[15] What this short tract actually seems to
represent is a record of the attempts of an unknown alchemist to deci-
pher and to reproduce the processes obscurely delivered in the corpus
A *Stein der Uhralten*, the first book to be published under the name of
Basil Valentine. *Stein der Uhralten*, better known as *Zwölff Schlüssel*
or *Twelve Keys*, presents an allegorical description of how to prepare
the Philosophers' Stone. These metaphorical descriptions were given
illustrative form in the 1602 second edition of the work that was
adorned with emblematic illustrations of the type that Jung interprets as
dream-symbolism and that Beretta scorns.[16] Thus, the *Offenbahrung*, as
an attempt to follow the earlier *Stein der Uhralten*, provides a showcase
example of the issues of replicability, experimentalism, and apparatus
design in chrysopoetic alchemy.

In the second key of *Stein der Uhralten*, "Valentine A" directed
allegorically that a precious water be made "cleverly and most carefully" from two fighters; then, switching metaphors in midstream, he
also called them an eagle "who makes his nest alone in the Alps" and an
"old dragon who has long had his habitation among the rocks."[17] In the
engraving that accompanied the second edition, these two allegories
are combined, with the eagle alighting on the sword of one fighter and
the dragon (portrayed as a snake) coiling itself around the sword of the
other (figure 3.1). The author of the *Offenbahrung*, under the heading
"meaning of the second key," began perfectly candidly with the
plaintext direction to "take saltpeter and sal ammoniac, equal parts . . . ,"
decoding Valentine's allegory into simple chemical terms.[18] The Alpine
eagle refers to the volatile sal ammoniac and the old dragon of the
rocks to potassium nitrate, which forms as an efflorescence on rock
walls, particularly near places where animals have lived. As for the
unspecified "clever and careful" means of extracting a water from these
dry salts, the anonymous author designed a special apparatus for the
operation. He prescribed a special earthenware retort with a tube projecting into its body and helpfully provided a cut-away illustration.
Indeed, this source seems, in fact, to be the first mention of a tubulated
retort in the chemical literature.[19] "Valentine B" then described in
detail how the retort is to be heated, how a glass receiver "as big as one

Figure 3.1
The second key of Basil Valentine, from *Practica cum duodecim clavibus*, pp. 377–431
in *Musaeum hermeticum* (Frankfurt, 1678), 396.

can get" is to be luted on and immersed in a basin of cold water, how
the salt mixture is then to be thrown into the red-hot retort through
the specially adapted tube and, finally, how an acidic liquor—the precious water of the two fighters—distills over.

In the third key, Valentine A spoke somewhat obscurely of casting a corpse into the salty sea and of a rooster who devours a fox so as
to prepare "the rose of our masters and the red dragon's blood."[20] The
Offenbahrung author instead calmly proceeded to the "use of the mineral water according to the third key" and prescribed the dissolution of
one part of gold in three parts of the acid distilled from the saltpeter
and sal ammoniac.[21] This solution then is distilled nearly to dryness,
fresh acid is poured on and distilled off and, finally, the gold is distilled
over into the receiver. Here, the author is again specific about the
proper apparatus needed for correct experimental reproduction:
"[N]ote that this distillation must be done in a short flask with a flat
bottom." Indeed, the process is very delicate and tricky, as the thermally sensitive gold salt is far more easily decomposed by the heat than
it is sublimed successfully.[22] The author thus prescribed a flat-bottomed

Kolbe—an unusual shape, to be sure—that would serve to keep the gold residue evenly and thinly distributed over a wide surface area, thus facilitating sublimation, and of a short height, which would allow sublimation at the lowest possible temperature, thus minimizing decomposition.

In other operations, this second author continued to describe specific pieces of equipment designed to facilitate the success of the difficult operations of *Stein der Uhralten*. In one place, he specified the use of an "earthenware bottle from Waldenburg" and elsewhere designed a special apparatus of copper and glass intended to produce Valentine A's "feurige Weingeist" (or "fiery spirit of wine") by condensing the "fiery mercury" released when spirit of wine is set alight.[23] Clearly, these are the notes of a man who had deciphered Valentine and had reduced his allegories to practice after considerable laboratory trials.

Unfortunately, after the sixth key of *Stein der Uhralten*, the halfway point to the Stone, the instructions of *Offenbahrung* become considerably less detailed; no clear decipherment of keys 7 through 10 is offered. The reader is left with the instructions to seal up all the prepared matters hermetically and to heat gradually for an extended period. The balance of the text recites the standard descriptions of what the maturing Philosophers' Stone *should* look like—common knowledge for any seventeenth-century alchemist. Perhaps our student of Valentine A went as far as he could but reached a point after which he could not proceed successfully. However, this outcome in no way changes the conclusion that he did expect Valentine A's directions to be reproducible; he was not put off by metaphorical disguise but clearly expected the obscure *Stein der Uhralten* to contain the rough, though cryptic, outlines of an experimental practice. Indeed, he was not disappointed until halfway into the scheme, at which point, after so much success, he reasonably believed that the balance of the text contained an equal measure of experimental alchemy but that he was yet unable to reduce it to practice.

Another example comes from Eirenaeus Philalethes, the extraordinarily popular and influential chrysopoetic author and romantic disguise of George Starkey. William Newman's recent study of Starkey-Philalethes has already shown how several successive alchemical practitioners endeavored to decode the often-enigmatic Philalethes tracts, particularly in regard to the meaning of the all-important "doves of Diana" crucial for preparing the Philosophers' Mercury. Some, including Hertodt von Todtenfeldt and Carolus de Maets, were successful in this attempt, correctly identifying the doves as metallic silver, as its true meaning is revealed to be in private documents of Starkey that were unknown to them.[24] Such endeavors in the second half of the seventeenth century nicely corroborate the similar activities we have documented between the "Valentine A" of the *Stein der Uhralten* and the "Valentine B" of the *Offenbahrung*.

In several places in the Philalethes tracts, Starkey showed his own keen appreciation of the problems of replication. For example, consider the crucial topic of applying the correct degree of heat to the vessel in which the Stone is to be formed. Starkey devotes considerable space in most of his treatises to detailed descriptions of how a furnace is to be built to provide a constant source of heat over long periods. It was generally acknowledged that the developing Stone required continuous and constant heating for a year or more and that this in turn necessitated a careful consideration of the principles of furnace construction and management. In one place, he provided precise measurements for the clearances, sizes, and positions of vents; the thickness of insulation; and other details needed for his operations to succeed.[25] Furthermore, in *Ripley Reviv'd* (1678), a commentary on George Ripley's fifteenth-century treatise, *Compound of Alchymie*, Starkey showed an acute understanding of the problems involved in following experimental procedures set down two centuries earlier under different technological conditions:[26]

> [D]o not believe that Philosophers did formerly use our Art of Furnaces, but made them of Brick, or Earth, with Earthen Covers ... this Earthen Cover was not so reflective of heat, as our Iron Covers are.... Therefore, in thy Furnace let thy Cover or Top be luted with good Loam every-where, at the least half an inch thick, so shalt thou be sure not to have too scalding a heat in the concavity of thy Nest, which otherwise thou wouldst have.

These covers of which Philalethes (Starkey) speaks were hemispherical shells perforated with one or more holes through which the long necks of the digesting flasks could protrude. After a flask had been set in a sand or ash bath in the furnace, such a cover would be fitted over it to conserve the heat. But Philalethes warned the practitioner that the iron covers of the seventeenth century heat the flask differently than did the earthen ones of the fifteenth and, thus, would change the results. Ripley had prescribed that the experimenting alchemist judge

the "measure of Firing" by the rule, "[L]et never thy Glass be hotter
then thou mayst feel/And suffer still in thy bare hand to hold": that is,
the operator is ordered to check the temperature of the flask by grab-
bing hold of the protruding neck of the flask. The heat is not to be
increased beyond that allowing him to rest his hand on the neck of the
flask. However, Philalethes here explains that the modern iron covers
—unlike Ripley's clay covers—reflect heat back onto the enclosed
flasks more efficiently, so they will heat the body of the flasks more
strongly, even though the necks remain relatively cool. Thus, the
practitioner using Ripley's rule by feeling the exposed necks will not
stop increasing the heat soon enough and will allow "too scalding a
heat" so that the temperature of the matter in the flask will be too high
to provide the desired results.[27]

By way of summary, first Thomas Norton not only showed a
recognition of the differences in manufacture, quality, and design of
apparatus but gave his readers advice on how to choose good items so
as to profit from their laboratory endeavors. In the case of Valentines
A and B, a later alchemist not only deciphered the allegorical text of
his predecessor but reduced it to workable practice and designed
specialized apparatus for that purpose. Finally, Philalethes showed
an appreciation of how changes in material technology would affect
reproducibility and then suggested steps to obviate the problem.
Clearly these alchemists were neither obsessive nor blind and were
most certainly involved in daily practical experimentation that they
expected to be reproducible.

However, one further factor is key to this chapter's attempted
refutation of the belief in a nonexperimental or only marginally experimental alchemy. The Jungian interpretation has owed much of its
acceptance and continuance to the fact that it advances an explanation
of the origins of the notoriously extravagant imagery of alchemical
texts. The very frequent references to hermaphrodites, flowers, drag-
ons, kings, queens, and a multifarious menagerie of real and mythical
creatures involved in everything from birth and marriage to incest and
death have been a chief locus for arguing the "otherness" of alchemy.
Indeed, one notable author, following Jung's lead, has declared rather
incautiously that "clearly these picturesque symbols have nothing to do
with chemical realities or with rational theories of transmutation."[28]
These figures have thus played a key role in allowing alchemy to be
disjoined from chemistry and seem to countenance the formulation of
the rather far-fetched Jungian notion of "irruptions of the uncon-

scious." Indeed, the promise of providing an explanation of alchemical imagery has probably been the chief preservative of the Jungian interpretation over so many years, even though Jung's system does little more than explain the apparently inexplicable by means of something yet more inexplicable, even though tricked out in pseudoscientific language. Thus, it is crucial for any convincing dismissal of the Jungian interpretation to advance a more plausible origin for these alchemical images.

Consider, then, one often-recurring image in accounts of attempts to make the Philosophers' Stone: Hermes' Tree, the Tree of the Philosophers, or the Tree of the Hesperides. This image occurs regularly in both written and emblematic format, and many of these emblematic images are regularly reprinted in modern books of alchemical illustrations. For example, in figure 3.2, a late fifteenth-century manuscript shows the tree sprouting up, rather grotesquely, out of the body of a dead man.[29] This specific image was, in fact, used by Jung in his exposition of the psychological interpretation of alchemy.[30] Another figure of the tree (of later date) appears in Michael Maier's *Atalanta Fugiens* (figure 3.3), where we see the tree enclosed within a small circular edifice that is almost too small for such a plant.[31] Many other images of naturalistic trees can be found in alchemical literature, both in illustrative and textual form.

Among the school of chrysopoeians whom we may call the *mercurialists*, who believed that the Philosophers' Mercury was somehow to be produced out of prepared quicksilver and gold, this tree image is absolutely crucial.[32] For these mercurialists, the tree signifies the vegetation of the "seed of gold." Employing an agricultural trope, part of their theoretical framework dictated that the propagative principle within gold had to be stirred up or awakened and then "sown" within the proper matrix to provide a generous harvest of new gold via the transmutatory Philosophers' Stone. George Ripley referred to this image on several occasions, sometimes as Hermes Tree, which must eventually be burnt to ashes and then revived; on other occasions, he likened his Philosopher's tree to a hawthorn.[33] Speaking of the Philosophical Mercury—the mineral water with which the "gold plant" is watered—Ripley wrote, "Therwith dyd Hermes moysture hys Tre:/ Wythyn hys Glas he made to grow upryght,/Wyth Flowers dyscoloryd bewtyosely to syght."[34]

Jean Collesson, a mercurialist of the first half of the seventeenth century, wrote that the power of the true Philosophers' Mercury lies in

Figure 3.2
Illustration from Ashburnham MS BL1166, f. 16, Laurentian Library, Florence.

Figure 3.3
Emblem 9 from Michael Maier, *Atalanta fugiens* (Oppenheim, 1617), 45.

its ability to "make gold vegetate and germinate" and set this down as a test of alchemical students' prepared materials. If the students' Mercury does not make gold "vegetate to sight," it is not the true Mercury of the Philosophers; it is a sign that they are not on the right path and must retrace their steps or give up the search.[35] At the risk of multiplying references *ad infinitum*, we consider again Philalethes, one of whose most valuable assets was his habit of collating and summarizing similiar directions or tropes found in earlier authors. Noting the multitude of authors who write of Hermes' Tree, he wrote in his exposition of Ripley's *Compound of Alchymie*:[36]

> This tree of ours some have compared to one thing, and some to another; some to a Cypress or Fir-Tree, which indeed may seem to resemble it; others to Haw-Thorn Trees, ... others to shrubs and

Bushes, others to thick Woods … others have called it their Coral, which is indeed the fittest comparison.

This repeated image, like any of the others employed in traditional chrysopoetic alchemy, could trace its ubiquity and permanence to three causes. First, it could arise from a purely or predominantly literary tradition (i.e., its use passed on from one writer to the next as a stock image); a later writer copies or develops the tropes encountered in earlier exponents of the tradition. In a related sense, the image might be a mere consequence of traditional philosophical or theoretical assumptions. Specifically, practitioners seeking the seed of gold would, in fact, associate with that construct certain related agricultural images, such as a plant or tree. A second possibility is that which Jungians suggest: such recurrent images in alchemy originate in manifestations of the collective unconscious. Both the images of alchemy and the images in the dreams of Jung's putative patients are drawn from a common cistern of racial or universal unconscious imagery. A third possibility is that some alchemical images recur because they are based on actual experimental results that themselves recur reproducibly. A fifteenth-, sixteenth-, seventeenth-, or even twentieth-century alchemist combining the same ingredients in the same way should achieve the same results. I wish to argue for the contribution of this last source and suggest that at least *some* of the repeated imagery in alchemy arises because several different alchemists all saw the same experimental results in their attempts to approach the Stone. The same visual results, particularly if strange or evocative in appearance, could naturally lead to the same metaphorical *Decknamen* for allegorical expression or to the codification of an apt metaphor.

Working from this hypothesis, I believed that the modern replication of alchemical processes derived from the literature of mercuralist chrysopoeians might well provide physical phenomena that could elucidate, or at least render more plausible, the images recorded in their texts. Therefore, I compared the corresponding texts of various mercurialists, particularly Ripley, Collesson, Alexander von Suchten, Gaston "Claveus" Duclo, Philalethes, and others, in conjunction with a private alchemical communication between George Starkey and Robert Boyle known as the *Key*.[37] These I reduced to a chemical protocol following the methodology either clearly stated or darkly hinted. After a fairly lengthy process involving various materials and numerous distillations, I obtained an "animated" Mercury, which was supposedly the necessary

Figure 3.4
The "tree" produced in the laboratory inside a glass flask ("egg"), height ca. 3 cm.

"mineral water" that mercurialists required for the "moistening of the seed of gold."

Then, after interpreting Starkey's *Marrow of Alchemy* for advice regarding exact proportions and method of mixture and digestion, I used this material along with gold to prepare a mixture that was sealed in a "glass egg" and heated. The mixture soon swelled and bubbled, rising like leavened dough, recalling (perhaps not unwarrantably) the numerous references to fermentation and leavening in mercurialist literature. Then it became more pasty and liquid and covered with warty excrescences, again perhaps accurately recalling the "moorish low bog" that "Toads keep."[38] After several days of heating, the metallic lump took on a completely new appearance, as illustrated in figure 3.4. Some today might call this a *dendritic fractal*, but I think that most onlookers would refer to it first as a *tree*.

This remarkable product is, exactly as Eirenaeus Philalethes wrote, a tree with "Shoots and Springs without any thing that may be properly likened to Leaves," and aptly called *coral* by him.[39] As the mixture contains gold, here, as Collesson wrote, "gold vegetates to sight" and, as Ripley said, is "made to grow upryght." Maier, too, who stated that the tree is "enclosed in a house of glass, full of dew," seems to describe just such a sight, with the glistening droplets of quicksilver clinging to the glassy walls of the flask.[40] Ripley's "Flowers dyscoloryd bewtyosely" appear on further heating when the tips of the branches take on various colors, owing to the formation of mercury oxides. Even the popular illustration of the tree used by Jung now makes new sense, especially if the dead man is not Adam (as has long been presumed without evidence) but rather a representation of the body of gold, "pierced" by the Philosophical Mercury—the flying arrow suggesting its quickness and volatility—and now growing into a tree. The pithy aphorisms of this manuscript do not allow for a firm conclusion that it describes a true mercurialist approach (which leads to the chemical tree), but an early reader of the Laurentian library copy has glossed the text, writing that "*nullus mercurius sumatur quam mineralis,*" ("let no mercury be taken except mineral mercury"), which is an axiom among all the other mercurialists cited here. Boyle himself obtained such a sight when carrying out the process partially delivered to him by George Starkey in 1651, for he too wrote allusively of obtaining by a special process on gold certain "very pretty vegetations."[41]

Thus, we come to the surprising turn that these very same repeated images—which led Jung to his psychological interpretation of alchemy and led the Enlightenment writers and, more recently, discontinuity partisans to their rejection of alchemy as serious experimentalism—may actually be (in at least some cases) not only artifacts of, but *arguments in favor of* the reality and reproducibility of experimental programs carried out by Stone-seeking alchemists. Of course, I agree that the choice of image is closely bound up with a variety of cultural factors, philosophical, theological, artistic, experiential, and so forth, but all I contend is that the admittedly culturally influenced metaphorical clothing, no matter how externally bizarre, may (in more than a few cases) cover a solid body of repeated and repeatable observations of laboratory results.[42] The common source for these sometimes extravagant images is neither the unconscious psychic state nor the merely literary but rather the experimental. One needs no invocation of so shaky a notion as the "collective unconscious" or its "irruption" or "projection" when simple observation of reproducible chemical phenomena in the physical world suffices.

Consequently, we can now see that alchemy as a whole is not to be distanced from the practical work of the laboratory. Some alchemists, including the chrysopoeians, were highly solicitous in their manual work, carefully employing or designing specialized apparatus for specific goals. They followed the records (though secretive and allusive) of their predecessors and attempted to reduce them to experimental protocols.[43] Further, some of the central imagery of the alchemical tradition has been shown to be explicable, and even rational (though metaphorical), in terms of actual physical phenomena presented to the sight of alchemical workers by empirical, experimental laboratory operations.

The "otherness" of alchemical texts rests, then, more on their modes of expression than on their modes of laboratory work. We are not justified in disconnecting "alchemy" from "chemistry" on the basis of a radically differing valuation of or involvement in laboratory experimentation. Indeed, if we cannot make such a disconnection, we must look even more closely at the whole of early alchemical/chemical thought and practice and at the evolving role and method of laboratory practice over a long period of "chymical" history, free from the obscuring shadows of untenable interpretations of alchemy.

NOTES

1. Regarding the Jungian, discontinuity, and other interpretations of alchemy, see Lawrence M. Principe and William R. Newman, "Some Problems in the Historiography of Alchemy," forthcoming in *Archimedes*. On the problem of distinguishing "alchemy" from "chemistry," see Newman and Principe, "Alchemy vs. Chemistry: The Etymological Origins of a Historiographic Mistake," *Early Science and Medicine* 3(1998): 32–65.

2. See Richard Noll, *The Jung Cult: Origins of a Charismatic Movement* (Princeton: Princeton University Press, 1994); and *The Aryan Christ* (New York: Random House, 1997).

3. B. J. T. Dobbs, *Foundations of Newton's Alchemy* (Cambridge, UK: Cambridge University Press, 1975), 26–36.

4. B. J. T. Dobbs, *Janus Faces of Genius*, (Cambridge, UK: Cambridge University Press, 1991). See also her "From the Secrecy of Alchemy to the Openness of Chemistry," 75–94 in Tore Frangmayr, ed., *Solomon's House Revisited* (Canton, MA: Science History Publications, 1990), 76.

5. Lawrence M. Principe, "Robert Boyle's Alchemical Secrecy: Codes, Ciphers, and Concealments," *Ambix* 39(1992): 63–74; and *The Aspiring Adept: Robert Boyle and His Alchemical Quest* (Princeton, NJ: Princeton University Press, 1998). Also Steven Shapin and Simon Schaffer, *Leviathan and the Air-Pump* (Princeton: Princeton University Press, 1985), 39, 56–59, 70–72, 78, 335–336.

6. Brian Vickers, "Alchemie als verbale Kunst," in Jürgen Mittelstrass and Günter Stock, eds., *Chemie und Geisteswissenschaften: Versuch einer Annäherung* (Berlin: Akademie Verlag, 1992), 17–34.

7. Marco Beretta, *The Enlightenment of Matter* (Canton, MA: Science History Publications, 1993), 331.

8. Ibid.

9. Georg Ernst Stahl, *Philosophical Principles of Universal Chemistry*, trans. by Peter Shaw (London, 1730), 396, 401–408.

10. Thomas Norton, "Ordinall of Alchymie," in Elias Ashmole, ed., *Theatrum Chemicum Britannicum* (hereinafter TCB), (London, 1652; reprinted in 1967, New York: Johnson Reprint Co.), 92.

11. Ibid., 97–98; illustrated on p. 103 in a slightly reworked seventeenth-century engraving.

12. Ibid., 94–96.

13. The complete Valentinian corpus appears in *Basilii Valentini Chymische Schrifften* (Hamburg, 1677; reprinted in 1976, Hildesheim: Verlag Dr. H. A. Gerstenberg). The most important recent addition to the Valentine literature is Claus Priesner, "Johann Thoelde und die Schriften des Basilius Valentinus," 107–118, in Christoph Meinel, ed., *Die Alchemie in der europäischen Kultur- und Wissenschaftseschichte*, Wolfenbütteler Forschungen, vol. 32, (Wiesbaden: Otto Harrassowitz, 1986).

14. The first edition was Erfurt, 1624; thereafter, it was published as part of the "expanded" *Letztes Testament*; see Valentine, *Schrifften*, vol. 2, 319–338.

15. Friedrich Roth-Scholtz, *Deutsches Theatrum Chemicum*, (Nurnberg, 1727), vol. 1, 663: "*Die Offenbahrung der verborgenen Handgriffe aber/ sind nicht wehrt/ daß sie unter die Schrifften Basilii Valentini erzehlet werden/ die auch mit seinen übrigen Schrifften im geringsten nicht überein kommet.*"

16. The first edition is extremely rare: *Ein kurtz Summarischer Tractat, Fratris Basilii Valentini Benedicter Ordens, vom dem grossen Stein der Uhralten* (Eisleben, 1599). The second edition (Leipzig, 1602) is expanded both by emblematic figures and by a *Klare Repetition oder Wiederholung*. The work is also found in the collected *Chymischen Schrifften*, vol. 1, 1–74.

17. Valentine, *Schrifften*, 31–32.

18. Ibid., vol. 2, 322–323.

19. Priesner, "Basilius Valentinus und die Labortechnik um 1600," *VCH Berichte zur Wissenschaftseschichte* 20 (1997): 1–14, on pp. 2–3.

20. Ibid., vol. 1, 34–35.

21. Ibid., vol. 2, 324–325.

22. Regarding the volatilization of gold chloride, see Thomas Kirke Rose, "The Dissociation of Chloride of Gold," *Journal of the Chemical Society* 67(1895): 881–904. I have successfully repeated Valentine B's procedure; the ammonium salts carried from the "water of the fighters" facilitate the sublimation of the gold salt. Robert Boyle also reproduced the procedure of Valentine's *Twelve Keys*; see Principe, "The Gold Process: Directions in the Study of Robert Boyle's Alchemy," in Z. R. W. M. van Martels, ed., *Alchemy Revisited* (Leiden: E. J. Brill, 1990), 200–205.

23. Valentine, *Schrifften*, vol. 2, 326, 338.

24. William R. Newman, *Gehennical Fire: The Lives of George Starkey, an American Alchemist in the Scientific Revolution* (Cambridge, MA: Harvard University Press, 1994), 125–133, esp. 128, 132–133.

25. Eirenaeus Philalethes, *Ripley Reviv'd* (London, 1678), 36–39.

26. Ibid., 215–216.

27. George Ripley, *The Compound of Alchymie*, 107–193; in *TCB*, p. 138.

28. Dobbs, *Foundations*, 32.

29. Ashburnham MS, BL1166, f. 16, Laurentian Library, Florence; a copy exists at the British Library, Sloane MS 1316, on fol. 12. Urszula Szulakowska, "The Tree of Aristotle: Images of the Philosophers' Stone and their Transference in Alchemy from the Fifteenth to the Twentieth Centuries," *Ambix* 33(1986): 53–77, discusses this image and points out how alchemical images have often been wholly divorced from their context. She rightly argues for the importance of an art historical appreciation of such figures to understand them fully.

30. Carl G. Jung, *Psychology and Alchemy* (London: Routledge and Kegan Paul, 1980), 256.

31. Michael Maier, *Atalanta fugiens* (Oppenheim, 1617), 45.

32. I am at work on an extended project to delineate and define the major schools of thought within the broad topic of "alchemy." An early facet of this project is published as "Diversity in Alchemy: The Case of Gaston 'Claveus' DuClo, a Scholastic Mercurialist Chrysopoeian," in Allen G. Debus and Michael Walton, eds., *Reading the Book of Nature: The Other Side of the Scientific Revolution* (Kirksville, MO: Sixteenth Century Press, 1998), 181–200.

33. Ripley, *Compound*, 114, 125, 170.

34. Ibid., 141.

35. Jean Collesson, *Idea perfecta philosophiae hermeticae* in *Theatrum Chemicum*, vol. 6 (Strasburg, 1659–1661; reprinted in 1981, Torino: Bottega d'Erasmo) 143–162, on pp. 146, 149.

36. Philalethes, *Ripley Reviv'd*, 65.

37. Newman, "Newton's *Clavis* as Starkey's *Key*." *Isis* 78(1987): 564–574.

38. Philalethes, *Ripley Reviv'd*, 65.

39. Ibid.

40. Maier, *Atalanta*, 45.

41. Robert Boyle, *Producibleness of Chymical Principles* (London, 1680), 220; see also Principe, *Aspiring Adept*, 174–175, and Dobbs, *Foundations*, 186–187. The growing of such trees eventually emerged from secrecy, and an account of the practice was published in fuller form in several later texts—for example, George Wilson's *A Compleat Course of Chymistry*, 3rd ed. (London, 1709), 393–394, and *Coelum philosophorum* (Dresden and Leipzig, 1739), 8–9, 110. A similar metallic tree was investigated (among more common varieties of the well-known "Arbre de Diane") by Wilhelm Homberg in 1692 as reported in his "Reflexions sur différéntes végétations métalliques," *Histoire de l'Académie Royale des Sciences*, vol. 10 (Paris, 1730–1733), 171–179.

42. In conjunction with this experimental demonstration, see Newman's brilliant unraveling of the bizarre dreamlike sequence of Philalethes' *Ripley Reviv'd* in *Gehennical Fire*, 118–133, which clearly demonstrates the experimental activities decodable from the extravagant imagery commonly adduced as supporting evidence for Jungian views.

43. Such endeavors will be showcased further in the study of the laboratory notebooks of George Starkey being carried out by Principe and Newman.

II

FROM HALES TO THE CHEMICAL REVOLUTION

At the beginning of the seventeenth century, chemistry and alchemy were nearly synonymous terms. During the first half of that century, teachers of chemistry gradually distinguished their subject both from the esoteric goals of the alchemists and the practical medical purposes of the pharmacists. They defined their aim as the separation of substances into their simplest constituents, or principles. They identified these constituents with various combinations of the elements of Aristotle and the principles of Paracelsus. In the third edition of his popular *Elements of Chymistry*, Nicolas Lemery attacked alchemy as a worthless, fraudulent pursuit, redefining it, in the process, by limiting it to such activities as the transmutation of base metals into gold or the search for the Philosopher's Stone, activities that had been rejected by the emerging chemistry.[1]

From then onward, chemists sought to distance themselves from an increasingly discredited alchemical practice. Nevertheless, their own practice continued to be based on apparatus, operations, and instruments that they had once shared with traditional alchemy. Except for gradual refinements in design and materials, the distillation flasks, furnaces, balances, and other vessels described in the practice of alchemy in part I, as well as the laboratory space in which they were employed, remained largely as they had been during the preceding centuries.

With these traditional material resources, chemists were able, during the first two-thirds of the eighteenth century, to build a flourishing investigative science, oriented largely around the study of acids, bases, and neutral salts. As the number of known acids and bases grew, the number of their known and unknown combinations multiplied, providing ever-expanding opportunities for further discoveries.

Even as they deployed traditional methods and apparatus with growing success, eighteenth-century chemists began to introduce novel apparatus, instruments, and methods as they turned to the study of phenomena that had lain beyond the horizons of their predecessors and particularly as they made quantitative measurement more central to their practice. The seven chapters in this part treat the most important of these additions to the chemical repertoire and present examples of smaller innovations that were, by the end of the century, cumulatively extending the reach and improving the quality of chemical experimentation.

The most conspicuous new methods and apparatus were associated with the collection and measurement of what were known by the end of the century as *gases*. Chemists of the seventeenth and early eighteenth century had been aware that air and various vapors were often absorbed or given off in the course of a chemical operation but, having no way to capture them, they could not account for their part in the chemical changes observed. This technical obstacle was suddenly overcome in 1728, when Stephen Hales invented the device later called the *pneumatic trough*. The introduction of this apparatus into chemical laboratories later in the century enabled chemists to handle a new dimension in the production and interpretation of chemical phenomena.

In reviewing the history of these changes, however, Maurice Crosland also cautions against attributing too large a role to the experimental apparatus itself. Hales's device was a turning point, but its manipulation alone could not lead to the identification of the gaseous state. Far more difficult than the capture and measurement of whatever was given off or absorbed was, in Crosland's view, the mental transition necessary to understand the invisible, extremely tenuous, and vastly expansible matter captured in the pneumatic trough as species of what eventually came to be called *gases*.

The chemist who finally defined the gaseous state at the end of the century was Antoine Lavoisier. Frederic L. Holmes shows that the experiments through which Lavoisier constructed his new theoretical structure were carried out mainly with relatively simple apparatus, based primarily on combinations of Hales's pneumatic devices and traditional chemical equipment. By now, however, the design of the specific forms of the apparatus, and the methods employed, became crucial to success, because his reasoning required Lavoisier to obtain reliable, if not necessarily precise, quantitative measurements of gases as well as of liquids or solids.

By late in the eighteenth century, the investigation of gases demanded instruments with capabilities exceeding those of the simple pneumatic trough. Trevor H. Levere discusses the advent and refinement of two such instruments invented for particular purposes. The eudiometer allowed chemists to carry out a reaction within a closed chamber and to determine the chemical change quantitatively by means of the change in volume. Gasometers enabled them to measure the quantities of gases supplied to a chemical operation in which those gases are consumed. Levere examines the question of the degree of accuracy of the early eudiometers and gasometers and follows their further development into the early nineteenth century.

Next to the volumetric measurement of gases, the most significant new form of measurement that emerged in chemistry during the eighteenth century was that of heat. Jan Golinski shows that the introduction of the thermometer into chemistry raised problems about the nature of what was measured that somewhat parallel the question of what was collected and measured in Hales's pneumatic apparatus. The early assumption that thermometers directly measure the quantity of heat present was shattered by the discovery of latent heat. When Lavoisier and Laplace invented the ice calorimeter to measure the heat released in a physical or chemical change, the thermometer was reduced to a subordinate, although still essential, role in controlling the circumstances under which the operation was carried out. Golinski's essay provides an illuminating illustration of the ways in which the meaning of a measurement can change and become problematic even when the measuring instrument itself has become relatively stable.

Searching for other means to quantify chemical properties, Lavoisier and some of his contemporaries devoted much attention during the last quarter of the century to the hydrometer, an instrument long used commercially to determine the specific gravity of liquids. Bernadette Bensaude-Vincent shows that the great hopes that they placed on the method were not fully realized. Once again, the meaning of measurements that were themselves simple to make proved problematic, posing (in this case) obstacles too formidable to be overcome fully.

Concentration on the new instruments and methods for measuring gases, heat, or the densities of liquids and on the problematic nature of such conspicuous innovations should not lead us to overlook the many small, qualitative improvements that were taking place as chemistry became increasingly dynamic during the late eighteenth century. William A. Smeaton guides us through some of the changes introduced by one of the more inventive chemists of the time, Guyton de Morveau. These include multiple uses for the newly discovered metal, platinum, and the introduction of ground-glass stoppers. As the expanding range of chemical experimentation increased the demand for convenient, reliable, leak-proof apparatus usable over a broader range of temperature and in other conditions, such incremental improvements became more common and more essential. The chemical laboratory, which had remained relatively stable for more than two centuries, was giving way by the end of the eighteenth century to a rapidly evolving workplace.

NOTE

1. See William Newman and Lawrence Principe, "Alchemy vs. Chemistry: The Etymological Origins of a Historiographic Mistake," *Early Science and Medicine* 3(1998): 32–65.

4

"SLIPPERY SUBSTANCES": SOME PRACTICAL AND CONCEPTUAL PROBLEMS IN THE UNDERSTANDING OF GASES IN THE PRE-LAVOISIER ERA
Maurice Crosland

A British newspaper headline in 1995 reported that two American physicists had been awarded the Nobel prize for revealing "almost nothing."[1] The journalist was referring to the neutrino, a subatomic particle with no mass, and the tau particle, which exists for virtually no time at all. It requires all the sophistication of science on the eve of the twenty-first century to explain the importance of these particles to the physicist's understanding of the universe. The extreme sophistication of modern physics may be contrasted with the simplicity of seventeenth-century knowledge of the natural world. Yet perhaps some aspects are common to the science of both periods, as Robert Boyle had occasion to comment that most men of science, "not finding the air to be a visible body," tend to ignore it, because "what is invisible, they think to be next degree to nothing."[2]

A preliminary step in understanding gases was, therefore, to consider air as a material substance. Stephen Hales deserves great credit for suggesting in 1727 that chemists should begin to take air more seriously:[3] "May we not with good reason adopt this now fixt, now volatile Proteus among the chymical principles ... notwithstanding it has hitherto been overlooked and rejected by Chymists..."

Gases were to assume crucial importance in the eighteenth century. It was as if chemists had created a whole new realm. In the seventeenth century, gases had been almost unknown, whereas by the nineteenth century, their existence was largely taken for granted, a point of view still prevailing in our time.

After J. B. van Helmont (1579–1644), born in Brussels, the main *dramatis personae* in the early history of gases were British: Robert Boyle and John Mayow in the late seventeenth century, Stephen Hales in the early eighteenth century, followed in close succession by Joseph Black, Henry Cavendish, and Joseph Priestley. Special attention is paid here to the latter as the person who arguably made the greatest contribution to the knowledge of gases. Yet paradoxically, his own understanding was limited. He was much stronger on practice than on theory. Other eighteenth-century characters also are mentioned: William Brownrigg and the Swedish apothecary C. W. Scheele. This list is by no means exhaustive, but it covers most of the main actors (figure 4.1).[4]

Boyle and Mayow both worked in Oxford and London, and both were early members of the Royal Society, founded in 1660. Hales too was a member two generations later, when so much science was performed under the spell of the recently deceased Isaac Newton. Black was exceptional among the chemists in that he held a university chair (at Edinburgh). Cavendish and Priestley worked in the best tradition of British amateurism, being largely self-taught in chemistry, yet succeeding in becoming adept in the handling of gases. It was only in the eighteenth century that the concept of the gaseous state gradually emerged, even then—with what might appear to us who have the benefit of hindsight—very seriously confused with atmospheric air. It was understandable that Boyle and Hales should have spoken of "air," but the practice of describing gases as airs continued into the second half of the eighteenth century with the other British pneumatic chemists.

GAS

The early history of gases is related in some ways to the history of certain chemical operations and, in particular, to distillation. In distillation, a liquid is converted by heat into a vapor; the vapor then condenses in a receiver. However, in some operations, a chemical reaction would take place, and sometimes (our modern knowledge tells us) a gas would be produced. Operators did not understand this. Anything we would call a *gas* would be regarded as irrelevant, no more than a will-of-the-wisp. What concerned them most was the immediate practical consequence of any buildup of pressure in the system: the retort might explode. The apparatus would be damaged beyond repair, and the operator himself might well have been injured. Therefore, operators adopted the practice of making a hole near the receiver to allow any "spirits" to escape. This expedient saved the apparatus and allowed the distillation to be successfully concluded. What was lost seemed of no importance. As one later writer expressed it:[5]

> Very far from suspecting what they might have gained from this work, they preferred to safeguard their apparatus. They were happier to abandon a product the value of which they did not understand than to risk losing the fruit of the different operations they were carrying out.

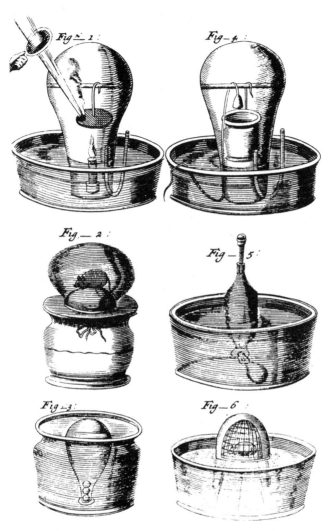

Figure 4.1
Various experiments by Mayow, mainly on combustion and respiration in a confined volume of air. From *Medical-Physical Works* (London: Alembic Club reprint,
1674). (For a discussion of the experiments illustrated, see J. R. Partington, *History
of Chemistry*, vol. 2, London, 1961, 594–604.)

If sixteenth-century "chemists" had been questioned about any possible loss, on reflection they might have spoken of a "spirit" or even a "wild spirit." This brings us to the famous definition of van Helmont: "I call this Spirit, unknown hitherto, by the new name of Gas, which *can neither be constrained by Vessels*, nor reduced into a visible body . . ."[6]

This idea of associating gas with potential explosions continued well into the eighteenth century. Hales refers to "this much neglected volatile Hermes, who has so often escaped through . . . burst receivers, in a guise of a subtile spirit, a mere flatulent explosive matter."[7]

Even more relevant to my argument than explosibility is the idea of incoercibility. The great *Encyclopédie*, edited by Diderot and D'Alembert, stated in relation to gases that "their incoercibility will remove them from our researches for a long time to come."[8]

Several interesting points may be seen here. The first is the use of the future tense. This is not, as with Hales, a mere historical statement. On the contrary, as with Helmont, incoercibility is seen as such an essential property of gases that it would be very difficult, if not impossible, to collect them. This has an obvious epistemological implication. If we cannot collect gases to examine them, perhaps we may never fully understand them.

A second point about these remarks regarding incoercibility is that they were written around the same time as Black's experiments which led to the discovery of "fixed air" (carbon dioxide). We may note that Black was able to announce his discovery without the use of any pneumatic apparatus. Instead, he used a balance to detect a loss in weight in the calcination of magnesia. His explanation of the loss in weight in terms of the loss of "the volatile parts" echoes the words of Helmont, because Black said that they "cannot, it seems, be retained in vessels, under a visible form."[9] For his discovery, he did not need to collect the gas. Later, however, as a professor at Edinburgh, he enlivened his lectures by several experimental demonstrations, one of which involved simple pneumatic apparatus. Even here, however, his apparatus seems to have been based more on caution than on efficiency. The apparatus he used made no attempt to retain all the gas evolved, as the generating vessel contained a hole that acted partly as a safety valve and partly as a means of introducing reagents.[10]

After completing his doctoral thesis, Black performed very little research. Rather, he directed his creativity toward pleasing his students. As his editor, Robison, commented, "Dr. Black seems to have turned his whole attention to rendering his lectures as popular and profitable as possible by a neat exhibition of experiments. He multiplied these without any new views. . ."[11]

Here is a reminder that apparatus is sometimes more useful for didactic purposes than for research. In the history of eighteenth-century natural philosophy, this is well illustrated by the case of Desarguliers. We are reminded of the distinction between the context of discovery and the context of justification.

By the 1770s, it had become a fairly commonplace practice to set up apparatus to prepare gases. This is how the French chemist Lassone expressed it: "Chemistry has finally succeeded in rendering sensible and palpable those principles which had always been regarded as incoercible."[12] This achievement deserves to be noted as a major advance.

"Air"

The history of gases is anything but simple, and clearly more than one tradition prevailed. The late seventeenth-century work of Robert Boyle, particularly his studies of air, seem to lie outside the tradition of gases as essentially incoercible, partly because Boyle's starting point usually was not the product of a chemical reaction but simply atmospheric air. Indeed, Boyle's studies of air belong more to natural philosophy (or "physics," to use an anachronistic label) than to what he called "chymistry." This is how Boyle defines air: "By the air I commonly understand that thin, fluid, diaphanous, compressible and dilatable body, in which we breathe and wherein we move, which envelopes the earth on all sides . . ."[13]

Note that, apart from mentioning air's obvious transparency, Boyle emphasizes its compressibility and ability to expand. This is the Boyle of "Boyle's law."[14] Although this law is now interpreted as describing the behavior of gases generally (however approximately), for Boyle it was simply about air. Any gas for Boyle would have been air, possibly impure air. When Boyle spoke of "true air," he did not mean atmospheric air as opposed to one of the gases to be later discovered. He meant permanent air, air that did not condense on standing because it was really, for example, water vapor.

This is a problem that continued to exercise Stephen Hales two generations later. Hales is credited with the invention of the pneumatic trough, normally regarded as simply a method of collecting gases. Yet we should remember that for Hales, all gases were no more than air. Also, his trough was only indirectly a means of collecting "air." It was

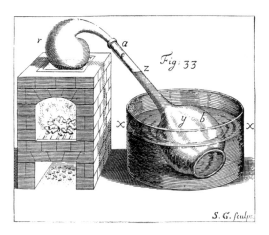

Figure 4.2
Hales's apparatus for measuring the volume of "air" evolved by heating a substance in a retort. From *Vegetable Staticks* (London, 1727).

essentially a device for washing the "air" evolved in the reaction vessel.[15] In several cases, he concluded that the "air" evolved was only partly permanent air: "A good part of the air thus raised from several bodies by the force of fire was apt gradually to lose its elasticity in standing [over water] several days" (figure 4.2).[16]

The change in the level of the water in the collecting vessel might be due to the evolution of something additional in the reaction, such as "sulphurous fumes." Partly to solve the problem of the loss of elasticity of the air, Hales introduced his pneumatic trough. The mouth of his retort was placed under a large inverted receiver full of water and suspended in a basin, "so that as the air which was raised in distillation passed though the water ... a good part of the acid spirit and sulphurous fumes were by this means intercepted and retained in the water" (figure 4.3).[17]

In other words Hales was using his basin as a wash bottle. This intention did not prevent his successors, and notably Priestley, from using it simply as a pneumatic trough. The most important and permanent advance offered by Hales's device was to demonstrate the ease of separating the generator from the collector.

My thesis is that the question of apparatus for the collection of gases must be put in perspective. I suggest that certain basic concepts were even more important than the apparatus in the early stages. I have already mentioned the prejudice that, when natural philosophers and

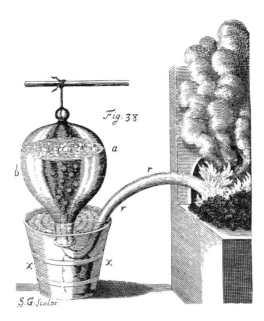

Figure 4.3
Hales's collection of "air" by the displacement of water. From *Vegetable Staticks*
(London, 1727).

chemists were studying solids and liquids, they tended to ignore air—
the nearest most came before the second half of the eighteenth century to identifying the gaseous state. Material substances were considered to be either solid or liquid. To go beyond these tangible realms
would have been to invoke the mysterious world of spirits. For many,
this would have represented a regression from the physical to the
metaphysical.

Although Boyle studied air, he did so in a very limited way. For
him, air was something with properties of compressibility, like a pile of
wool or a metal spring, something with physical rather than chemical
characteristics. Only atmospheric air existed, possibly in various states
of purity, as Boyle allowed for the atmosphere to be permeated with
various exhalations. Even for some later investigators confronted with
the decidedly poisonous properties of carbon monoxide, this was not
recognized as a new substance but was simply "[atmospheric] air
infected with the fumes of burning charcoal."

When people started investigating different samples of air, they
were not surprised to find differences. After all, since the very beginnings of chemistry, samples of a like substance from different sources

were never exactly the same. To cite a famous example, "alum of Tolfa" was particularly prized as a mordant in dyeing, presumably because it was very pure. Alum from other sources might be much less useful to the dyer, being contaminated with various extraneous substances. Thus it was easy to think of a continuum in terms of states of purity rather than as a sharp discontinuity that implied that an air from a different source was essentially different from atmospheric air. The various gases were so similar physically that appreciating their similarities was easier than appreciating their differences. Priestley even began his first book on "airs" by remarking how convenient the term *air* was to denote substances that had so many common properties, such as transparency and elasticity.[18]

THE GASEOUS STATE

It was an enormous advance to contemplate the possibility of an entire new state of matter—the gaseous state—parallel to and no less real than the solid and liquid states. This might even be claimed as one of the greatest (but largely unappreciated) discoveries of the second half of the eighteenth century. Yet, many obstacles had to be overcome. In the early eighteenth century, air was one of the four Aristotelian elements but, in the chemistry of Stahl and G. F. Rouelle, it was also an instrument of chemical change, even if it was not as powerful as fire or water.[19] Both emphasized the role of solvents; although water played the most prominent part, air could also be considered in this light. Boyle had spoken of atmospheric air as "that great receptacle ... of celestial and terrestrial effluviums"[20] and, if we think of air as containing a variety of impurities, we may appreciate the difficulty in the mid-eighteenth century of thinking of any new type of air as chemically distinct from ordinary air. Much easier was thinking of it as ordinary air with something added to it to give it slightly different chemical properties. It required a conceptual leap to consider air as an entire state of matter rather than as a single substance.

    According to this new view, atmospheric air was not privileged; it was merely the longest-known example of the gaseous state. After a few decades, research would show that, contrary to intuitive belief, it was not a simple substance but a mixture of at least two distinct substances in the gaseous state. Yet, up to the 1750s, "air" existed only in the singular. Finally, Joseph Black discovered that "air" had a plural, as he demonstrated, in addition to ordinary atmospheric air there was another "air" that he called *fixed air*, contained in magnesia alba.[21] By the time Cavendish had gone on to discover "inflammable air" (1766), we might think that all chemists understood that air was not a single substance but rather a state of matter, of which there were several distinct species.

The temptation is to take the work of Black as a turning point in the history of gases, as it must seem today. However, this is not how it was seen by most of his contemporaries, many of whom were more concerned with his new interpretation of the nature of alkalis and, insofar as they thought about air at all, could not escape from the view of a single air. The problem was made worse by Black's unwillingness to mark his discovery with a new name. He was content to call the new gas (our carbon dioxide) "fixed air—a word already familiar in philosophy."[22] Black was merely reflecting the fact, well-known in the Newtonian tradition, that many substances contained air in a fixed state, able to be released by heat.[23] This was the tradition in which Hales had worked, but Black had gone much further than Hales in revealing that atmospheric air and fixed air were chemically distinct. The pity is that many chemists failed to appreciate this. Thus, Baumé in 1773 insisted on the existence of only one kind of air.[24] For him and for many others, it was one of the four Aristotelian elements and, if it sometimes appeared different, it was because it could enter into an infinite number of combinations. Thus, he insisted that fixed air was not a separate species:[25]

> ... Fixed air ... is ordinary air but which is charged with foreign substances which it holds in solution, air which can often be purified and brought back to the state of pure air, similar to that of the atmosphere, by passing the fixed air through different liquors suitable for filtering the air and retaining the foreign substances which alter its purity.

It would be all too easy to dismiss Baumé as an obscurantist and turn for enlightenment to someone such as the young Lavoisier. Yet, in 1774, Lavoisier remarked that it was not certain whether fixed air "is a substance essentially different from the air or whether it is ordinary air, to which something has been added or taken away."[26]

However, in this chapter, we are more concerned with investigators before Lavoisier, such as Priestley. It is significant that Priestley's predecessor, Stephen Hales, rather than his contemporary Joseph Black, was his main inspiration.

Perhaps Priestley's grounding in natural philosophy rather than chemistry led him to believe that he was dealing with ordinary air in his experiments or, at most, with some modification of atmospheric air. In 1772, he prepared what he called "inflammable air" (in this case, coal gas) by heating coal in a gun barrel. When he shook the "air" with water for a long time, he reported that it was converted into common air.[27] He also claimed that this method of shaking with water was an infallible method "to restore any kind of noxious air."[28]

This account helps to explain why Priestley was content to use the term *air* to describe gases. When he prepared another "air" by heating charcoal in atmospheric air, thus obtaining a mixture of carbon dioxide and carbon monoxide, he described the product as "air infected with the fumes of burning charcoal,"[29] because for him it was only air (i.e., atmospheric air contaminated with impurities), even though these so-called impurities might be lethal. For a gas to have different properties is enough for us to appreciate that it is an entirely different substance. This concept is something that Priestley came to appreciate only slowly, and inconsistencies permeate his writing throughout his career as a pneumatic chemist. Therefore, we cannot say that the early Priestley walked in the footsteps of Hales and then, in some sudden conversion, came to have an essentially modern understanding of his various gases. His understanding developed slowly and in a confused way.

PHLOGISTON AND THEORY

At this point, something must be said about the phlogiston theory. In the first edition of his *Dictionary of Chemistry*, Macquer wrote, "Most gases ... seem to be nothing but phlogiston which is pure or nearly pure, which escapes from bodies without being in a state of ignition."[30] Of course, this was pure speculation, but it is a reminder that in the eighteenth century, gases were not the only mysterious invisible fluids. According to the theory of phlogiston, this mysterious "flame stuff," or phlogiston, was supposed to be given off in combustion. It too, at first, seemed to be intangible. Chemists were prepared to believe in its existence because it seemed to provide a rational explanation of combustion, yet they did not expect to be able to collect it. As one defender of the theory wrote:[31]

You do not surely expect that chemistry should be able to present you with a handful of phlogiston, separated from an inflammable body; you may just as reasonably demand a handful of magnetism, gravity or electricity to be extracted from a magnetic, weighty or electric body; there are powers of Nature which cannot otherwise become the objects of sense than by the effects they produce, and of this kind is phlogiston.

Here is a reminder of an even greater number of invisible and subtle fluids in the natural philosophy of the eighteenth century. Gravity was often explained as an "ether."[32] Also, an "electrical ether" was discussed, as was something similar to explain magnetism. The most favored theory of heat in the eighteenth century was that it was an imponderable fluid. Lavoisier was to speak of heat as a fluid, in fact "the most subtle, the most elastic,"[33] and, he implied, the most rare of all fluids, which he later would call *calorique*.

In his early speculations on "airs," Priestley tended to think of them as variations of atmospheric air, but soon he brought in phlogiston as an explanation of differences of properties. Most famously, he referred to the air that allowed a mouse to live longer than in ordinary air as "dephlogisticated air."[34] Depriving such air of more than its normal share of phlogiston would explain why it took longer for the air to be used up, or "phlogisticated." The gas, later called *nitrogen*, was "phlogisticated air" for Priestley. Thus, the atmosphere consisted essentially of a mixture of phlogisticated and dephlogisticated air.[35] In explaining how to make dephlogisticated air, Priestley said that great care should be taken to exclude all phlogistic matters, as they are "exactly plus and minus to each other."[36] One cannot help thinking of the analogy with Franklin's theory of electric charge; as Priestley had made a special study of electricity before he had turned to pneumatic chemistry, it is likely that such an analogy was not too far from Priestley's mind. We may compare the English chemist's views with those of his Dutch contemporary, Ingen Housz, who spoke of air departing from its simple nature "by the addition or subtraction of something."[37] To explain the properties of light inflammable air (hydrogen) as phlogiston was easy, as something that burned entirely and left no residue apart from a few drops of water was a good candidate for being phlogiston itself, or phlogiston plus water, the latter alternative being favored by Priestley.[38]

The difficulty experienced by our ancestors in grasping the concept of a gaseous state is illustrated by their tendency to think that air could have chemical or physiological properties only if it held tiny solid particles in suspension. We have evidence of this in the seventeenth century with Mayow's "nitro-aerial particles."[39] Again, from at least the sixteenth century, it was well-known that people could be poisoned by breathing the air in certain mines; it raised no question of postulating a poisonous gas. The only known poisons were solids or liquids, so they explained the deaths by postulating the existence of tiny arsenical particles in the atmosphere.[40] Even Joseph Black in the eighteenth century conjectured whether the peculiar properties of "fixed air" might be due to "an exceedingly subtile powder."[41]

It is important to try to see into the mind of the experimenters of the eighteenth century. In the case of Hales's work on airs, that endeavor is not very difficult. Interestingly, Hales ignored simple qualitative experiments, which might have revealed that he was dealing with different gases, in favor of a rigidly quantitative approach. This obvious fact may not have been made sufficiently explicit by some historians. Hales was so convinced that he was dealing with a single substance—atmospheric air—that his overriding concern was to measure the quantity of air evolved in different experiments. We now know that his experiments produced gases later called *hydrogen, oxygen, coal gas*, and so on, but he never thought of examining the various "airs" because his mind was made up in advance. Having prepared these gases, he measured their volumes and then threw them away unexamined. His mind-set was at least as important as his apparatus. In the history of chemical apparatus, he is a pioneer, but one blinded, as some later Enlightenment writers said, by a deep conceptual prejudice. We have no less a figure than Priestley, who spoke of his predecessors' deep prejudice that air was one thing.[42]

When the full early history of gases comes to be written, it will include chapters on mineral waters, mines, caves, and other natural sources of exhalations, not to mention a survey of meteorological conditions. If one were making a study of vocabulary, something would have to be said about "exhalations," "smokes," "steams," "damps," "vapors," "mofettes," "spirits," and so on. For present purposes, however, mentioning the mineral water dimension should be sufficient, because this was a matter of popular interest in the eighteenth century as part of the expanding market for medical remedies. This was certainly the starting point of William Brownrigg, whose "Experimental enquiry into the mineral elastic spirit, or air, contained in Spa water" was published by the Royal Society in 1765.[43] He related the sharp taste of Spa and Pyrmont waters to the "air" they contained. His work was a source of inspiration for Priestley who, early in his career as a pneumatic chemist, published a pamphlet on the preparation of artificial Pyrmont water, entitled, *Directions for Impregnating Water with Fixed Air* (1772). As far as apparatus is concerned, Brownrigg's great contribution was the beehive shelf, which was adopted by Priestley and many subsequent experimenters (figure 4.4).

Apparatus

Although some branches of science have depended heavily on the invention, construction, and use of sophisticated apparatus, this is not true of pneumatic chemistry in the pre-Lavoisier era. Gases could be prepared and collected with the very simplest apparatus, as the writings of Joseph Priestley amply testify. I have argued elsewhere that, although no doubt important religious and philosophical dimensions are found in Priestley's work, we should not overlook the mundane and severely practical aspect of finance.[44] In the first part of his scientific career, he had invested a substantial proportion of his income in expensive books and electrical apparatus to write about natural philosophy, his *History of Electricity* being particularly successful. However, he suffered from major financial problems, and one of the things that attracted him to pneumatic chemistry was that its study involved so little expense. In his memoirs, he refers constantly to the problems posed by his expenses.[45] At the end of his first publication on gases, he congratulated himself that "the apparatus with which the principal of the preceding experiments were made is exceedingly simple and cheap" (figure 4.5).[46]

In the interests of economy, Priestley made full use of kitchen utensils: an "earthenware trough" ... "such a one as is commonly used for washing linen,"[47] "pots and dishes of various sizes," including "common tea dishes," a kettle, a basin, candles, and even a red-hot poker and the kitchen fireplace. The household also supplied other objects, such as "a tall beer glass" and a tobacco pipe. Priestley was able to adapt objects made for other purposes, including a gun barrel. Even the humblest household utensil could be adapted to reveal a whole new world. Encouraged by his early successes, Priestley exclaimed, "By working in a tub of water ... we may perhaps discover principles of more extensive influence than even that of gravity itself."[48]

Although Priestley used mainly simple apparatus, he did not confine himself entirely to household objects. He constructed a few

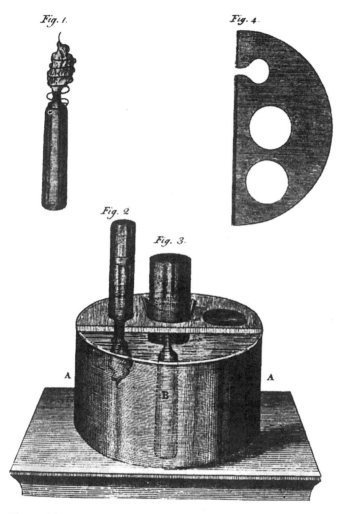

Figure 4.4
Brownrigg's beehive shelf. From *Philosophical Transactions* 55(1765).

Figure 4.5
Priestley's laboratory, showing very simple apparatus. From *Experiments and Observations on Different Kinds of Air* (London, 1774).

pieces of apparatus himself (e.g., by perforating a tin plate with many holes to enable mice to have access to outside air, or by making a wire stand) (figure 4.6). That is not to say that he had no contact with instrument makers. In particular, Mr. W. Parker of Fleet Street, London, manufacturer of cut glass and maker of philosophical instruments, was happy to provide him with glassware free of charge.[49] This included several phials with expensive ground-glass stoppers. After Priestley went to America in 1794, the son of W. Parker offered to send him any philosophical apparatus he might require. Much earthenware equipment, mainly retorts, was supplied to Priestley by his friend, Josiah Wedgwood. Rubber tubing was nonexistent in the eighteenth century; when Priestley wanted a flexible tube, he used one made of leather or otherwise of glass or earthenware.

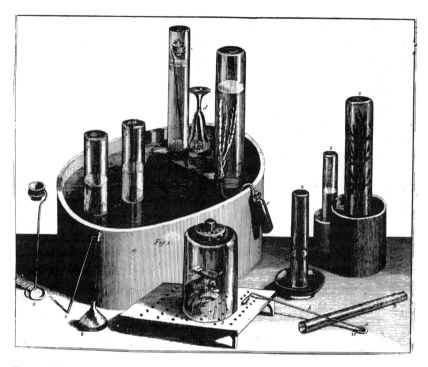

Figure 4.6
Priestley's examination of different "airs." From *Experiments and Observations on Different Kinds of Air* (London, 1774).

Most experimenters of the time used glass vessels, such as flasks, to collect gases. They had the great advantage that one could witness the progressive displacement of water in the vessel by the gas and then carry out further experiments on the gas, such as combustion, within the collecting vessel itself. The gas could not be lost through the sides of the container. The same cannot be said about animal bladders, which proved attractive to a few investigators. They were used very occasionally by Priestley (who used bladders to weigh gases),[50] but probably the most famous instance of collecting gas in a bladder was that of the Swedish apothecary Scheele. In his book, *On Air and Fire*, published in 1777, he described the preparation of "fire air" (oxygen) by heating saltpeter with oil of vitriol. Red fumes of nitric acid were first produced, but later Scheele connected to the retort a bladder containing some milk of lime, which absorbed any remaining red fumes. The bladder gradually filled with a gas, which then was trans-

Figure 4.7
Scheele's collection of "fire air" (oxygen) in a bladder. From *Von der Luft und dem Feuer* (Upsala, 1777).

ferred to a bottle filled with water, when the water was displaced by the gas. Scheele collected other gases in bladders (figure 4.7). This was usually fairly satisfactory, especially if the bladder was moist. Inflammable air (hydrogen), however, penetrated all bladders, whether wet or dry. When large hydrogen balloons were later made by the French, the fabric had to be treated with varnish to prevent escape of the hydrogen.

Yet, we should not assume that experimental success depended mainly on apparatus. An easy approach would be to study apparatus in greater detail than that presented here and to ignore the undoubted experimental skill of such people as Scheele, Cavendish, and Priestley. Thus Cavendish, for example, was able skillfully to transfer a gas from one bottle to another by pouring upward through a funnel under water (figure 4.8).[51] The manual dexterity of an operator when dealing with what Black termed "these slippery substances"[52] could sometimes compensate for simple apparatus.

Nor should we neglect entirely the reagents. I would argue that the chemicals used were sometimes of even greater importance than was the apparatus, remembering that the purity of reagents could still not be taken for granted in the eighteenth century.[53] Thus, when Priestley found that he could obtain a new air, in which a candle would burn very brightly, by heating *mercurius calcinatus per se* (our oxide of mercury, obtained by heating mercury in air at an appropriate temperature), he was suspicious about the purity of the reagent, as it had

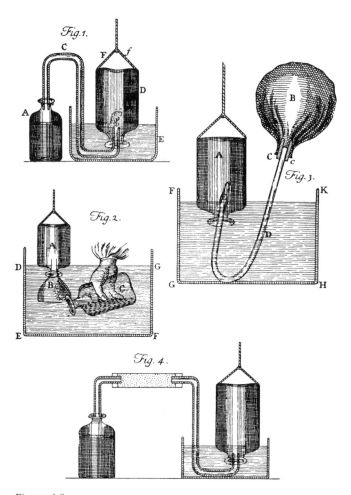

Figure 4.8
Cavendish's apparatus for the manipulation of "airs." From *Philosophical Transactions* 56(1766).

been "bought at a common apothecary's" and might have had nitric acid used in its preparation. His friend, Waltire, however, kindly supplied him with some *mercurius calcinatus*, which he assured him was genuine. Not completely satisfied even then, however, Priestley took advantage of a visit to Paris a few months later in 1774 to ask for a sample of *mercurius calcinatus* "prepared by M. Cadet, of the genuineness of which there could not possibly be any suspicion."[54] Having satisfied himself that no trace of nitric acid was present, he could eliminate the hypothesis that the bright burning of the candle was caused by the presence of nitrous particles.

PRIESTLEY AND LAVOISIER

In discussing Priestley, we must admit that we have given special attention to one aspect of his work that might be considered to show him in a poor light. We should, therefore, remember that Priestley was the most successful of the line of British pneumatic chemists, which included Hales, Black, and Cavendish. Priestley is credited with the isolation of numerous gases that, reported under their modern names, would include oxygen, nitric oxide, nitrous oxide, nitrogen dioxide, carbon monoxide, sulfur dioxide, ammonia, and hydrogen chloride. (In the case of the latter two gases, which are extremely soluble in water, Priestley deserves credit for using the method of collection over mercury.) Partington is happy to credit him with the "discovery" of all these gases.[55]

More recent historians exercise greater caution in speaking of "discoveries." Lavoisier's oxygen, for example, was only "dephlogisticated air" for Priestley. Yet, his experimental investigation of this gas is usually accepted as a step along the path to Lavoisier's oxygen, which was to become the centerpiece of the new chemistry. Also, although at times Priestley interpreted these gases as variants of atmospheric air, at other times he pointed out the many differences between these gases and common air. In any case, many of his contemporaries, including Cavendish, understood that these gases were fairly distinct. Priestley's reputation is liable to suffer from his transparent honesty and his willingness to commit his early speculations to print alongside descriptions of his experiments. Most other chemists were more discrete, so only by having access to their private laboratory notebooks does the historian have any chance of reconstructing their early ideas.[56] This study is not intended to belittle Priestley but simply to take advantage of his frankness in sharing his thoughts with the reader.

We could argue that Priestley was led astray by the language he used or, at the very least, that the terms he used were anything but helpful in achieving a better understanding of his experiments. The expression *inflammable air* does not have the same status as its translation as *coal gas* or *hydrogen*. These latter terms necessarily denote distinct substances, whereas inflammable air could be interpreted (and often was by Priestley) as (atmospheric) air with the property of inflammability. This property had little more status than, say, a color (i.e., it could be changed without losing the air). Thus, Priestley claimed to be able to "deprive inflammable air of all its inflammability."[57]

Priestley, like Black before him, was unwilling, despite all his pioneering work, to "introduce new terms, or change the significance of old ones."[58] Although this remark was made at the beginning of his research, when his knowledge was understandably slight, his resolution to continue to use the term *air* to describe all his gases persisted to the end of his experimental career.

Some blame must attach to Priestley for continuing to use the term *air* even after he understood that he was dealing with gases very different from common air. Even toward the end of his career, Priestley struck a conservative note in explicitly preferring the traditional term *air*, as used by his British predecessors, to the word *gas*, which he thought "unnecessary."[59] Yet in France, Lavoisier too had continued to speak of gases as *airs* in the 1780s. Thus, in 1783, 4 years after he had coined the term *oxygine*, he spoke of *air nitreux*, *air inflammable*, and even *air vital*.[60] In 1784, his contribution to the nomenclature of gases was to refer to "fixed air" (sometimes known as *acide crayeux*) as *acide de charbon*, thus emphasizing both its composition and its acidic nature but playing down its gaseous state.[61] In March 1787, 3 weeks before his famous memoir on the reform of chemical nomenclature, he was still prepared to speak of *air vital*, although he had also been using the synonym *oxygène* for some time.[62]

Seemingly, Lavoisier's program for the foundation of a new chemistry, as set out in his *Traité élémentaire* of 1789, prompted him to abandon the term *air* in favor of *gas*. In the very first chapter of his textbook, on "the formation of elastic aeriform fluids," he wrote, "Henceforth I shall express these elastic aeriform fluids by the generic term *gas*."[63]

Had Lavoisier by any chance been reading Helmont? The answer is clearly given by Lavoisier himself. He was following the precedent of his former colleague in the Academy of Sciences, P. J. Macquer.[64] Macquer was the author of an influential *Dictionary of Chemistry*. In the first edition, published anonymously in 1766, Macquer had included a short entry under the heading "Gas." This referred to the invisible volatile parts of certain bodies that cannot be retained. In the second edition of the *Dictionary*, published 12 years later (1778), the author greatly expanded the article "Gas" from two paragraphs to 100 pages. Incoercibility was no longer emphasized. Instead, gases were characterized as invisible and elastic, an obvious property of atmospheric air.

Macquer concluded that "the air itself should be regarded as a true gas—the simplest and purest of all gases."[65] He admitted that the various gases recently prepared by Priestley could be called air:

> However, as some of these aeriform substances have nothing in common with the air, it is better to use the term gaz (*sic*) for all of them, which is the term van Helmont and other chemists before Hales had given in general to expansible, volatile substances which could not be retained in ordinary apparatus ...

He went on to toy with the alternative name of spirit (*esprit*) but, realizing its ambiguity, came down in favor of *gas* "because it is a barbarous name, which means nothing in our language" and, therefore, could not cause confusion. Interestingly, Black had rejected the term *gas* precisely on the literary grounds that it was a barbarous term.[66] Macquer argued that to use the term *air* to describe gases was seriously misleading,[67] and Lavoisier probably took this to heart.

The new chemistry of Lavoisier was sometimes described as the "antiphlogistic theory," a polemical term introduced by Kirwan in 1787.[68] Yet, although this term was sometimes used in France, Fourcroy tells us that it was mainly used by "foreigners" (*les étrangers*).[69] Fourcroy himself preferred to describe the new theory as "the pneumatic theory" (*la doctrine pneumatique*), which emphasises how much the theory owed to the greater understanding of gases in the second half of the eighteenth century. Definite advances had taken place in apparatus during that century, advances of an evolutionary kind. But it would hardly be an exaggeration to describe *conceptual changes* as revolutionary.

In the end, of course, we do not have to choose between apparatus and ideas: Both were necessary. Nevertheless, the emphasis in this chapter has been on the overriding importance of the mind of the experimenter. This applied in the earliest instance to taking air seriously and recognizing it as a material substance. A major step was to advance from "air" to "airs." For a long time, when people could think of only one air, any new form of air was simply regarded as impure (atmospheric) air. To offer advice to scientists of the past in the light of later knowledge is not usual. Yet, if Black only had used a name that proclaimed to all the essential difference between his "fixed air" and ordinary air, his work might have been recognized by his contemporaries as more of a landmark.

Similarly understandable is that the early Priestley should have thought of many of his new gases as simply different forms of common air. Yet, when he came to appreciate more their essential differences, he might, with greater confidence, have used a name different from *air*, particularly as the term *gas* was available and waiting in the wings. As late as 1786, Priestley was still postponing the time when he would be ready to present a general theory of "airs."[70] He persisted to the end in his ad hoc explanations of gaseous phenomena. If everyone can agree that tools are important to the worker, we may argue the case for the tool of language. This was, of course, an argument that, immediately after Priestley's many contributions to pneumatic chemistry, was to be used by Lavoisier and his colleagues in their *Method of Chemical Nomenclature*.[71]

NOTES

Note on illustrations. Unfortunately it is difficult to illustrate some of the major points of this chapter, such as Helmont's view of *gas* as incoercible, Boyle's exclusively physical approach to air, or Black's failure to use pneumatic apparatus for his discovery of "fixed air," all of these features being negative. It is obviously easier to illustrate apparatus than ideas.

1. *The Guardian*, 12 October 1995.

2. Robert Boyle, *Works*, vol. 5, T. Birch, ed. (London, 1744), 178 (author's italics).

3. *Vegetable Staticks* (London, 1727), 179–180.

4. A brief history of gases is given by A. F. Fourcroy in his *Système des Connaissances Chimiques*, vol. 1 (Paris, 1801), 27–36. According to Fourcroy and most subsequent authors, the French entered the story in a major way only with the work of Lavoisier. For a compact history in English, see J. R. Partington, *Short History of Chemistry*, 2nd ed. (London: Macmillan, 1951), chap. 6.

5. Sigaud de la Fond, *Essai sur Différentes Espèces d'Air Fixe* (Paris, 1785), xiv.

6. J. B. van Helmont, *Oriatrike* (London, 1663), 106 (author's italics).

7. Hales, *loc. cit.*, 180.

8. The original reads, "leur incoercibilité les soustreira longtemps à nos recherches." *Encyclopédie*, vol. 7, 1757, 520.

9. Joseph Black, *Experiments upon Magnesia Alba* (1756; reprinted Edinburgh: Alembic Club, Reprint, 1944), 16.

10. D. McKie, "Thos. Cochrane's Notes of Black's Chemical Lectures," *Annals of Science* 1(1936): 105. See also D. Macbride, *Experimental Essays* (London, 1764), 52.

11. John Robison to James Watt, 25 February 1800, *Partners in Science*, Eric Robinson and Douglas McKie, eds. (London, 1970), 339. See also Thomas Cochrane, *Notes from Doctor Black's Lectures on Chemistry 1767/8*, Douglas McKie, ed. (Wilmslow: Imperial Chemical Industries, 1966).

12. Lassone, *Mémoires de l'Académie Royale des Sciences* (Paris, 1776).

13. Boyle, "The General History of the Air," in *Works*, T. Birch, ed. vol. 5, 107.

14. The law traditionally ascribed to Boyle actually depended on several other independent investigators. See I. B. Cohen, *Nature* 204(1964): 618.

15. John Parascandola and Aaron Ihde, "History of the Pneumatic Trough," *Isis* 60(1969): 351–361 (355).

16. Hales, *op. cit.*, 105.

17. Ibid.

18. Priestley, *Experiments and Observations on Different Kinds of Air*, vol. 1, (London, 1774), xxii.

19. See, for example, Rhoda Rappaport, "Rouelle and Stahl—the phlogistic revolution in France," *Chymia* 7(1961): 73–102 (84–85, 78–79).

20. Boyle, *Works*, vol. 3, 463.

21. *Experiments upon Magnesia Alba*, 30.

22. Ibid.

23. Isaac Newton had speculated about the air contained in bodies in his *Opticks*, 4th ed., (London, 1730; reprinted in 1931, *Query* 31: 396.

24. A. Baumé, *Chimie, Expérimentale et Raisonnée*, vol. 3 (Paris, 1773), appendix on fixed air, 694.

25. Ibid., 697.

26. Lavoisier, *Opuscules Physiques et Chimiques* (Paris, 1774), 320.

27. Priestley, *Philosophical Transactions of the Royal Society of London* 62(1772): 147–264 (180).

28. Ibid., 200.

29. J. Priestley, *Experiments and Observations on Different Kinds of Air*, vol. 1 (1774), 129. This description is even adopted as the title of a section.

30. P. J. Macquer, "Gas," in *Dictionnaire de Chymie* (Paris, 1776).

31. R. Watson, *Chemical Essays*, vol. 1, 2nd ed. (London, 1782), 167.

32. G. N. Cantor and M. J. S. Hodge, *Conceptions of Ether: Studies in the History of Ether Theories* (Cambridge, UK: Cambridge University Press, 1971).

33. Lavoisier, "Mémoire sur le combustion en générale," *Mémoires de l'Académie des Sciences*, Paris, 1777. Lavoisier, *Oeuvres*, vol. 2, 228.

34. J. Priestley, *Philosophical Transactions of the Royal Society of London* 65(1775): 387.

35. Priestley, *Experiments and Observations on Natural Philosophy*, London, vol. 3 (1786), 406.

36. Priestley, *Experiments and Observations on Different Kinds of Air*, vol. 2 (1775), 59–60.

37. Ingen Housz, *Expériences sur les Végétaux*, vol. 1, 2nd ed. (Paris, 1787–89), 140.

38. J. Priestley, *Experiments and Observations on Natural Philosophy*, vol. 3 (1786), 406.

39. John Mayow, *Medico-Physical Works* (1674; reprinted 1957, Edinburgh: Alembic Club), 75 ff.

40. See, for example, P. J. Macquer, "Mofettes," in *Dictionnaire de Chymie* (Paris, 1766).

41. Black, *op. cit.*, (ref. 9), 30.

42. J. Priestley, *Experiments and Observations on Different Kinds of Air*, vol. 2 (1775), 30.

43. W. Brownriqg, *Philosophical Transactions of the Royal Society of London* 55(1765): 218.

44. M. Crosland, "Priestley Memorial Lecture: A practical perspective on Joseph Priestley as a pneumatic chemist," *British Journal for the History of Science* 16(1983): 223–238.

45. J. Priestley, *Autobiography of Joseph Priestley*, with Introduction by Jack Lindsay (Bath: Adams & Dart, 1970). See, for example, pages 116, 120, 126.

46. J. Priestley, *Philosophical Transactions of the Royal Society of London* 62(1772): 250.

47. J. Priestley, *Works*, vol. 1, J. T. Rutt, ed., (London: 1817–32), 77.

48. J. Priestley, *Experiments and Observations on Different Kinds of Air*, vol. 2 (London, 1774–77), viii.

49. Priestley, *Works*, vol. 1, 216.

50. Priestley, *Experiments and Observations on Different Kinds of Air*, vol. 2 (1775), 93–94.

51. Cavendish, "Three papers containing experiments on factitious air," *Philosophical Transactions of the Royal Society of London* 56(1766): 141–184.

52. J. Black, *Lectures on the Elements of Chemistry*, vol. 2, John Robison, ed., (Edinburgh: 1803), 102.

53. Regarding the subject of purity of reagents, see M. Crosland, "Changes in chemical concepts and language in the 17th century," *Science in Context* 9(1996): 225–240.

54. Priestley, *Experiments and Observations on Different Kinds of Air*, vol. 2 (1775), 35–36.

55. Partington, *Short History*, 148.

56. See, for example, F. L. Holmes, *Lavoisier and the Chemistry of Life* (Madison: University of Wisconsin Press, 1985). The Paris Academy of Sciences had instituted a system of sealed notes, in which a savant could make any claim, however reckless, confident in the knowledge that it would not be made public unless the author wanted it in the light of subsequent evidence. Priestley had no access to such a system.

57. J. Priestley, *Philosophical Transactions of the Royal Society of London* 62(1772): 180.

58. Ibid., 148.

59. J. Priestley, *Experiments and Observations on Different Kinds of Air, and Other Branches of Natural Philosophy*, vol. 1, (Birmingham, 1790), 9.

60. M. Daumas, *Lavoisier, Théoricien et Expérimentateur*, (Paris: Presses Universitaires de France, 1955), 50, 51.

61. Ibid., 54.

62. Ibid., 61.

63. Lavoisier, *Elements of Chemistry* (Edinburgh, 1790), 15.

64. Ibid., 50.

65. P. J. Macquer, *Dictionnaire de Chymie*, 2nd ed., vol. 1 (Paris, 1778), 536.

66. D. McKie, *loc. cit.*, 110.

67. Macquer, *Dictionnaire de Chymie*, 635–636.

68. Richard Kirwan, *An Essay on Phlogiston and the Constitution of Acids*, 2nd ed. (London, 1789), 7.

69. Fourcroy, *Système des Connaissances Chimiques*, 36–49 (47).

70. *Experiments and Observations on Natural Philosophy*, vol. 3 (1786), 401.

71. De Moreau, Lavoisier, Berthollet, and Fourcroy, *Méthode de Nomenclature Chimique* (Paris, 1787).

BIBLIOGRAPHY

Badash L., "Joseph Priestley's apparatus for pneumatic chemistry," *Journal of the History of Medicine* 19(1964): 139–155.

Crosland M. P., "The development of the concept of the gaseous state as a third state of matter," *Proceedings of the Tenth International Congress of the History of Science* 2(1962): 851–854.

Eklund J., "Of a spirit in the water: Some early ideas on the aerial dimension," *Isis* 67(1972): 527–550.

McKie D., "Priestley's laboratory and library and other of his effects," *Notes and Records of the Royal Society of London.* 12(1956): 114–136.

Parascandola J., and A. J. Ihde, "History of the pneumatic trough," *Isis* 60(1969): 351–361.

Partington J. R., *A History of Chemistry*, vol. 3 (London: Macmillan, 1962).

5

MEASURING GASES AND MEASURING GOODNESS
Trevor H. Levere

As Maurice Crosland shows (chapter 4), gases, although occasionally chemically characterized from the turn of the sixteenth and seventeenth centuries, were not comprehensively incorporated into chemical theory until the second half of the eighteenth century. Most chemists thought of air as the matrix in which chemical reactions occurred, something that could be modified or contaminated, like a sponge with dirty water, although not in itself a proper subject for chemical investigation. However, the development of gasometry and eudiometry—the measurement and control of the volumes of gases and the measurement of their goodness for respiration, respectively—were specifically chemical, and the gasometers and eudiometers used by chemists in these operations were late eighteenth-century inventions. Accordingly, we concentrate on pneumatic apparatus from the work of Joseph Priestley through the early nineteenth century. Hales's work and the pneumatic trough (figure 5.1) rendered possible the isolation of different gases, which could then be investigated chemically. This was accomplished sometimes with a eudiometer, sometimes by using a gasometer to provide a controlled flow of a gas for a variety of different reactions, including the demonstration of the composition of water.[1]

Medical and chemical traditions converge in eudiometry. The medical tradition connecting air and health is ancient, related to humoral pathology, and very much alive in the eighteenth century. Thus, Robert Plot observed at the beginning of the century that "the Inhabitants of Fenny and Boggy countries, whose spirits are clogg'd with perpetual *Exhalations*, are generally of a . . . stupid, and unpleasant conversation."[2] The quality of an air in terms of health was of increasing concern as the eighteenth century advanced. Malaria and jail fever were just two of the diseases that were believed to be spread by "miasmata."

Public health reforms arose directly from such views of the role of air in transmitting disease. Sir John Pringle's *Observations on the Diseases*

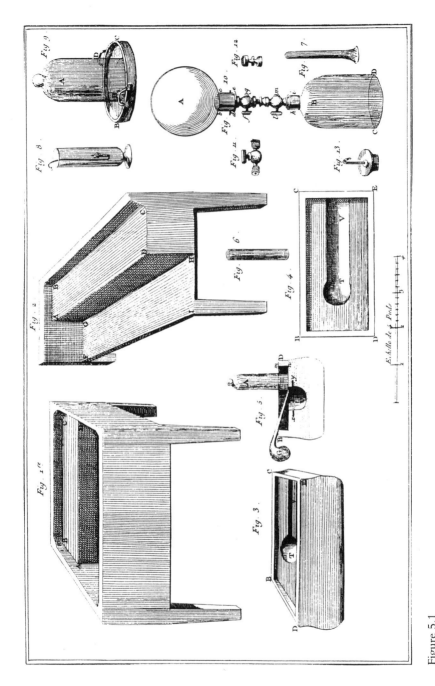

Figure 5.1
Hales's pneumatic trough. From A. L. Lavoisier, *Elements of Chemistry*, R. Kerr, transl. (Edinburgh: William Creech, 1790), plate 5.

of the Army (London, 1752), John Heysham's *An Account of Jail Fever or Typhus Carcerum as It Appeared at Carlisle in the Year 1781* (1782), and John Haygarth's *A Letter to Dr. Percival on the Prevention of Infectious Fever* (1801) all discussed ventilation as a factor in reducing the incidence of disease. A generally accepted belief connected air and health.

Consumption was a leading cause of death in the second half of the century. Public health rested on the collection of epidemiological data, and these generally took a quantitative form. The goodness of air might be an imponderable, but that did not mean that it could not be quantified, and eighteenth-century natural philosophers were good at quantification.[3] The question of how good an air was for health meant: How good was it for respiration? How well would it support life?

The answer was to come from chemistry. Once respiration was understood as a chemical process (and it was so understood by Priestley and Lavoisier, Senebier and Ingen Housz), researchers were able to imagine a chemical test that, by providing a laboratory analog to respiration, would show how good an air was. Further, once the existence of different species of air—Priestley's different *kinds* of air—had gained credence, the healthfulness and vitality conferred by air could be associated with a single constituent of atmospheric air. This was clearly the case in Lavoisier's new oxygen-centered theory. Now the goodness of air could be measured by the percentage of oxygen it contained. What was needed was a reaction that would identify oxygen qualitatively and a set of apparatus and procedures that would measure its percentage presence in the air. Measuring goodness would depend on measuring gases, so that eudiometry would become a branch of gasometry and would simultaneously involve qualitative and quantitative procedures. Eudiometers could double as gasometers.

Although gases were measured primarily by volume in the years around the time of Lavoisier's work, some researchers[4] often based their analyses on weight. As the weight of a gas was always much smaller than that of its containing vessel, the gravimetric examination of gases involved the determination of relatively small differences between relatively large weights. This meant using the newly developed precision balances. Without concentrating on the balance, we should note that precision balances were significant in gasometry and eudiometry, and their development in the second half of the eighteenth century was timely for these new practices.

In the 1750s, Joseph Black had given his results on magnesia alba and fixed air to the nearest grain when weighing 2 drams. That meant a sensitivity of approximately 1 part in 250.[5] The first great precision balance was made for Henry Cavendish by John Harrison, inventor of the marine chronometer. Cavendish was soon carrying out observations to one-tenth of a grain, a 10-fold increase in accuracy and sensitivity over Black's: We are up to 1 part in 2500.[6] Jesse Ramsden, arguably the finest instrument maker of the eighteenth century, made a balance that was used by Cavendish and others in the Royal Society of London in the 1780s and was sensitive to a hundredth of a grain, a further 10-fold increase in accuracy, to at least 1 part in 25,000.[7]

Modern estimates put the sensitivity of Lavoisier's great balance made by Fortin, formerly engineer to the king, at 1/400,000, an accuracy that cannot be beaten by the best mechanical balances today.[8] Michael Faraday, no slouch when it came to exact laboratory procedures, argued in 1829 that the chemist's "best balance" should be sufficiently delicate to weigh from 600 to 1000 grains and downward, "indicating distinctly and certainly materials equal to the 1/50,000 or 1/60,000 part of the weight in the scale";[9] that is, they should be able to measure differences in weight of the order of one-hundredth of a grain or better.[10] The only cases in which such precision was really important for chemists were those involving the weight of small volumes of a gas. In such cases, the containing vessel weighs far more than the gas, typically by a factor of 50 or more, so very small differences in weight are significant. However, there too, sources of error are many. Typical of late eighteenth-century measurements of air was Lavoisier's measurement of oxygen in the key experiment of the reversible oxidation of mercury. He heated mercury in atmospheric air in a closed vessel and found that the air had lost one-sixth of its bulk (we would expect just over one-fifth, or 21%); he then took 45 grains of the red matter produced (mercuric oxide), heated it, and collected 41.5 grains of mercury, and "7 or 8" cubic inches of oxygen.[11] The weights of mercury show approximately 6% error—the residue was too large. However, the measure of oxygen (7 or 8 cubic inches) was properly imprecise. Many chemists achieved similar results, and clearly measuring gases was not at first a very exact business.

In the early eighteenth century, Stephen Hales made observations about gases to which later chemists applied the understanding that different kinds of air were different chemical species. He devised an apparatus for separating gaseous products from reactants (figure 5.2) Among his observations was the notion that absorption took place when common air was mixed with the air obtained from a mixture of

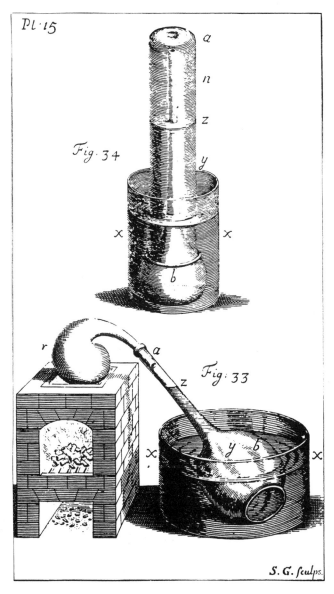

Figure 5.2
Apparatus for separating gaseous products from reactants. From Stephen Hales, *Statical Essays: Containing Vegetable Staticks ... Also a Specimen of an Attempt to Analyse the Air, by a Great Variety of Chymio-Statical Experiments* (London: Printed for W. Innys and R. Manby, 1733), plate 15.

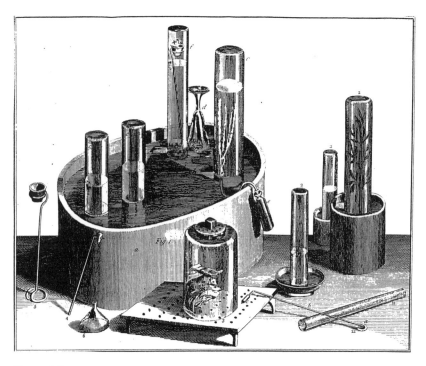

Figure 5.3
Pneumatic apparatus (i.e., eudiometer). From Joseph Priestley, *Experiments and Observations on Different Kinds of Air*, 2nd ed., vol. 1 (London: J. Johnson, 1775), frontispiece.

spirits of niter and Walton pyrites, a sulfide, and that in the process the airs, from being clear, became "a reddish turbid fume."[12] Joseph Priestley was struck by this experiment but was not sanguine about repeating it, as he thought that the result might be peculiar to the Walton pyrites.[13] Cavendish suggested that the red appearance of the mixture might be due to the spirits of niter only; Priestley, thus encouraged, dissolved various metals in the acid and obtained and named nitrous air (our nitric oxide). He already knew that some kinds of air supported respiration better than did others, a fact that could be quantified by asphyxiating more or less standard laboratory mice in standard volumes of different gases. His discovery of nitrous air, and the observation that, when mixed with atmospheric air over water, it turned brown and suffered a volume decrease, offered him a mouse-free route to measuring the goodness of air, because atmospheric air suffered a smaller diminution than vital air. He thus invented a eudiometer (figure 5.3).[14]

Priestley's method was straightforward, although not unproblematic, as the same proportional diminution could be obtained with an excess and with a much smaller part of oxygen in the test gas. "When I want to measure the goodness of any kind of air," he tells us, "I put two measures of it into a jar standing in water; and when I have marked upon the glass the exact place of the boundary of air and water, I put to it one measure of nitrous air; and after waiting a proper time, note the quantity of its diminution. If I be comparing two kinds of air that are nearly alike, after mixing them in a large jar, I transfer the mixture into a long glass tube, by which I can lengthen my scale to what degree I please."[15]

When working with exceedingly small quantities of air, Priestley filled a narrow tube containing a length of wire with gas and removed the wire in a jar of nitrous air; that air then took the place of the wire and combined with any good air in the tube, producing a measurable diminution. He obtained reasonably good results with these methods. Atmospheric air contained between one-fifth and one-sixth of an air that combined with nitrous air and was then absorbed; one-fifth (20%) was the maximum for that air.

The experiments were not particularly easy to perform, in spite of the simplicity of the apparatus. One can imagine determining the volume of the wire in the experiment with small quantities of gases, then extracting it from the tube, and allowing nitrous air to enter, without loss or excess. Priestley knew the importance of practice:[16]

> I would not have any person, who is altogether without experience, to imagine that he shall be able to select any of the following experiments, and immediately perform it, without difficulty or blundering. It is known to all persons who are conversant in experimental philosophy, that there are many little attentions and precautions necessary to be observed in the conducting of experiments, which cannot well be described in words, but which it is needless to describe, since practice will necessarily suggest them; though, like all other arts in which the hands and fingers are made use of, it is only *much practice* that can enable a person to go through complex experiments, of this or any other kind, with ease and readiness.

In the late 1770s and early 1780s, many modifications and supposed improvements were made over Priestley's original apparatus.[17] For some years, the best results were obtained by the Abbé Felice Fontana, who, since 1766, was natural philosopher at the court of the Grand Duke Pietro Leopoldo and in 1775 was appointed as direc-

tor of the Royal Cabinet of Natural Philosophy and Natural History of
Florence.[18] He was concerned with the goodness of air for respiration,
a question of chemical, medical, and environmental interest.[19] He
observed:[20]

> [T]he whole importance and difficulty now consist in finding con-
> venient and exact instruments, which will reliably measure for us the
> different degrees of goodness of the air that we breathe. In this alone
> everything consists; time and experience will do the rest.

Fontana described a series of different eudiometers (figures 5.4 and
5.5) working on the same principles. He filled two holders with nitrous
air and the air to be tested; opened the valve between them, thereby
allowing the gases to mix; added mercury to restore the volume; and
determined the increase in weight, which gave him a measure of the
volume decrease, itself a measure of goodness for respiration. Until
recently, none of these instruments was known to exist, but the recent
discovery of one in the Uffizi in Florence suggests that the instruments
were not just devised and described by Fontana but were, in fact, built.
What became known in England as "Fontana's eudiometer" (figure
5.6) used a glass measure with a brass cap fitted with a slide (figure
5.7).[21] The slide enabled one to seal the measure on a known volume
of gas and to release the gas at will. Careful use of this apparatus yielded
results "that very seldom differ by more than one fiftieth of a measure,"
an accuracy that one reviewer, Tiberius Cavallo, said "can hardly be
believed by persons who have not observed it." The way in which
Fontana used his apparatus was largely responsible for the constancy
and reproducibility of his results; note that accuracy meant precision
and reproducibility for Fontana and for Cavallo. Leaving the gases
mixed for a greater or lesser interval of time led to very different results,
so Fontana was at pains to specify and standardize his experimental
procedures.[22]

Although Fontana has deserved his reputation for accuracy and
ingenuity, he was not the first Italian to devise a eudiometer, nor did
he invent the name. That credit belongs to Marsilio Landriani,[23] who,
like Fontana, was concerned with what he saw as the necessary insalu-
brity of city air. He argued that measuring goodness was a step on the
way to the local amelioration of the air, so that even in cities one could
breathe a healthy dose. Landriani's instrument (figure 5.8) also worked
by mixing atmospheric air with nitrous air, but he did it over water

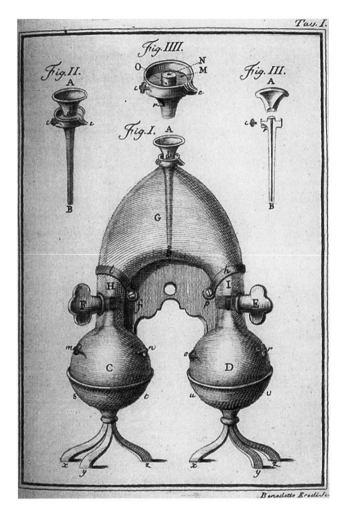

Figure 5.4
One of Fontana's eudiometers. From Felice Fontana, *Descrizione, ed Usi di Alcuni Stromenti per Misurare la Salubrità dell'Aria* (Firenze: Gaetano Cambiagi, 1775), plate 1. Courtesy the Museo di Storia della Scienza, Florence.

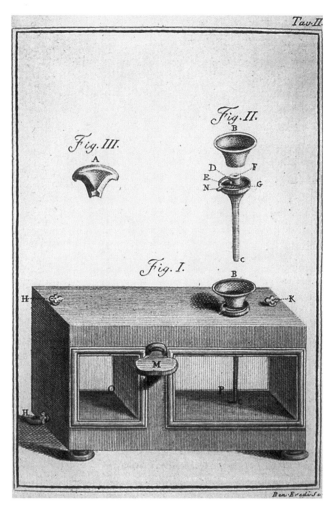

Figure 5.5
Another of Fontana's eudiometers. From Felice Fontana, *Descrizione, ed Usi di Alcuni Stromenti per Misurare la Salubrità dell'Aria* (Firenze: Gaetano Cambiagi, 1775), plate 2. Courtesy the Museo di Storia della Scienza, Florence.

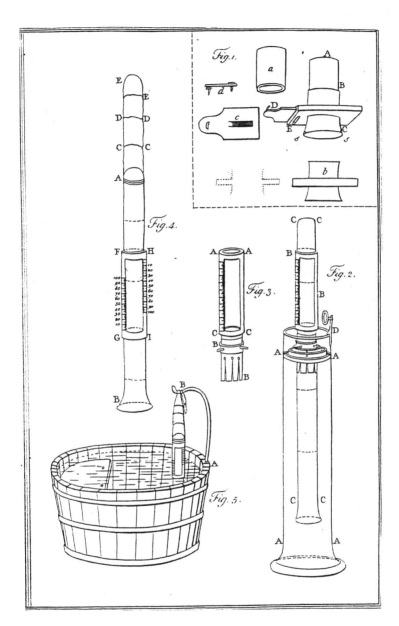

Figure 5.6
"Fontana's eudiometer." From Tiberius Cavallo, *A Treatise on the Nature and Properties of Air, and Other Permanently Elastic Fluids, to Which Is Prefixed an Introduction to Chymistry* (London: printed for the author, 1781), plate 3, figures 1–4. Courtesy the Dibner Institute.

Figure 5.7
Slide from Hauch's eudiometer, after Fontana. Instrument in the Hauch collection, Sorø Academy.

instead of mercury. In an appendix, he noted that he had scarcely finished his own research when he received a copy of Fontana's book; it was clear that Fontana had not had time, after seeing the description of Landriani's eudiometer, to revise expressions that might cast doubt on the latter's priority. So Landriani was gratified when Fontana wrote to him saying that he would make the priority clear and that no one would be able to take away Landriani's glory of being the original inventor.[24] However, it was Fontana's eudiometer that obtained the greatest fame outside Italy.

Henry Cavendish agreed in 1783 that Fontana's eudiometer was the most accurate to date, apart from the instrument that he himself had invented (figure 5.9).[25] He experimented to discover the best way of mixing the gases so that he obtained the greatest and most certain diminution in volume; this meant ascertaining the best rate of mixing, the best manner of agitating, the best time to allow for these operations, the order in which the gases were introduced, and other criteria besides. He devised a measure that was simple and effective: a glass cup

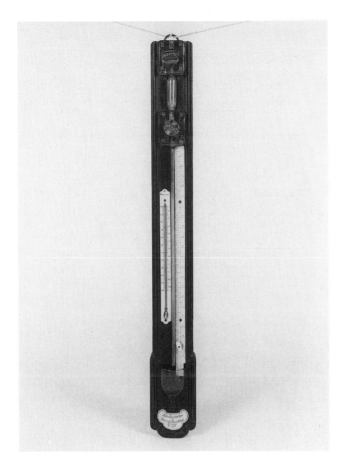

Figure 5.8
Landriani's eudiometer, at the Museo di Storia della Scienza, Florence. Courtesy
the Museo di Storia della Scienza.

with a mouth sealed by a brass plug.[26] Fontana had worked to one-
fiftieth of a measure; Cavendish regarded this as a very considerable
error "owing to more water sticking to the sides of the measure and
tube at one time than at another." So, unlike Priestley and Fontana, he
determined the quantities of air used by weighing them under water.
Most people assumed that it was more accurate to measure volumes
than weights. Cavendish proved otherwise. Most people also got poor
results from hydrostatic measurements; not so Cavendish.

His eudiometer is one of the simplest made, and it conforms to
Cavallo's general account of the history of the device:[27]

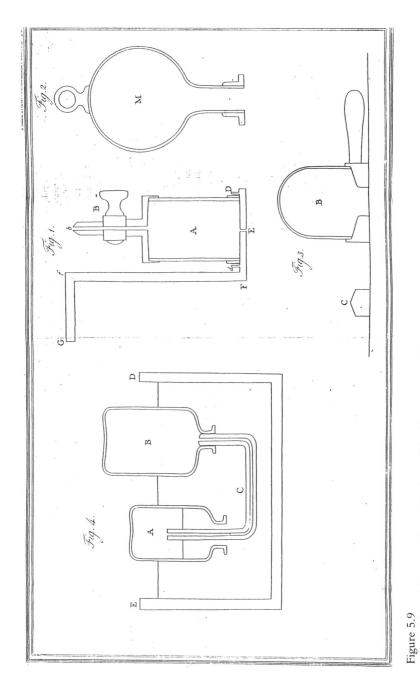

Figure 5.9

Cavendish's eudiometer. From H. Cavendish, "An Account of a New Eudiometer," *Philosophical Transactions of the Royal Society of London* 73(1783): 134 plate 2. Courtesy the Dibner Institute.

Figure 5.10
Top of Berzelius's spark eudiometer, showing wires sealed through glass (Berzelius Museum, Stockholm). After Cavendish's spark eudiometer (Royal Institution of Great Britain).

[A]t last those instruments seem to have undergone the same fate as barometers, viz. after various complicated constructions, the simplest of them has been found to answer the best. . . . Simple contrivances should always be preferred to more complicated ones; even when the latter seem to have some advantage in point of accuracy over the former, provided it be not very remarkable. Complex instruments are not only expensive, but they occasion very frequent mistakes, on account that the operator has generally many things to do, and to keep them in proper order, in which cases it is more easy to oversee or mistake some of them. . . .

Cavendish also, along with many contemporaries,[28] used another kind of eudiometer, a graduated cylinder inverted over a pneumatic trough and having two wires set through the glass near the sealed end, so that a spark could be made to pass through the gases (figure 5.10). He took pains to mark equal divisions on the cylinder by filling it successively with equal weights of mercury and engraving the levels of the meniscus with a diamond. Because each piece of glassware was hand-made and mouth-blown, rather than extruded, tubes and cylinders were never uniform, but not all of Cavendish's contemporaries worked with sufficient exactness to be much concerned about such a source of error. This eudiometer gave good quantitative results when known mixtures of gases were sparked and combined.

Cavendish, however, used a different set of apparatus for his work on the composition of water.[29] He had not set out with this goal. Rather, he began by investigating the phlogistication of common air, which took him to an examination of the result of sparking mixtures of an inflammable air (our hydrogen) with vital air, and that, in turn, took him to the composition of water. The discovery of the composition of water, like the discovery of oxygen, was achieved in several laboratories by different chemists at around the same time. Cavendish used cylinders and globes for the reaction, quickly determined the right mix for a smooth combustion rather than an explosion, and ignited the gases with an electric spark. The qualitative result was less obvious than it might seem, as problems arose from impure samples, from leakage, and from contamination by reaction vessels. Still, although skill was required and the results had to be interpreted. the qualitative result was relatively easily achieved. However, a problem arose in arriving at a quantitative result. How did one mix the gases smoothly and in sufficient quantities to give a measurable result? Storage, transfer, and measurement of the gases were all involved, and all raised problems.

Cavendish worked with a glass globe fitted with a brass stopcock, evacuated it using an air pump, closed the cock, then connected the globe via a bent glass tube to a reservoir of gas in an inverted jar over a pneumatic trough. The gas in the reservoir was a 1.9:1 mixture of inflammable air and dephlogisticated air. Cavendish filled the globe from the reservoir, sparked the gases, and repeated the process until almost all the gas in the reservoir had been converted to water. He measured the reactant gases by volume and measured their product by weight. The results were good but not exceptionally so. Cavendish here seemed to have been principally concerned with monitoring the qualitative transformation, noting measurements merely incidentally.

However, he could measure gases with great precision and with the simplest apparatus. His second paper, "Experiments on Air,"[30] looks more closely at what happens when a spark is passed through atmospheric air, which he described as "phlogistication of air by the electric spark." He took a glass tube, bent like an uppercase L at an angle of near 90 degrees, filled it with mercury, and inverted it in two glasses of the same fluid metal. He then introduced air into that tube, which was bent into an S, "by means of a small tube, such as is used for thermometers." When he wanted to introduce air repeatedly into the L tube, he used a modified S tube with a small-bore exit, a central ball, and a larger-bore main tube. He weighed the S tube at the beginning and end of each experiment, to see how much air had been forced out of it into the L tube.

Once the measured amount of air was in the L tube, he passed a spark through it.[31] He confirmed that common air consisted of one part dephlogisticated air (oxygen) and four parts phlogisticated air (nitrogen). He also confirmed that "the phlogisticated air was enabled, by means of the electrical spark, to unite to or form a chemical combination with the dephlogisticated air and was, thereby, reduced to nitrous acid;" this could be absorbed by an alkaline solution to form a solution of niter. The remaining phlogisticated air (nitrogen) was then sparked with more dephlogisticated air and was absorbed into an alkaline solution until the two principal constituents of atmospheric air had been completely removed and no further diminution took place. Then he removed the excess dephlogisticated air with liver of sulfur (a mixture of potassium polysulfides and sulfate), which left a small bubble of air unabsorbed, not more than 1/120 of the bulk of the phlogisticated air let up into the tube. Thus, "if there is any part of the phlogisticated air of our atmosphere which differs from the rest, and cannot be reduced to nitrous acid, we may safely conclude, that it is not more than 1/120 part of the whole."

Anyone who used Cavendish's apparatus and who looked carefully would have detected a very small residual bubble, but I suggest that it was so small that most observers would not have bothered about it; in any case, no one else in the eighteenth century recorded such an observation. Using small and very simple apparatus, Cavendish had not only correctly identified the proportions of oxygen to nitrogen in the atmosphere right, at close to 21 : 79[32] (others had done as much), but he had found a tiny residue, less than 1% of the whole by volume, and published his result instead of ignoring it. More than a century later, in 1894, William Ramsay and Lord Rayleigh announced their discovery of argon, the first inert gas; in Ramsay's investigation, the inert residue, after nitrogen and oxygen had been removed, was 1/80 of the original volume. Modern estimates are around 0.93%.[33] No one, to my knowledge, between Cavendish and Ramsay had detected the inert residue.[34]

Lavoisier also carried out research on gases, using simple apparatus, but then he transformed his experiments into demonstrations and, in the process, transformed his apparatus.[35] Frederic L. Holmes (chapter

6) discusses Lavoisier's apparatus and rightly emphasizes that Lavoisier's demonstration showpieces, including his spectacular gasometers, were not used in his key researches and were generally devised after the relevant part of those researches had been completed.

However, an important note is that Lavoisier, in collaboration with Meusnier, had devised a true gasometer, an instrument capable of controlling and measuring the uniform flow and volumes of the gases it dispensed. Gas flow was controlled by a mechanical balance. The instrument looked like the fruit of a marriage between industrial revolution engineering and physics. In Lavoisier's hands, aided by his collaborator Meusnier, it became what Lavoisier called "a precious instrument, both because of the large number of applications in which one can use it, and because there are some experiments that are almost impossible without it."[36]

Such special pieces of apparatus were expensive; chemistry, in approximating itself to physics, had, like physics, become big science. The two gasometers that Mégnié made in 1783 cost 636 livres (approximately 135,000 francs today). These instruments have not survived. Mégnié's improved gasometers, constructed between 10 September 1785 and 11 July 1787, are those described by Lavoisier in his Traité (1789) and are now in the Conservatoire des Arts et Métiers (Musée des Techniques) (figure 5.11). They cost 7,554 livres (more than 1.5 million francs today), a very high sum, owing principally to the finish and precision of every component, which had to be purpose-made and not mass-produced. For comparison with eighteenth-century values, Fourcroy's annual salary as a professor at the Jardin du Roi was 1,200 livres, or approximately one-sixth of the cost of the gasometers.[37] Lavoisier observed: "It is an inevitable effect of the state of perfection to which chemistry is beginning to approach. This requires expense and complicated apparatus: no doubt one should try to simplify them, but not at the expense of their convenience, and especially of their exactness."[38]

What sort of results did Lavoisier get with his elaborate gasometers? He was working volumetrically but had to combine gas volumes with the weights of solids. So he took the volumes, sometimes introducing a factor to account for estimated losses, corrected the volumes for temperature and pressure, and then converted these to weights via densities. He reported in his Traité of 1789 that atmospheric air contained 27 parts of oxygen to 73 parts of nitrogen[39]—a poor result, especially for oxygen. Because, as we have seen, Lavoisier

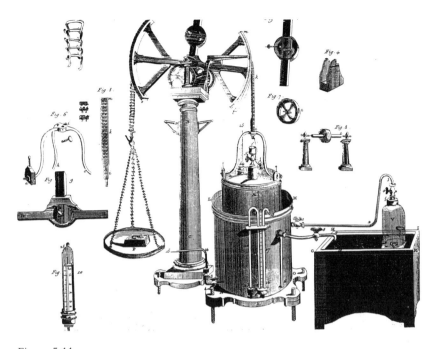

Figure 5.11
Gasometer. A. L. Lavoisier, *Elements of Chemistry*, R. Kerr, transl. (Edinburgh: William Creech, 1790), plate 8.

obtained a much lower and more accurate figure in earlier experiments, it is unfortunate that 27% is a widely reported value in the writings of Lavoisier's supporters.[40] Carbon dioxide came out at 72% oxygen to 28% charcoal by weight, just 3.6% in error. Water was 15.6% hydrogen to 84.4% oxygen, a pretty poor result again, as water is approximately 11% hydrogen by weight. Lavoisier may have had problems with lutes, leading to leakage.

When the conversion of a Dutch chemist, Martinus van Marum, to Lavoisier's chemistry occurred, it followed on his witnessing Lavoisier's demonstration of the composition of water. Being unable to afford to replicate Lavoisier's apparatus, he designed cheaper alternatives, including a vastly simplified gasometer, of which the hydrostatic control on pressure and rate of flow comes straight out of 'sGravesande's arsenal of apparatus. Van Marum's best gasometers (figure 5.12) cost less than one-sixth the price of Lavoisier's, and they were widely sought from Italy to the Baltic. Georges Parrot, rector of the new university in Dorpat (now Tartu, in Estonia) had them sent to

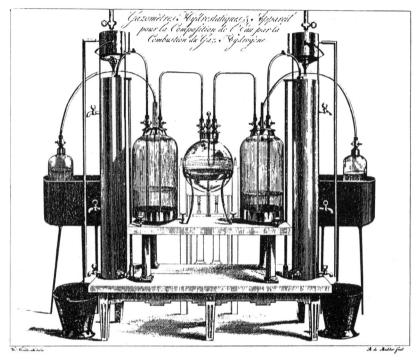

Figure 5.12
Hydrostatic gasometers in apparatus for composition of water. In Martinus van
Marum, *Verhandelingen uitgegeeven door Teyler's Tweede Genootschap*, 4 (Haarlem:
Tjeek Willink, 1798), plate 1. Courtesy Teyler's Museum.

him there; his son Friedrich, a medical student in the university, wrote
a prize-winning essay on gasometry.[41] He observed that the composition of water was the most striking phenomenon in the arsenal of the
new chemistry, so that chemists after Lavoisier were not so much led to
use "a gasometer in [his] senses, as to invent an apparatus for the composition of water. Both concepts, gasometer and apparatus for the
composition of water, became almost identical, and implicated in one
another" [p. 7]. Van Marum's large gasometer, with its hydrostatic
control, may have been much cheaper than Lavoisier's, but it was still
one of the most expensive items in a *cabinet de physique*. However, it
was accurate, far more so than its competitors, and therefore useful for
research; the smaller, simpler, and cheaper gasometer that van Marum
also had devised was useful only for demonstration lectures, lacking the
precision that had by now become necessary for research.[42]

Figure 5.13
Hauch's gasometers and part of apparatus for the composition of water. Courtesy
Sorø Academy.

No real competition challenged van Marum's hydrostatic gasometer, but various others attempted to construct gasometers in which
the pressure on the gases was maintained by a system of weights, as in
Lavoisier's instrument. Dumotiez in Paris invented one with weights
over pulleys controlling the pressure and advertised it in 1795. Adam
Wilhelm Hauch, a Danish soldier, courtier, diplomat, and amateur of
natural philosophy, devised a variant on Dumotiez's gasometer, with a
square gas holder suspended by weights over four pulleys and seated in
a square trough (figure 5.13). It was not easy to use, because if the walls
of the gas holder touched the walls of the trough, friction was a problem; but building it was cheap. Most gasometers using water for the
liquid in the trough followed the weight-and-pulley model, but for
gases soluble in water or reactive with water, what had become the
traditional design was unacceptable, because of the weight and the
expense of the amount of mercury needed. William Clayfield came up
with the answer, inspired by James Watt's device for containing factitious airs.[43] The inside of the trough was principally occupied by a
core, "allowing space enough between it and the sides of the cistern for
the jar, and for mercury to make it tight."[44] Clayfield's instrument was
made of glass.

William Hasledine Pepys, son of a cutler, who for a while had
followed that trade, made what Faraday called "an excellent instrument
of this sort," constructed of iron.[45] The movable vessel, steadied by a
vertical stem passing through a hole in a cross-bar, was counterpoised
by weights run over one or more pulleys. For some years, this gasometer was the standard in laboratory use. Brande described it in
1819.[46] Guyton de Morveau, in his 1812 article on the combustion of
diamonds, used and described a gasometer similar to Pepys's but, to
render it useful even in handling acid gases, had the vessels constructed
of porcelain.[47]

However, by the time Faraday was describing the uses of Pepys's
gasometer in 1829, it was about to enter the list of endangered species.
When Richard Griffin brought out his catalog of chemical apparatus in
1841, it was impressive in its range, but it did not include any gasometers. What was on his list was another item that had been around
for some years: "Gas Holders, Pepy's [sic] 2 gal. 20 s, 4 gal. 28 s each."[48]
George Cox's catalog of 1845[49] likewise listed Pepys's gas holders but
no gasometers. As Faraday noted [(1830) p. 344], "The gasometer has
to a great extent been superseded by another instrument, [the gas
holder], also the invention of Mr. Pepys, which can hardly be dispensed with in the laboratory."[50] This instrument (figure 5.14) had two
interconnected compartments, of which the lower one was closed
(except for stopcocks and tubes for conducting fluids and gases) and the
upper one was open, so that water could be poured into it or tipped
out. The gas holder was just what it claimed to be, and it replaced the
gasometer because of the expense of the latter and the accuracy of
alternative ways of measuring gases, either by weight or by volume.

We have seen that, before Lavoisier's invention of the gasometer,
Fontana had devised a eudiometer with which he obtained results
accurate to 2%. Cavendish, using even simpler apparatus and even
greater discipline, had improved on that. His eudiometric analysis of
atmospheric air in 1785 had given results to approximately 1%. His
investigation of the quantity of fixed air in marble, described in the
Philosophical Transactions for 1766, yielded results consistent to 0.1%,
and he was weighing comfortably to one-fourth of a grain; as has been
seen, he could weigh to one-tenth of a grain. If others could learn to
get results like Cavendish's, using fine balances but otherwise simple
and inexpensive apparatus, they would not need gasometers, whether
Lavoisier's enormously expensive ones, van Marum's cheaper ones, or
Dumotiez's, Hauch's, or Pepys's still cheaper ones. Simply having

Figure 5.14
Pepys's gas holder. From Berzelius museum, Stockholm.

the gas available and on tap, as it was in Pepys's gas holder, would be sufficient. The gas holder was to enjoy a much longer life as a species than had the eudiometer, and it was used throughout Europe. The gas holders in the Berzelius Museum in Stockholm, the Fondazione Scienza e Tecnica in Florence, and the Kekulé Laboratory in the History of Science Museum at the University of Ghent are clearly Pepys's instruments.

However, first a problem had to be solved. Others could not get results as good as Cavendish's; he was, by any standards, an extraordinary experimentalist. Perhaps apparatus could reduce the need for skill. Here, as before, it was Pepys to the rescue. In 1807, he read to the Royal Society of London an account of "A New Eudiometer,

Accompanied with Experiments, Elucidating its Application."[51] Pepys was entranced with the rapidity of the development of pneumatic chemistry, with its recent discoveries of many aeriform substances "till lately, confounded either with common air, or not even suspected to exist" [247, 248].

Cavendish had used nitrous air to analyze atmospheric air and, as Pepys noted, "by many laborious processes and comparative trials obtained results, the accuracy of which has been more distinctly perceived the more the science of chemistry has advanced." Dalton had offered some observations on the accuracy of nitrous air eudiometry.[52] Others had used different ways to estimate the amount of oxygen in atmospheric air. Some had tried the slow combustion of phosphorus and the decomposition of "fluid sulphuret of potash," but this process was slow and, like Guyton's,[53] attempts to accelerate it were likely to lead to the production of hydrogen sulfide, which interfered with the volumetric result. "The green sulphate of iron impregnated with nitrous gas, first discovered by Dr. PRIESTLEY, and recently used by Mr. DAVY for eudiometrical purposes" worked better than nitrous gas alone, because it did not combine with the other gases with which the nitrous gas was commonly found to be contaminated [Pepys, pp. 249, 250].

Thomas Charles Hope,[54] who had succeeded Joseph Black in Edinburgh, had invented a eudiometer consisting of a small bottle with a graduated glass tube fitted into its neck by grinding; also, a stopper was fitted at the base of the bottle. Hope filled the bottle with a solution of "alcaline sulphuret," filled the graduated tube with the air to be tested and fitted it to the bottle, then inverted and agitated the apparatus, opened the stopper under water, and measured absorption and thus goodness by the rise of fluid in the tube. William Henry of Manchester, who had studied under Black, substituted an elastic gum bottle for Hope's glass vessel.[55] Pepys modified Hope's and Henry's devices and produced a eudiometer (figure 5.15) consisting of a glass measure divided into a hundred parts; a small rubber (gum elastic) bottle, capable of holding approximately twice as much as the measure; a glass tube divided into thousandths, or tenths of the divisions in the measure; and a moveable cistern to be filled with water or mercury, with the graduated tube inserted through a hole in a cork at its base so that the tube could be moved in the cistern without any leakage of the water or mercury. Pepys filled the measure with gas over mercury and filled the elastic bottle with the reagent. He then connected measure

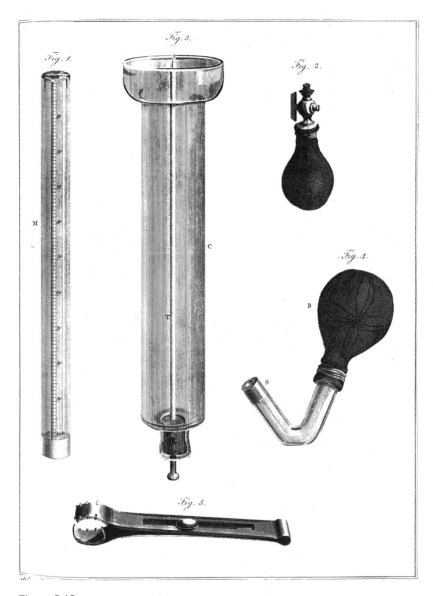

Figure 5.15
Pepys's eudiometer. From *Philosophical Transactions of the Royal Society of London* 97(1807), plate 15. Courtesy the Dibner Institute.

and bottle, squeezed fluid from the bottle into the measure, and repeated this until no more gas was absorbed by the fluid. He then separated the bottle from the measure under mercury and brought the mercury to the same level in measure and cistern. Hundredths could be read off on the measure and thousandths on the tube, using it like a vernier gauge. He gave his results to five parts in a thousand. For atmospheric air, Pepys found the oxygen content was 21.5% by volume; the modern value is 20.95%, so Pepys was out by 0.5%, a fairly good result. He thought highly of his apparatus: "Simple as [it] may appear, it is calculated to extend our knowledge of the different kinds of air, by the precision and accuracy which it enables us to obtain, and which solely constitutes the value of every experiment."[56]

Here was some of the best volumetric work to date, and a reasonable claim is that by 1807, eudiometry rendered gas analysis an exact science. However, Pepys's approach (which was based on Priestley's) was not destined to be the model for the mid–nineteenth century. We have seen that Cavendish in 1784 had used absorption with nitrous air but that he had also used an electric eudiometer. Electric eudiometers were to become dominant before the midcentury. Although Faraday in 1829 did describe nitric oxide, its combination with oxygen, and its subsequent absorption over water, he gave a detailed description only of the electric eudiometer. Similarly, in 1857, Bunsen's landmark book on *Gasometry*[57] described electric eudiometers and made extensive use of igniting gaseous mixtures by means of an electric spark. He generally worked with volumes, adjusted for temperature and pressure, and arrived at results accurate to four and sometimes to five significant figures, which is in a range between 0.1% and 0.01%. Volumetric work by the mid–nineteenth century had thus reached an accuracy better than Black had obtained gravimetrically and almost as good as Cavendish's. Gasometry had become an exact science, with legitimate claims to precision.

NOTES

1. The discovery of that composition used simpler and cheaper apparatus; see later and Frederic L. Holmes (chapter 6).

2. Robert Plot, *The Natural History of Oxford-Shire, Being an Essay Towards the Natural History of England* (Oxford: Leon Litchfield, 1705), 19.

3. J. L. Heilbron, "Weighing Imponderables and Other Quantitative Science Around 1800," in *Historical Studies in the Physical and Biological Sciences vol. 24, part 1* [supplement], (Berkeley: University of California Press, 1993).

4. Henry Cavendish was the most exact of these early practitioners.

5. One grain equals 64.8 milligrams, more or less, depending on whose grain is used.

6. John T. Stock, *Development of the Chemical Balance* (London: Her Majesty's Stationery Office, 1969), 11. Regarding the history of the chemical balance, see also P. D. Buchanan, *Quantitative Measurement and the Design of the Chemical Balance 1750–ca. 1900* (PhD thesis, University of London, 1982); and R. Multhauf, "On the Use of the Balance in Chemistry," *Proceedings of the American Philosophical Society* 106(1962): 210–218.

7. Stock, *Development of the Chemical Balance*, 11–14.

8. Maurice Daumas, *Les Instruments Scientifiques aux XVIIe et XVIIIe Siècles* (Paris: Presses Universitaires de France, 1953), 366; and *Lavoisier Théoricien et Expérimentateur* (Paris: Presses Universitaires de France, 1955), 134.

9. Michael Faraday, *Chemical Manipulation: Being Instructions to Students in Chemistry on the Methods of Performing Experiments of Demonstration, or of Research, with Accuracy and Success* (London: John Murray, 1829), 25. (I have used the 1829 edition, which is a corrected version of the 1827 first edition.)

10. That is only one-sixth or one-seventh of the sensitivity of Fortin's balance for Lavoisier 50 years earlier.

11. Antoine Laurent Lavoisier, *Elements of Chemistry in a New Systematic Order, Containing All the Modern Discoveries*, Robert Kerr, transl. (Edinburgh: William Creech, 1790), 34–35. Compare this with the 27% oxygen that he reports later in the same volume (see later).

12. Stephen Hales, *Statical Essays: Containing Vegetable Staticks . . . Also a Specimen of an Attempt to Analyse the Air, by a great Variety of Chymio-Statical Experiments* (London: W. Innys and R. Manby, 1733), vol. 1, 224; vol. 2, 280.

13. Joseph Priestley, "Observations on Different Kinds of Air," *Philosophical Transactions of the Royal Society of London* 62(1772): 147–252 at 210.

14. Most people experienced difficulty in getting readily reproducible results using nitric oxide, and other substances were used around the turn of the century, including sulfuret of lime (calcium sulfide). See, for example, Thomas Charles Hope, *Nicholson's Journal of Natural Philosophy, Chemistry and the Arts* 6(1803): 210, and a related effort by Humphry Davy, "An Account of a New Eudiometer," *Journal of the Royal Institution* 1(1801): 45–48.

15. Priestley, *Experiments and Observations on Different Kinds of Air*, vol. 2 (London: J. Johnson, 1775), 20–21.

16. Ibid., 6–7.

17. Priestley's apparatus was derived from that of Hales and others. A review of different eudiometers appears in Tiberius Cavallo, *A Treatise on the Nature and*

Properties of Air, and Other Permanently Elastic Fluids, to Which is Prefixed, an Intro-duction to Chymistry (London: printed for the author, 1781). See also Ionanne Andrea Scherer, *Eudiometria sive Methodus Aeris Atmosphaerici Puritatem Salubri-tatemque Examinandi* (Vienna: Typis Iosephi Nobilis de Kurzbeck, 1782).

18. Felice Fontana, *Descrizione, ed Usi di Alcuni Stromenti per Misurare la Salubrità dell'Aria* (Firenze: Gaetano Cambiagi, 1775); and Ferdinando Abbri, *Science de l'Air: Studi su Felice Fontana* (Cosenza: Edizioni Brenner, 1991). See also Marsilio Landriani, *Ricerche Fisiche intorno alla Salubrita dell'Aria* (Milan: il Polifils, 1775; reprinted in 1995, *Descrizioni ed Usi di Alcuni Stromenti per Misurare la Salubrità dell'Aria: Con un Articolo di Alessandro Volta sull'Eudiometria*, edited with an introduction by Marco Beretta, Firenze: Giunti).

19. P. K. Knoefel, "Famine and Fever in Tuscany: Eighteenth-Century Concern with the Environment," *Physis* 21(1979): 7–35.

20. Fontana, *Descrizione*, ix.

21. Adam Wilhelm Hauch devised a eudiometer, using slides like those of Fontana's; a (broken) example is found in the Hauch collection in Sorø Academy in Denmark.

22. Cavallo, *A Treatise*, 329–333.

23. Landriani, *Ricerche Fisiche*, ed. Beretta (1995). In this edition, Beretta includes the text of Volta's edited article (1783) on eudiometry for the Italian translation of the second edition of Macquer's *Dictionnaire de Chimie* (Paris: Didot, 1778). See also Scherer, *Eudiometria*, 55, regarding Landriani and Fontana.

24. Landriani (1995), 98.

25. H. Cavendish, "An Account of a New Eudiometer," *Philosophical Transactions* 73(1783): 106–35.

26. Faraday [*Chemical Manipulation*, 2nd ed. (London: John Murray, 1830), 329] said that Cavendish used a measure with a flat slider, similar to Fontana's. I have not found this in Cavendish but am unwilling to impute error to Faraday.

27. Cavallo, *A Treatise*, 315, 353–354.

28. To take just two examples: Martinus van Marum used electric eudiometers in Haarlem (T. H. Levere, "Martinus van Marum and the Introduction of Lavoisier's Chemistry in the Netherlands," in R. J. Forbes, ed., *Martinus van Marum: Life and Work*, vol. 1 (Haarlem: Tjeenk Willink, 1969), 158–286 at 206–207); also, several examples of electric eudiometers from the late eighteenth century are located in the Museo di Storia della Scienza in Florence. For a discussion of Volta's electric eudiometer, see William A. Smeaton, chapter 9.

29. H. Cavendish, "Experiments on Air: First Paper," *Philosophical Transactions of the Royal Society of London* 74(1784): 119–153. Reprinted in 1926, Edinburgh: *Alembic Club Reprints (no. 3)*.

30. H. Cavendish, "Experiments on Air," *Philosophical Transactions of the Royal Society of London* 75(1785): 372–384. Reprinted in 1926, Edinburgh: *Alembic Club Reprints (no. 3)*.

31. Cavendish, *Alembic Club Reprints (no. 3)*, 42:

It is scarcely necessary to inform any one used to electrical experiments, that in order to force an electrical spark through the tube, it was necessary, not to make a communication between the tube and the conductor, but to place an insulated ball at such a distance from the conductor as to receive a spark from it, and to make a communication between that ball and the quicksilver in one of the glasses, while the quicksilver in the other glass communicated with the ground.

32. Ferenc Szabadváry, *History of Analytical Chemistry* (1960, trans. 1966, reprinted Yverdon and Langhorne, PA: Gordon and Breach, 1992), 65, summarizes Cavendish's work: "After carrying out four hundred experiments at different localities and in varying weather conditions, he concluded that the atmosphere contained 20.84% dephlogisticated air (oxygen)."

33. Mary Elvira Weeks, *The Discovery of the Chemical Elements*, 6th ed. (Easton, PA: Journal of Chemical Education, 1956), 784–785. Erwin N. Hiebert, "Historical Remarks on the Discovery of Argon: The First Noble Gas," in Herbert H. Hyman, ed., *Noble Gas Compounds* (Chicago: University of Chicago Press, 1963), 1–20. Ramsay knew about Cavendish's discovery at some time in the 1880s, having read about it in George Wilson, *The Life of the Honourable Henry Cavendish* (London: Cavendish Society, 1851). He seems to have forgotten about it in 1894 [see Morris William Travers, *A Life of Sir William Ramsay* (London: E. Arnold, 1956), 100].

34. I am not, of course, suggesting that Cavendish in any sense discovered argon; I am asserting that here, as in his measurement of the gravitational constant, he was a most scrupulous and dependable observer.

35. Lavoisier used mostly the pneumatic apparatus of Hales and Priestley in his initial researches (Frederic L. Holmes, chapter 6; and Lavoisier MS Académie des Sciences, laboratory notebooks for March 1774–February 1776, f6; for November 1786 to end 1787, f167r and loose-leaf insert). Lavoisier's notebooks also show that he modified, added, and used increasingly sophisticated apparatus. He first improved the pneumatic trough, went on to invent (with suggestions from Meusnier) a pneumatic device that was essentially a pneumatic trough with taps for the entry and egress of gases; [notebook 1 ff. 34v & 51v] and, by 1783, had two of them built, for handling two gases [Daumas, *Lavoisier*, 47–54]. Two gasometers would be needed to demonstrate the formation of water by the combustion of hydrogen. Meusnier made suggestions for transforming them into true gasometers. Between 1783 and 1785, Mégnié tackled their construction, and that assignment was completed by the end of 1785 (Daumas, *Lavoisier*, 132–156). Further improvements, conceived by Meusnier, were incorporated in new gasometers built by Mégnié and presented to the Académie in March 1788. Meusnier's collaboration was important and is admirably indicated in volume 4 of Lavoisier's correspondence [*Oeuvres de*

Lavoisier. Correspondance . . . , Fascicule 4, 1784–1786, Michelle Goupil, ed. (Paris: Belin, 1986), Annexe 3, 299–304].

36. *Oeuvres de Lavoisier*, vol. 1, J. B. Dumas, ed. (Paris, Imprimerie Impériale, 1862), 267.

37. Maurice Daumas, "Les Appareils d'Expérimentation de Lavoisier," *Chymia* 3(1950): 45–62 at 54; Daumas, *Lavoisier*, 132–156; Bruno Jacomy, "Le Laboratoire de Lavoisier," *La Revue Musée des Arts et Métiers*, 6(1994): 17–22 at 20; T. H. Levere, *Chemists and Chemistry in Nature and Society 1770–1878* (Brookfield, VT: Variorum, 1994), 4: 323. Apart from the time spent on the construction of the later gasometers by his workers, Mégnié himself had spent 92 days on the job.

38. *Oeuvres de Lavoisier*, vol. 1, 267.

39. Lavoisier, *Elements of Chemistry*, 38. Compare this with his one-sixth oxygen earlier in the same volume (see note 11).

40. See, for example, A. F. Fourcroy, *Système des Connaissances Chimiques, et de leurs Applications aux Phénomènes de la Nature et de l'Art* vol. 1 (Paris: Baudouin, 1801–2), 156. I am grateful to Dr. W. A. Smeaton for pointing this out.

41. Friedrich Parrot, *Ueber Gasometrie nebst einigen Versuchen über die Verschiebbarkeit der Gase* (Dorpat, 1811).

42. F. Hauch, 46–52. Martinus van Marum, *Description de Quelques Appareils Chimiques Nouveaux ou Perfectionnés de la Fondation Teylerienne, et des Expériences faites avec ces Appareils (Verhandelingen Uitgegeeven door Teyler's Tweede Genootschap*, vol. 10) (Haarlem: Tjeek Willink 1798), chaps. 1 and 2.

43. Thomas Beddoes and James Watt, *Considerations on the Medicinal Use, and on the Production of Factitious Airs*, 5 parts (Bristol: Printed by Bulgin and Rosser for J. Johnson, London, 1794–1796). Watt's device was essentially a simple gas holder, with two outlets on the top and one at the side of the base; each outlet could be closed with a cork.

44. Faraday, *Chemical Manipulation*, 2nd ed., 342; and William Clayfield, "Description of a Mercurial Air-Holder, Suggested by an Inspection of Mr. Watt's Machine for Containing Factitious Airs," in H. Davy, *Researches, Chemical and Philosophical; Chiefly Concerning Nitrous Oxide, or Dephlogisticated Nitrous Air, and Its Respiration* (London: Biggs and Cottle, 1800), 573–575.

45. Faraday, *Chemical Manipulation*, 2nd ed., 342; William Hasledine Pepys, "Description of a Mercurial Gazometer Constructed by Mr. W. H. Pepys," *Philosophical Magazine* 5(1799): 154; and William Allen and W. H. Pepys, "On the Changes produced in Atmospheric Air, and Oxygen Gas, by Respiration," *Philosophical Transactions of the Royal Society* 98(1808): 249.

46. William Thomas Brande, *A Manual of Chemistry: Containing the Principle Facts of the Science, Arranged in the Order in Which They Are Discussed and Illustrated in the Lectures at the Royal Institution of Great Britain* (London: John Murray, 1819), 81–82.

47. Louis Bernard Guyton de Morveau, "Nouvelles Expériences sur la Combustion du Diamant et Autres Substances Charbonneueses en Vaisseaux Clos," *Annales de Chimie* 84(1812): 20–33.

48. Richard Griffin, *Descriptive Catalogue of Chemical Apparatus, Chemical Reagents, Cabinets and Collections of Rocks, Minerals, and Fossils; Models of Crystals; Models of Chemical Manufactories; Apparatus for Electrography, Glass-Blowing, and Blowpipe Analysis: and Every Other Convenience for the Pursuit of Experimental Science. With the Prices Affixed at which the Articles are Sold by Richard Griffin & Co. Glasgow* (Glasgow: 1841). For information about Griffin, see Brian Gee and W. H. Brock, "The Case of John Joseph Griffin. From Artisan-Chemist and Author-Instructor to Business-Leader," *Ambix* 38(1991): 29–62; *The Fontana/Norton History of Chemistry* (London: Fontana, 1992), 185–192; and F. Kraissl, "A History of the Chemical Apparatus Industry," *Journal of Chemical Education* 10(1933): 519–523. See also J. J. Griffin, *Chemical Recreations* (Glasgow: 1834).

49. George Cox, *Cox's Condensed Catalogue of Chemical, Philosophical, Optical, and Mathematical Instruments, Apparatus, etc.* (London: 1845).

50. W. H. Pepys. "Description of a new Gas-holder," *Philosophical Magazine* 13(1802):153.

51. W. H. Pepys, "A New Eudiometer, Accompanied with Experiments Elucidating Its Application," *Philosophical Transactions of the Royal Society of London* 97(1807): 247–259.

52. John Dalton, *Philosophical Magazine* 28.

53. For Guyton's eudiometer, see William A. Smeaton, chapter 9.

54. Hope, *Nicholson's Journal* 4.

55. William Henry, *The Elements of Experimental Chemistry*, 8th ed., vol. 1 (London: 1818), 149.

56. W. H. Pepys, "A New Eudiometer, Accompanied with Experiments, Elucidating Its Applications." *Philosophical Transactions* 97(1807): 247–259, at 259.

57. Robert Bunsen, *Gasometry Comprising the Physical and Chemical Properties of Gases,* H. E. Roscoe, transl. (London: Walton and Maberly, 1857), published simultaneously with the German edition. Note that Bunsen's gasometers were simply graduated storage jars over mercury and thus were even simpler than the gas holders that had supplanted the gasometers which controlled flow as well as measured volume.

6

THE EVOLUTION OF LAVOISIER'S CHEMICAL APPARATUS
Frederic L. Holmes

Historians of the Chemical Revolution have recently emphasized the novel instruments and methods with which Lavoisier established and demonstrated the tenets of the new chemistry. "What distinguished his instruments," Arthur Donovan has written, "in addition to their great cost, was their innovative design and the precision of the quantitative data they made available."[1] Suggesting that "much of Lavoisier's fortune was probably spent on the best scientific apparatus that money could buy," William Brock has characterized some of his apparatus as "unique and so complex that his followers were forced to simplify his experimental procedures and demonstrations in order to verify their validity."[2] The very expense and complexity of his apparatus became, according to Jan Golinski, a major issue between Lavoisier and those who opposed his views. "To Priestley and his followers," Golinski writes, "expenditure on this scale was not only undesirable but reprehensible, because it foreclosed the possibility of Lavoisier's experiments being replicated by others who lacked his wealth."[3]

These impressions are drawn largely from three or four of Lavoisier's most elaborate instruments: the calorimeter, the gasometer, the precision balance, and the apparatus for the synthesis of water. These instruments were involved in some of his most spectacular experimental demonstrations, three of which are showpieces of the series of plates at the end of his *Traité élémentaire de chimie* depicting the apparatus of a chemical laboratory, and they are prominent in the collection of Lavoisier's surviving apparatus displayed in Le Conservatoire des Arts et Metiers in Paris. They serve readily as icons of Lavoisier's chemistry. They were not, however, the instruments on which he built the core of his quantitative methodology or his revolutionary challenge to the existing theoretical structure of chemistry.

The calorimeter, the Baroque-appearing gasometer, and the large glass vessel in which he burned a stream of inflammable air in oxygen were all products of the period, beginning in 1780, in which Lavoisier and his collaborators consolidated the theoretical structure whose main outlines he had developed with simpler apparatus during the 1770s. The famous precision balances designed for him by Fortin and Mégnié were probably delivered to him in 1787 or later, when he had already performed most of the measurements on which the new system of chemistry rested. The equipment with which Lavoisier carried out the experiments basic to the general theory of combustion that he proposed in 1777—and on which he relied even in the 1780s for the majority of his chemical investigations—consisted of improvements of what Maurice Daumas has called "classical apparatus within the reach of any chemist."[4]

Within the definition of classical apparatus, Daumas included not only the traditional furnaces, retorts, receivers, and diverse vessels on which Lavoisier depended, just as generations of chemists before him had, but the various forms of vessels, inverted over water or mercury, that were used to collect and measure airs released or absorbed in chemical operations. Invented by Stephen Hales in 1728, these "pneumatic" vessels had become indispensable, in the years just preceding Lavoisier's entry into the field, for the study of the various airs recently discovered. Joseph Priestley described the apparatus that he used for experiments on air as "nothing more than that of Dr. Hales, Dr. Brownrigg, and Mr. Cavendish, diversified, and made a little more simple." Typically, Priestley made do with readily available objects, such as ordinary jars inverted over a wooden or earthen trough. When he needed to produce an air by heating a substance, he placed it in a gun barrel connected by a clay pipe stem to the vessel in which he collected the released air.[5]

The investigation of the processes that fix or release airs that Lavoisier began in October 1772, required him also to use apparatus descended from that of Hales. Whereas Priestley had been able to meet his needs with simplified renditions of Hales's apparatus, Lavoisier's experimental program pressed him toward more robust equipment. After observing weight gains in the combustion of phosphorus and sulfur, Lavoisier believed that the most decisive experiment he could try to demonstrate that the same phenomenon occurred in the calcination of metals was "the reduction of lead calx in the apparatus of Hales." The version of the Hales apparatus Lavoisier used was a modification devised by his teacher, Guillaume François Rouelle, which combined two forms of the experimental setups described in Hales's *Vegetable Staticks*. As reconstructed in Henry Guerlac's *Lavoisier—The*

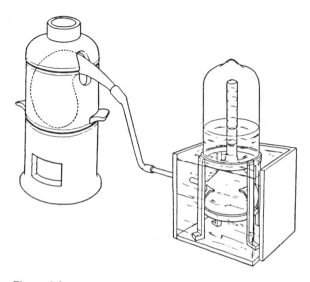

Figure 6.1

G. F. Rouelle's combination of Hales's two devices. From Henry Guerlac, *Lavoisier—The Crucial Year.*

Crucial Year, the Rouellian apparatus resembled that in figure 6.1. The substance to be heated was inserted into an ordinary glass chemical retort, which then was placed in a reverbatory furnace with close-fitting walls that enabled the operator to heat the vessel all around. The disengaged air passed through tightly luted connections into a receiver opening into the upper part of a glass jar inverted over water. The descent of the water measured the volume of air produced.[6]

When Lavoisier heated a mixture of the lead ore minium with charcoal until the retort began to redden, the water descended rapidly. After cooling, the air released amounted to 462 cubic inches, or at least a thousand times as much as the volume of the litharge employed. In this experiment, Lavoisier did not appear concerned to establish exact quantities. To show that a "prodigious quantity of air" was released was enough to establish his initial argument about the calcination and reduction of metals.[7]

When Lavoisier began, in February 1773, an extended study of the processes that fix or release air, his first step was to try to complete his demonstration that air is fixed in the calcination of metals and is released in the reduction of their calxes. On February 22, he attempted to calcine lead in the Hales apparatus. He quickly ran into trouble. Perhaps because he knew that he would have to apply prolonged, intense heat, he chose a clay retort. He pushed the fire "strongly for four hours," turning the iron support bars bright red. The air expanded in the inverted vessel, owing to the heating. When the water level should have begun rising owing to absorption of air by the lead, however, it continued to fall. Lavoisier suspected, correctly, as it turned out, that air was entering through a crack in the retort. After two more experiments failed for the same reason, he "had to resort to glass retorts." Trying tin in place of lead, he tried to keep the fire constant but allowed it to "augment in spite of myself, which made me fear that the retort had collapsed, and that is, in fact, what had happened."[8] Persuaded by these four successive failures that a Hales apparatus put together from ordinary chemical equipment could not withstand the conditions required for his calcination experiments, Lavoisier began to seek alternatives.

Pressed for time, because he was anxious to announce his new theory of combustion and calcination at the public meeting of the Academy of Sciences on April 21, Lavoisier decided at the end of March that he might be able to "do some of the same experiments in a simple way with the burning glass." This was the great Tschirnhausen lens that he had used, along with his more senior colleagues, to study the burning of diamonds during the previous summer. In his hurry, Lavoisier put together a simple form of Hales's apparatus with easily obtainable objects: a supply of glass jars ranging from 5 to 7 inches in diameter, an ordinary wash basin for the pneumatic trough, and a pedestal (normally used in restaurants to hold fruits) for a support on which he could place the material to be heated under the inverted jar above the water in which the latter was immersed. With this makeshift apparatus, he performed, between March 29 and March 31, five calcinations and two more reductions of lead ore. All the calcinations and the first reduction gave inconclusive results. Only the second reduction experiment yielded a moderate success.[9] For the presentation of his theory at the Academy, Lavoisier put the most optimistic face he could on the meager experimental support he had been able to muster,[10] but he soon realized that the burning glass method was, in principle, inadequate for accurate quantitative experiments. The amount of metal or calx that could be heated was so small, in proportion to the volume of the jar, that the volume of the air disengaged or absorbed in the operation was "spread through such an extended space that the differences must be little sensible."[11]

Forced to return to heating the metals and calxes in a retort placed in a furnace, so that he could operate on quantities large enough to produce volumes of air large enough to be measured accurately in the Hales apparatus, Lavoisier now went beyond the standard chemical equipment, by having made for his use a special retort able to withstand the heat to which he must subject it. A craftsman fabricated the retort from four pieces of iron soldered together with copper.[12] The retort is shown in figure 6.2. In adapting this centerpiece of traditional chemistry to a more demanding need, Lavoisier departed as little as possible from the standard form of the apparatus. He merely had it reproduced in a material more resistant to heat and to leaks than the materials customarily employed.

Lavoisier made some changes also in the remainder of the Hales apparatus to adapt it to the larger-scale operation he contemplated. The inverted jar he used was some two and a half feet tall. To the receiver at the bottom of the trough holding the water, he attached a long, vertical metal pipe to carry the air to the top of the jar. At first, he raised the water into the jar by sucking the air out through a long syphon. The elevation was so great, however, that he eventually attached a simple air pump to the apparatus. By the time Lavoisier had fully developed this apparatus, it appeared more complex than what was common in a contemporary chemical laboratory. All its components, however, were only modifications or new combinations of familiar equipment. He described his apparatus accurately as one "whose idea came originally from M. Hales, which has since then been amended by the late M. Rouelle, and to which I have myself made a few changes and additions related to the circumstances."[13]

Meanwhile, Lavoisier was also devising some auxiliary apparatus intended to handle the air released in the operations he planned to perform. Here, too, he did not begin anew but adapted designs made by his contemporaries. Because he was convinced at the time that the air absorbed or disengaged in all the operations he was studying was the fixed air of Joseph Black, Lavoisier contrived this apparatus with the special properties of fixed air in mind. The immediate inspiration for it was an apparatus consisting of two bottles connected by a stopcock, a device that his colleague, Jean-Baptiste Bucquet, had recently constructed to generate and collect fixed air. During May 1773, Lavoisier sketched in his laboratory notebook three variations of such an apparatus (figure 6.3). The first enabled him to collect, in the inverted

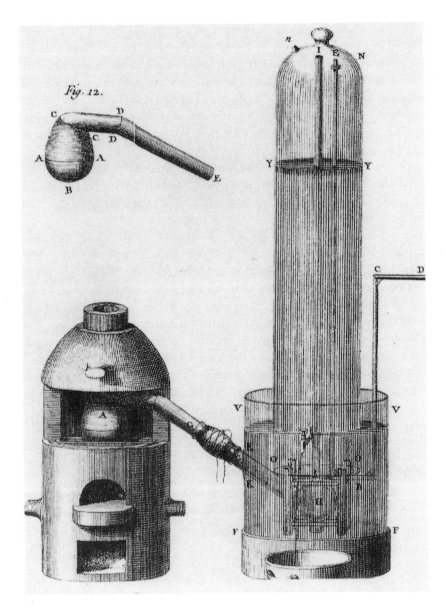

Figure 6.2
Apparatus used by Lavoisier for reduction of minium. Upper left, retort fashioned from four pieces of iron. Lower and right, retort in place in furnace, and connected to modified Hales apparatus. From Lavoisier, *Opuscules*, Plate II, figs 10, 12.

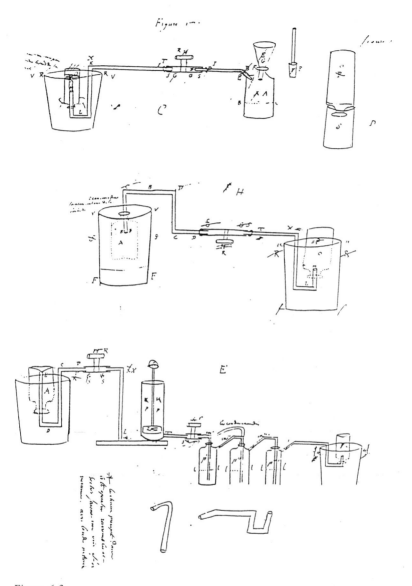

Figure 6.3
Lavoisier's sketches of apparatus designed especially to handle the air released in various operations. From Lavoisier, Cahier lab. 1, facing pp. 33, 35, 37.

bottle at the left, fixed air generated in the flask at the right by the action of an acid on chalk. The second was designed for the special purpose of testing a theory that fixed air could be converted to ordinary air by cooling it to the temperature of melting ice. The third, and ultimately most important, apparatus (shown at the bottom of figure 6.3) enabled Lavoisier to pass air collected in the inverted bottle at the left successively through three bottles containing lime water and to collect whatever air was not absorbed in these solutions in the inverted bottle at the right.[14] Despite its overall appearance of complexity, this apparatus was easily assembled from readily available components.

I have described elsewhere, in detail, how Lavoisier used various combinations of this apparatus, with some further adaptations to special circumstances, to perform during the summer of 1773 the experiments that make up most of the contents of his first publication concerning the absorption and fixation of elastic fluids, the well-known *Opuscules Physiques et Chymiques*. In the course of solving his technical problems, learning the power and limits of his evolving techniques, he was also forging, piece by piece, what has become known as the "balance sheet method" through which he eventually transformed the experimental practices of chemistry.[15] Here I sketch briefly how the basic repertoire of apparatus and methods that was developed over this initial year of his experimental venture remained at its core through the entire two decades in which Lavoisier sustained that historic enterprise.

Because of his reputation for designing apparatus that was overly complex, it should be emphasized that, when his experimental circumstances permitted it, Lavoisier was as quick as any of his contemporaries to simplify his equipment and methods. He was as fully aware as anyone of the multiplication of sources of error with the increasing complexity of a chemical operation. An opportunity to simplify the apparatus he had designed for the reduction of lead ore in 1773 arose in 1775, when he turned to the reduction of mercury precipitate. His principal reason for doing so was that the possibility of reducing the mercury calx without charcoal offered the means to simplify the question at issue: to clarify whether the source of the air released in reductions with charcoal was the metallic calx or the charcoal. This question had remained insoluble for him as long as he had focused his experiments on the lead calx. He was also able, however, to simplify his apparatus, because the mercury calx was "reducible without addition at a very moderate degree of heat." Instead of using the redoubtable iron retort, he was able, therefore, to revert to a small glass retort placed in a reverbatory

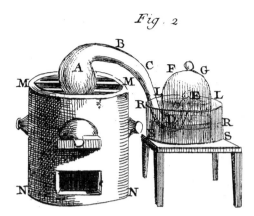

Figure 6.4
Apparatus used by Lavoisier for calcination and reduction of mercury. From Lavoisier, *Traité*, vol. 2, Plate IV, fig. 2.

furnace "proportioned to its size." The neck of the retort, some one foot long and four lines in diameter, he bent in several places with an enameling lamp, in such a manner that its extremity was directly engaged under a large jar filled with water and inverted in a tub. "This apparatus," he wrote, "as simple as it is, is all the more exact, because there is neither solder nor lutes, nor any passages across which air can be introduced or escape."[16] Having encountered many problems with all three of these features of his earlier apparatus, Lavoisier was well-prepared to appreciate the virtues of simplification.

With this simple and elegant version of the Hales apparatus, Lavoisier was able to show that the air released by the reduction without charcoal resembled common air, whereas that with charcoal was fixed air. One year later, with a similar apparatus modified to use mercury in place of the water to fill the inverted vessel, he performed the famous experiment on the analysis and synthesis of air that became a centerpiece in his new theory of combustion (figure 6.4).[17] The same type of apparatus served for two of the three experiments on which he based the beautiful paper of 1776 about the decomposition and recomposition of nitrous acid and from which he built his general theory of acidity. In this case, he carried out a different type of operation—the dissolution of mercury in the nitric acid—in the retort and collected a different kind of air (Priestley's nitrous air) with the same equipment (except for the furnace) that he had previously used to release "pure air" from the mercury calx.[18]

As he continued to expand his theory of acidity, just after announcing his general theory of combustion in 1777, Lavoisier encountered special problems that required him to adapt his basic procedures in different ways. In his memoir, *General Considerations on the Nature of Acids and the Principles of which They Are Composed*, he pointed out the following in 1778:

> This type of work demands a great variety of methods, and the procedures necessary to attain a combination vary according to the different substances on which one operates: for one case, one must have recourse to combustion, whether in atmospheric air, or pure air. . . . In other cases, simple exposure to the air suffices to produce the combination. . . . Finally, in most cases, one must have recourse to the art of affinities, and to employ the acidifying principle already engaged in another combination.

The special significance of this statement is that Lavoisier was not invoking the necessity for a great variety of instruments or of physical measuring procedures but of classical chemical operations. For the experiments he reported in this memoir, he could use the same repertoire of apparatus that had served in his previous experiments. He had to resort again to lutes and to connect several bottles in succession to the glass retort to collect samples of the airs produced by the action of nitrous acid on sugar; but his application of "the methods that modern chemistry furnishes us" had mainly to do with the exactitude with which he measured the quantities of nitrous acid required to decompose the sugar that disengaged the aeriform fluid in successive portions. He did so by applying one of the most traditional quantitative methods of the chemistry of his century: the determination, by neutralization with alkali, of the portion of the original quantity of acid not consumed in the decomposition of the sugar.[19]

During the spring of 1784, Lavoisier performed a series of experiments intended to determine the proportions of carbon and oxygen in fixed air. He encountered a great deal of trouble. He carried out the combustion on charcoal prepared in various ways and on a wax candle in an effort to obtain consistent results. Along the way, he came to the conclusion that the combustion included not only the formation of fixed air but of water from the combustion of inflammable air contained as an impurity. His calculations became more complicated than in the combustion experiments of the previous decades. In the design of his apparatus and in the operation itself, however, Lavoisier displayed his continued drive for simplification. In all these experiments, he "operated in a pneumatico-chemical apparatus over mercury." He was able to ignite the charcoal within the inverted pneumatic jar itself by passing a curved red-hot iron rod through the mercury. The rod first ignited a piece of phosphorus, which kindled some tinder, which heated the charcoal enough to start it burning in the atmosphere of pure oxygen. To absorb the fixed air formed in the combustion, he did not pass the air remaining afterward through a series of external bottles but managed to introduce liquid caustic alkali through the mercury into the inverted vessel itself. Through these dexterous maneuvers, he eliminated the need for separate connecting vessels with their potential for leaks. "There was," he remarked in reporting the experiments, "no loss of weight when one operates in glass vessels hermetically sealed." It was because of the simplicity of the experimental arrangement that he could be confident, when he did encounter a deficit in his balance sheet, that it represented some other process occurring within the vessel—in this case the formation of water—rather than material escaping through an imperfect seal.[20]

Among the refinements Lavoisier made in his pneumatical-chemical apparatus sometime during this period was to substitute a carved marble basin for the wooden or earthen tubs into which he and others had previously placed the inverted jars. The marble provided a stable foundation, sculpted in a way that facilitated inserting objects through the mercury, while minimizing the total quantity of mercury required. This customized basin undoubtedly cost Lavoisier a good deal more than what it replaced, but the elegant simplicity of the resulting apparatus is clearly evident in the illustration included in his *Traité* (figure 6.5).

After 1785, Lavoisier concentrated his experimentation on the determination of the proportions of carbon, inflammable air, and oxygen in plant substances. For this work, he continued to rely mainly on variations of the same apparatus and methods that he had been developing ever since 1772. The combustion of spirit of wine he carried out within the usual jar inverted over mercury but connected it to an additional pneumatic jar over water, from which he could replenish periodically the oxygen consumed by the spirit of wine. In the 1787 fermentation experiments and in the troublesome efforts to analyze sugar that occupied him during much of 1787 and 1788, he returned to

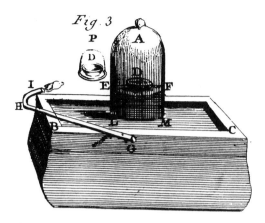

Figure 6.5
Carved marble basin substituted by Lavoisier for earthen or wooden tubs traditionally used to hold the inverted jars of pneumatical-chemical apparatus. From Lavoisier, *Traité*, vol. 2, Plate IV, fig. 3.

the series of limewater bottles, ending with a pneumatic vessel that he had designed originally in the spring of 1773 to collect the products of the operation.[21]

Through nearly two decades, Lavoisier pursued the problems of combustion and calcination and the determination of the combining proportions of substances, with a repertoire of relatively simple apparatus and with methods clearly descended from combinations of ordinary chemical equipment and the pneumatic vessel invented by Stephen Hales. In the light of this continuity, we can see from a different perspective those conspicuous instruments and apparatus of Lavoisier that were very dissimilar to earlier equipment. They departed more sharply from prior practices because they were designed to confront problems that had not arisen in prior chemical investigations. The calorimeter was designed to measure the quantity of heat released in a chemical or physical process. Four years earlier, Lavoisier had not thought it possible to do so. The vessel in which Lavoisier and Laplace burned inflammable air in oxygen and the gasometers that measured the quantities of the two airs consumed were novel, because a combustion in which both the combustible and the oxygen were aeriform (though the product was liquid) presented a unique problem for chemical analysis. Lavoisier did not resort to complicated and unique apparatus fortuitously or to make chemical experimentation inaccessible to those less wealthy than he. He did so when confronted with problems that required new solutions.

Lavoisier did not himself design the apparatus that has received most attention as marking his departure from previous chemical methods. The basic idea for the calorimeter is generally attributed to Laplace. The gasometers (and a special apparatus designed to burn oils) were the handiwork of mathematician and engineer Jean-Baptiste Meusnier, who collaborated with Lavoisier in the early 1780s.[22] On his own, Lavoisier tended not to invent new apparatus but to adapt existing apparatus to evolving purposes. A comparison between Lavoisier and Meusnier reminds one of Claude Lévi-Strauss's comparison between the bricoleur and the engineer:[23]

> The bricoleur is able to execute a great number of diversified tasks, but in contrast to the engineer, he does not subordinate each one to the acquisition of primary materials and tools conceived and procured for the specific needs of his project.... The rule of his game is always to manage with the "means at hand," that is, with an ensemble of materials and tools that is at each moment finite and heterogeneous, because its composition ... is the contingent result of all the occasions that have been presented to renew and enrich the stock, or to sustain it with the residues of previous ... constructions.

To say that Lavoisier regularly improvised by adapting existing apparatus and methods to new tasks is not to detract from his originality. That is how most innovations in the techniques of scientific experimentation originate.

Daumas has emphasized that Lavoisier's experimental success relied not on instruments of precision alone but to an equal extent on his personal skill as an experimental operator.[24] His manual dexterity and observant eye were as important as the design of his apparatus, and the adroitness with which he could execute complex maneuvers, sometimes with great rapidity, enabled him often to make do with relatively simple apparatus. These were skills that Lavoisier learned only with time and experience. In the first few months of the experimental venture with airs that he began in February 1773, most of the operations that he attempted went wrong in some major or minor way. Only gradually did he learn to appreciate the capacity and the limits of his methods. By the 1780s, he had acquired a surety with his methodological repertoire that no one else could match. The claims of his opponents to be unable to replicate Lavoisier's results sometimes had less to do with the alleged complexity of his methods than with his opponents' relative inexperience with them.

This chapter has emphasized the continuities linking Lavoisier's apparatus and methods with those of his predecessors. In other places, I have argued that he viewed his efforts to determine the composition of acids and bases as a continuation of the process by which his predecessors had determined the acids and bases composing the neutral salts.[25] I have also asserted that Lavoisier was not the initiator of the new chemical nomenclature but that he captured for his purposes a nomenclature reform that was already taking place.[26] In another place, I have suggested that in the famous statement of February 1773 predicting a revolution in physics and chemistry, Lavoisier may not have viewed himself as the initiator of a coming revolution but as a participant in a revolution that had already begun.[27]

Other Lavoisier scholars have tended to depict starker contrasts between Lavoisier and the chemistry that came before him. Some have portrayed the new nomenclature as a total break with the past or as the basis of a "new chemical practice." Some have seen his quantitative arguments as a new standard of chemical knowledge or as a new rhetorical form of science. Others have located the rupture in a new structure of chemical composition or in the complex new instruments and apparatus with which he changed the landscape of the chemical laboratory. The tendency to seek a single "key" to the Chemical Revolution, whatever it may be, oversimplifies a multidimensional historical process.

The overall result of Lavoisier's activity was a true revolution but not one that can be reduced to one or two events or features lifted from his scientific odyssey. In a recent review of the first volume of Janet Browne's masterful biography of Charles Darwin,[28] Stephen Jay Gould has commented that the explanation of Darwin's great achievement "can only be achieved through dense narrative. No shortcuts exist; the answer lies in a particular concatenation of details—and these must be elucidated and integrated descriptively." The same thing can be said about the explanation of Lavoisier's Chemical Revolution. The revolutionary effect of his work lies not in any single aspect of his thought, his laboratory methods, his rhetoric, or his ability to organize the ranks of his followers, but in the way they fitted together to form a whole more powerful than its parts. Only by following the trail of his activity through its full development can we understand in depth why he, more than any other individual, reshaped the chemistry of his time.

NOTES

1. Arthur Donovan, *Antoine Lavoisier: Science, Administration, and Revolution* (Oxford: Blackwell, 1993), 277.

2. William H. Brock, *The Norton History of Chemistry* (New York: W. W. Norton, 1992), 95.

3. Jan Golinski, *Science as Public Culture: Chemistry and Enlightenment in Britain, 1760–1820* (Cambridge, UK: Cambridge University Press, 1992), 138.

4. Maurice Daumas, *Lavoisier: Théoricien et Expérimentateur* (Paris: Presses Universitaires de France, 1955), 129.

5. Joseph Priestley, *Experiments and Observations on Different Kinds of Air*, vol. 1 (Birmingham: Thomas Pearson, 1790), 12–20.

6. A. L. Lavoisier, "Sur la cause de laugmentation de pesanteur quaquierent les metaux et quelques autres substances par la calcination," transcribed in C. E. Perrin, "Lavoisier's Thoughts on Calcination and Combustion, 1772–1773," *Isis* 77(1986): 664 and Henry Guerlac, *Lavoisier—The Crucial Year: The Background and Origin of His First Experiments on Combustion in 1772* (Ithaca: Cornell University Press, 1961), 31.

7. Lavoisier, "Sur la cause," p. 669; Lavoisier, "The Sealed Note of November 1, 1772," in Guerlac, *Lavoisier—The Crucial Year*, 227–228.

8. "Cahier de Laboratoire de Lavoisier," no. 1 (henceforth "Cahier lab. 1"), *Archives de l'Académie des Sciences*, Paris, 10–12.

9. Cahier lab. 1, 18–24.

10. A. L. Lavoisier, "Sur une nouvelle theorie de la calcination et de la reduction des substances metalliques sur la cause de laugmentation de poids quelles acquierent au feu et sur differens phenomenes qui appartiennent a l'air fixe," in René Fric, "Contribution à l'Étude de l'Évolution des Idées de Lavoisier sur la Nature de l'Air et sur la Calcination des Métaux," *Archives Internationales d'Histoire des Sciences* 12(1959): 155–162.

11. A. L. Lavoisier, *Opuscules physiques et chymiques* (Paris: Durand, 1774), 259.

12. Ibid., 263–265.

13. Ibid., 260–265.

14. Cahier lab. 1, opp. 33,35,37.

15. Frederic L. Holmes, *Antoine Lavoisier: the Next Crucial Year* (Princeton: Princeton University Press, 1998).

16. A. L. Lavoisier, "Mémoire sur la Nature du Principe qui se Combine avec les Métaux Pendant leur Calcination, et qui en Augmente le Poids," *Mémoires de l'Académie Royale des Sciences*, 1775 [publ. 1778], 520–526.

17. A. L. Lavoisier, "Expériences sur la Respiration des Animaux, et sur les Changements qui Arrive à l'Air en Passant par les Poumons," *Mémoires de l'Académie Royal des Sciences*, 1777 [publ. 1780], 185–194; and Lavoisier, *Traité Élémentaire de Chimie* (Paris: Cuchet, 1789), vol. 1, 33–41; vol. 2, plate IV, figure 2.

18. A. L. Lavoisier, "Mémoire sur l'Existence de l'Air dans l'Acide Nitreux," *Oeuvres de Lavoisier*, vol. 2 (Paris: Imprimerie Impériale, 1857), 129–138.

19. A. L. Lavoisier, "Considérations Générales sur la Nature des Acides, et sur les Principes dont ils sont Composés," *Mémoires de l'Académie des Sciences*, 1778 [publ. 1781], 535–547 (quote, p. 538).

20. A. L. Lavoisier, "Mémoire sur la Formation de l'Acide, Nommé Air Fixe ou Acide Crayeux, & que je Designerai Désormais sous le nom d'Acide du Charbon," *Mémoires de l'Académie des Sciences*, 1781 [publ. 1784], 448–467.

21. For detailed descriptions of these experiments, see Frederic L. Holmes, *Lavoisier and the Chemistry of Life: An Exploration of Scientific Creativity* (Madison: University of Wisconsin Press, 1985), 261–409.

22. Daumas, *Lavoisier*, 145–146.

23. Claude Lévi-Strauss, *La Pensée Sauvage* (Paris: Plon, 1962), 27.

24. Daumas, *Lavoisier*, 129–131.

25. Frederic L. Holmes, *Eighteenth-Century Chemistry as an Investigative Enterprise* (Berkeley: Office for History of Science and Technology, 1989), 107.

26. Frederic L. Holmes, "Beyond the Boundaries," in Bernadette Bensaude-Vincent and Ferdinando Abbri, eds., *Lavoisier in European Context* (Canton, MA: Science History Publications, 1995), 271.

27. Holmes, *Lavoisier: The Next Crucial Year*, pp. 150–151.

28. Stephen Jay Gould, "Why Darwin?" *The New York Review of Books* 43 (April 4, 1996): 12.

7

"The Chemist's Balance for Fluids": Hydrometers and Their Multiple Identities, 1770–1810

Bernadette Bensaude-Vincent

The hydrometer, or pèse-liqueur, was not an innovation of the Chemical Revolution. Its origin can be dated back to ancient Alexandria. The standard reference is to Hypathia, a female Platonist philosopher living in Alexandria in the fifth century AD. However, an accurate earlier description has been found in a Latin poem of Remnius, a Roman of the second century AD. Remnius stressed the analogy between the use of this instrument and the method used by Archimedes to determine the proportion of silver in the crown of King Hiero. Remarkably, he stated the axiomatic principle lying behind the use of this instrument: the existence of a *pondus naturae* (i.e., nature gave a weight to all corporeal bodies and ascribed a specific weight to each element).[1]

Something similar to a hydrometer was represented by Libavius in his *Alchymia* (1597), but it was Robert Boyle who coined the English term *hydrometer* and described the instrument in 1675.[2] He extended the use of his hydrometer in *Medicina Hydrostatica* (1690) with a substantial appendix recommending the hydrometer to assay metallic coins, by counterbalancing a liquid of a known density against a metallic sample. Although hardly mentioned in Maurice Daumas's *Scientific Instruments of the Seventeenth and Eighteenth Centuries and Their Makers*, hydrometry was mainly developed through this period. Toward the end of the eighteenth century, a great number of instruments of various types were built in England, in France, and in Italy for various purposes. Hydrometers were used for the analysis of mineral waters and to titrate alcohols and proportion the price and the taxes to the strength of the spirits. They were also used in breweries to measure the specific gravity of sugar solutions and in the making of saltpeter to determine the proportion of salt in the solution of mother-liquors and, consequently, to decide the appropriate moment for evaporation. As many academic scientists became involved in public administration in late Enlightenment France,[3] efforts were made to improve the hydrometers and to find reliable methods for comparisons between the data available from different instruments used in different places and circumstances.

This revived interest can be illustrated through a comparison of the entries "aréomètre" in the *Encyclopédie de Diderot*, vol. 1, 1751, and in vol. 2 of the *Dictionnaire de Chimie of the Encyclopédie Méthodique* (1792). In 1751, this instrument merited no more than four columns, versus thirteen in 1792, to which should be added sixteen columns in the entry "alcool," the major part of this article being devoted to the hydrometer. Only three kinds of hydrometers were described in 1751, as compared to seven in 1792. Most striking is the contrast between the styles of description. In 1751, the hydrometer was presented as the embodiment of a general law—Archimedes' law. It was aimed not at measuring but at illustrating a general rule: that a body—such as a boat or a floating island—did not need to be made out of materials less dense than the water to float. In 1792, the article "aréomètre" still began with a general definition of specific weight as opposed to absolute weight, and the treatises of physics still presented the hydrometer as an illustration of Archimedes' law.[4] What mattered then, however, was how to determine practically the specific gravity of liquids. In other words, the 1751 hydrometer was mainly intended to render visible, to *faire voir*, a natural phenomenon. The hydrometer was, by 1792, an instrument intended to identify liquors accurately by measurement of their density. The revival of interest thus corresponded to changing functions: from illustration of a general principle to measurement in specific contexts; from demonstration to control. This changing function deserves attention from three different perspectives.

From the chemical viewpoint, the hydrometer that Lavoisier named "the chemists' balance for fluids" was a specific illustration of the methods of weighing (figure 7.1). In this respect, Lavoisier echoed Macquer, who identified two ways of measuring specific gravity: the hydrostatic balance for solids and the hydrometer for liquids.[5]

The hydrometer used to determine the specific gravity of individual liquors could be considered even more crucial than the simplest balance when chemists intended to identify individual substances, because the ordinary balance gave only a general property, not a property specific to the substance. The hydrometer should thus play a key role in the development of quantitative chemistry. One main purpose of this chapter is to examine the hydrometer's place and functions in chemical investigations and to try to understand why it was superseded by the balance in the chemical revolution.

The chemical laboratory was by no means the only "natural space" for the hydrometer, as it was an indispensable piece of equip-

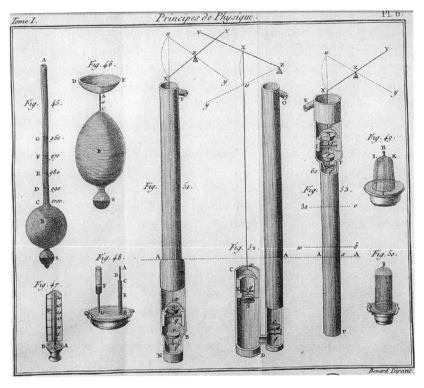

Figure 7.1
Hydrometers at end of eighteenth century, illustrated in Mathurin Jacques Brisson, *Traité élémentaire ou principes de physique*, 2d. Ed. (Paris: Bossange, 1797), vol. 1, plate 6.

ment for the field mineralogist. It was also at home in many manufactures and in the marketplace or wine shops. In addition to these practical functions, the hydrometer had legal functions for the excise on alcohols. The connection between metrics and taxes dated back to the origins of metrics in ancient Egypt.[6] Because it embodies the key notion of proportion, which is the origin of both rationalism and social justice, the hydrometer is a paradigmatic example of the dual functions of metric techniques and, therefore, allows us to appreciate the state of the art in the eighteenth century.

Clearly, the interest in hydrometric techniques exemplified the quantifying spirit that developed at the end of the eighteenth century through a common concern shared by experimental scientists and bureaucratic states.[7] "The quantifying spirit was at work in quantifying spirits," as Simon Schaffer so succinctly states.[8] So the case of the hydrometer may provide a better understanding of this general concern for quantification. Should it be described as an extension of the geometric spirit already at work in some areas of science into the broader field of culture, including administration and industrial practices? Does it suggest a more interactive view between science and society, similar to the balance characterized as a mediator between the cognitive and the economic area, between nature and society?[9] Does it belong to the same culture of precision?

Finally, hydrometers are also interesting from a broader epistemological point of view. Boyle used his instrument to define a universal criterion for distinguishing pure gold from its adulterations. As Simon Schaffer has pointed out, this concern with hydrometry and coin assaying was the onset of a process of construction of a universal view of nature bound by laws—departing from the natural historical view that Nature was locally variable and heterogeneous—that Schaffer has followed through the works of the British "founders" of modern science.[10] Considering the efforts made by French scientists in the late eighteenth century to design hydrometers of universal applicability, can we similarly describe the Enlightenment concern for hydrometry as part of a general movement from particularism to universalism?

Through a selection of French memoirs, this chapter first emphasizes the variety of cultures of the hydrometer, embedded in their social contexts of uses. Then considering the attempts made to improve its performance, it examines how the hydrometer became a subject of scientific investigation. Finally, it shows that such efforts resulted neither in the making of a universal instrument nor even in the triumph of one optimal instrument.

WINE OR WATER?

In fact, two terms were in competition in the eighteenth century to refer to the family of instruments intended for weighing liquors: The word *hydrometer* was more familiar in the English language, and *areometer* was more familiar in French. Was this distinction a simple nuance due to different cultural contexts or did it express the use of two different instruments?

According to M. Bories, a French scientist from Montpellier, the hydrometer and the areometer were two different instruments exemplifying two different methods for determining the specific gravity of a liquor.[11]

The areometer maintained a fixed volume and measured different weights. The hydrometer was based on a fixed weight with variations of volume. The areometer was derived from Fahrenheit's so-called hydrometer, made up of a bulb and a vertical stem going through the bulb. The lower extremity of the bulb was loaded with weights to sink the bulb into the liquid under trial, and the upper and longer extremity ended with a small pan used to add weights until the level of the liquid reached a single mark on the stem. From the number of weights added in the upper pan, one could infer comparisons of densities between various liquors.

The hydrometer, on the contrary, had a graduated scale on its stem, allowing a direct reading of the result. In other words, the comparisons have been incorporated into the instrument. Ironically, the areometer seems to be an English invention, of Robert Boyle[1], improved by Fahrenheit, whereas the hydrometer is better known as the achievement of two French chemists, Montigny and Baumé.

According to Bories, the areometer was a good instrument for physics laboratories but unfit for commercial use, because of the number of operations required for a comparison. The hydrometer was obviously more convenient because it required only one simple operation by its user.

However, Bories's distinction between the hydrometer and areometer was never adopted by his contemporaries. For instance, when describing the two kinds of instruments, Lavoisier called them "areometer with weights" ("aréomètre à poids") and "areometer with stem" ("aréomètre à tige").[12] Bories himself confessed that the words *hydrometer* and *areometer* were commonly used as synonyms:[13]

> I believed that it is necessary to distinguish the hydrometer from the areometer, although these two words are generally accepted as synonyms. Both are derived from the Greek. The word areometer signifies, in strict usage, the measurement of thin, porous, spongy bodies, a measure of their volume: whereas the hydrometer is meant to measure water, or in general, any liquid, a pèse-liqueur.

The word *areometer* derived from the Greek *araios*, meaning "subtle," suggesting an instrument meant to measure degree of subtlety, which sounds somewhat odd. If one remembers, however, that the word *alcohol* first referred to all substances reduced to a state of fine powder and then referred to extremely light and volatile liquors,[14] a plausible conclusion is that the word *aréomètre* may have emerged from the earliest uses of this instrument for determining the degree of alcohol in liquors, whereas the hydrometer was intended to weigh water.

If this origin is correct, what should be noticed is that the instrument has been named after its applications rather than according to a general definition of the physical quantity that it measures (i.e., the specific gravity or density). Remarkably, although the terms *areometer* and *hydrometer* sound more academic than does *pèse-liqueur*, which belongs to common language, the latter is a more general term than the former two, which are deeply embedded in their local contexts of application.

Is the differential terminology between French and English languages linked with two different practices: measuring the density of spirits of wine in France and of mineral waters in Britain? It is far from certain that the contrast can be explained in terms of national cultures, because hydrometers were used for measuring many other liquors in both countries. The United Kingdom, for instance, experienced a great demand for hydrometers for breweries.[15] In France, areometers played a key role in the campaign launched in 1775 by the Régie des Poudres et Salpêtres to increase the production of gunpowder.[16] However, the applications to wine distillation and mineral waters are analyzed here in further detail as exemplifying two different cultures' use of the same instrument.

ANALYZING MINERAL WATERS

As a young geologist working under the supervision of Guettard, Lavoisier had to carry out many analyses. He started to use the hydrometer as early as 1765 in his study of gypsum and then in analyzing a hundred mineral waters in the East of France.[17] Lavoisier presented his own method for analyzing mineral waters. He first discarded the traditional qualitative method by chemical reactants. The quantitative method by evaporation he criticized as so unreliable that correct analytical results appeared as a miracle. Lavoisier's method was grounded in an elementary gravimetric consideration: a salt dissolved in water increases its weight. He then designed a hydrometer and became so interested in this matter that he devoted a whole memoir to this technique in 1768.

Before describing his own instrument, Lavoisier gave a critical review of the other instruments available for the same purpose. Homberg's instrument lacked accuracy. It might be suitable for "trade and society," but investigating mineral waters required the utmost accuracy. Interestingly, Lavoisier's review assigned a specific space to each kind of instrument. Most hydrometers are inconvenient for ordinary use because they are too heavy and too fragile to be portable. The hydrostatic balance is too complex for popular uses in the marketplace.[18]

As a member of the "Ferme Générale" in charge of collecting the excise on alcohols, Lavoisier belonged to the sphere of society's users. As a member of the Academy, he belonged to the "Republic of Science." In his memoir on the areometer, he carefully distinguished both spheres, describing two kinds of instruments—one for the physicist's use and another for commerce. Such is the contrast between their respective needs that it required an inversion of the principles of the instrument: fixed volume for the physicist, fixed weight with a graduated stem for the people.

One interesting feature of the hydrometer is that a more sophisticated instrument was necessary for commercial use than for scientific purposes. The less skilled the user, the more complex the instrument should be. In fact, Lavoisier never managed to build an areometer practical for commercial use. He described a complex and painstaking procedure for graduating the stem, with repeated accurate weighings of small quantities of mercury and a table providing the specific gravity according to the volume of liquid displaced. Lavoisier confessed that only a "skilled artist" would be able to make such an instrument, and he mentioned the name and address of one.

Lavoisier was more successful with the areometer for academic use. His main preoccupation was to "limit the range of possible errors," but he insisted that he had to combine "sensibility and convenience," to build an instrument suitable for all kinds of liquids from water to alcohols, including acids. This project of a universal, multipurpose hydrometer required a general investigation of the performance of an areometer. Lavoisier identified volume and shape as the decisive factors and grounded the design of his instrument in mechanical and geometrical considerations concerning the position of the center of gravity.

Although he claimed to improve on the hydrometer designed by de Parcieux, a member of the Paris Academy of Science, the reference seems rather rhetorical, as his own instrument was more directly inspired by Fahrenheit's areometer. The reasons he invoked for his choice were opposite to those in Bories's comparison between areometer and hydrometer. To Lavoisier, measuring weights obviously simplified the calculations associated with the act of measuring volumes:

"Other things being equal, this equality of volume is a very real advantage; it avoids many reductions and calculations in practice, and thereby simplifies the operations."[19]

Lavoisier's memoir thus exemplified a nascent divorce between scientific and commercial instruments based on the alternative between gravimetry and volumetry. Physicists and chemists were more confident about weighing than about measuring volumes. Mathurin Brisson, for instance, argued that the stem hydrometer showed only the existence of a difference of density between liquids; only Fahrenheit's method allowed an accurate estimate of the difference (figure 7.2).[20] Reciprocally, in his review of the art of distillation, Demachy noted that Lavoisier's instrument could be used only in the "cabinets de curieux."[21]

Even among scientific instruments, however, Lavoisier's improved version of Fahrenheit's areometer was not the favorite. Despite his reputation as a member of the prestigious Paris Academy of Science, Lavoisier was much less successful than was Antoine Baumé, a pharmacist-chemist. Baumé's method consisted of a hydrometer with a stem graduated by reference to a saturated solution of salt and a large table of data, analogous to Bergman's table of affinities, summarizing the results of hundreds of measurements performed by varying the experimental parameters: the weight of the liquor; its volume referred to the same weight of pure water; the variation of volume due to the mixture of two liquors; and the variations of densities according to temperatures. Thanks to this database, Baumé's instrument allowed more flexible uses. Finally, even for its initial function—analyzing mineral waters—Lavoisier's hydrometer proved insufficient. The standard method adopted toward the end of the century was the one developed by Torbern Bergman in 1778, which combined the procedures rejected by Lavoisier with hydrometry.[22]

TITRATING ALCOHOLS

For titrating alcohols, Baumé's hydrometer, imitated and popularized by Cartier, became the standard method in France in the 1770s. It was widely adopted even by Lavoisier's colleagues, the "fermiers généraux." This choice was the result of an intense rivalry between various instruments designed and built in response to a fierce competition in the wine trade.

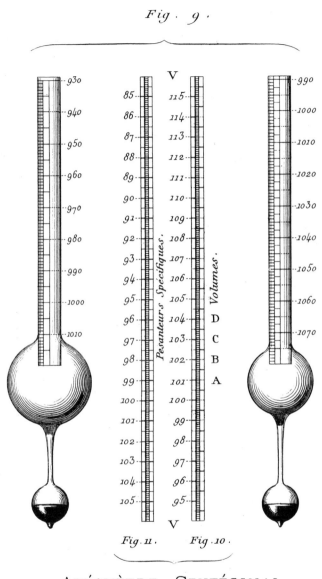

Fig. 9.

V

Pesanteurs Spécifiques.   *Volumes.*

Fig. 11.   *Fig. 10.*

V

ARÉOMÈTRE   CENTÉSIMAL

Figure 7.2

Areometers used by Antoine Lavoisier to measure specific gravities of waters. 1000 marks level of immersion in distilled water. Remainder of scales give relative volume displaced in waters of different density. Lavoisier provided tables for conversion of volume displaced to specific gravities. (Left) Areometer for use in liquids less dense than distilled water and (right) for denser liquids. (Center) Scales marked on strips of paper during calibration operation. From *Oeuvres de Lavoisier* (Paris: Imprimerie Impériale, 1865), Vol. 3, Plate VIII.

This situation can be illustrated by the commercial rivalry between Languedoc and Catalonia in the 1770s, which prompted the Etats du Languedoc to secure reliable methods for determining the degree of alcohol and for detecting frauds.[23] Prizes were proposed by the Société Royale des Sciences to inventors of such methods. The prizes were awarded in 1771 by the Abbé Poncelet and in 1772 by Bories, a medical doctor.[24] Both offered a hydrometer as the best method, while criticizing more conventional methods in use.

The reason the hydrometer was so crucial for wine distillers was that the prices of alcoholic beverages for the customer and the amount of excise charged by the state were established according to the relative proportions of water and alcohol in the liquor. Before the use of one standard hydrometer was established by legal acts, an agreement had been reached on the basis of a nomenclature of alcohols referring to a complex combination of three parameters: the number of distillations, the proportion of water and alcohol, and the nature of the test used. The standard reference was a liquor, in common use, named after the Holland Test ("Preuve de Hollande"), which was obtained after the second distillation. All other degrees of alcohol were defined by reference to the "Preuve de Hollande".[25]

What were the conventional methods for titrating alcohols that the hydrometer was supposed to replace? Four tests were mainly used by distillers and wine traders, all of which rested on skill and experience.[26]

The Holland proof: In this routine quality test, the distiller simply shook a small quantity of the liquor in a test tube and evaluated the degree from the number and velocity of the bubbles generated by the injection of air within the liquor.

The gunpowder test: Itself an improvement on Geoffroy's test of "ustion," the gunpowder test consisted in inflaming the liquor. The addition of gunpowder provided a very fast method by which to identify the "trois-six": if the gunpowder was burnt, the titer was good; if not, the titer was too low.

The tartaric salt method: Described by Boerhaave to rectify the spirit of wine, this method tested the concentration of water precipitated while the alcohol remained on the surface.

The drop of olive oil test: Specific gravity was not measured by this test, but rather it was used as a visual indicator. The titer of the liquor was too low if the oil floated on the surface and was correct only if the oil sank in the liquor.

By the 1760s, all these methods were rejected on grounds that they were mainly qualitative and too dependent on a variety of circumstances that were not under control. The general agreement reached among authors concerned with distillation was to substitute for all these conventional methods one single quantitative indicator: specific gravity.[27]

> The uncertainty of these methods does not escape the notice of the scientist who cannot be too scrupulous, of the farmer who wants to be sure that it is all alcohol, or of the merchant who may wish to fool the buyer but who does not want the manufacturer to impose on him. The pèse-liqueur satisfies all three.

Apparently, the best way to determine the specific gravity of liquors was no longer a matter of debate in the 1770s. The question was closed in favor of the hydrometer. The use of an ordinary balance to weigh the same volume of different liquors in the same flask was simply discarded as being too dependent on the performance of the balance. The hydrostatic balance, "instrument plus digne du physicien," was discarded for the same reason. An alternative method of measuring specific gravity with the help of small glass "bubbles" labeled as *philosophical beads* was simply ignored by the French chemists. This method, invented by Alexander Wilson in 1751, was used in Britain. The beads were supplied in sets that covered a wide range of specific gravities. Each bead was calibrated and patented for a specific value at a particular temperature, and the bead floating on the liquid under test indicated its specific gravity.[28]

Behind the consensus regarding the hydrometer, a question remained, however: Which one should be preferred: that of Homberg or that of Fahrenheit? A multitude of rival instruments existed, and each claimed to improve on the others. In 1773, Demachy examined the comparative advantages of 11 rival instruments available at that time in France. Except for the "common pèse-liqueur," all were named after their inventors: Fahrenheit, Homberg, Musschenbroek, Cartier, de Parcieux, Lavoisier, Demachy, Baumé, Brisson, and Gouvernain. In their efforts to design a precise, reliable, flexible, and convenient hydrometer, the chemists concerned with wine distillation clearly identified the parameters to control and the theoretical principles involved in this gravimetric technique.

Like all techniques of measurement, hydrometry is necessarily relative. The strength of a liquor is always determined by comparison with a standard. The specificity stressed by Bories and Poncelet, however, was that to be adaptable to all kinds of liquors, a hydrometer needs not one but two standard units, two marks corresponding to the extreme densities. On the stem of the hydrometer, the lower mark would indicate the density of distilled water, and the upper mark would show the density of the most highly rectified alcohol. The quality of the instrument thus depended first on the preparation of pure distilled water and on the determination of its specific gravity, which was to provide one reference point. Lavoisier worked for some time on this problem and published his results in a memoir: "Pesanteur absolue de l'eau distillée. Poids du pied cube d'eau et contenance de la pinte de Paris."[29]

The main challenge for the second reference point was to ensure that one had prepared the "most highly rectified alcohol." The method of rectification consisted in removing the "phlegme" (the aqueous constituent) and the "oil" from the spirit of wine, a complex process that required approximately 6 months. Interestingly, the lengthy description of the process by the two prize winners of the competition in Languedoc made use of the conventional qualitative tests, such as the gunpowder test that they strongly criticized. Individually, each method was considered as erroneous—Poncelet even used the gunpowder as a proof of impurity—but a suitable combination of all of them provided a reliable estimation of purity. Significantly, despite the variability of graduations from one instrument maker to another, all were graduated from 0 to 10. The hydrometer was thus one of the earliest applications of a decimal system, some decades before its generalization in the metric system.

Because hydrometers had to be used in all seasons and under different climatic conditions, the second problem to solve was how to control the influence of temperature on the densities. Using a thermometer in conjunction with the hydrometer was indispensable (preferably a thermometer containing alcohol rather than mercury) to compare similar variations of densities between the two instruments. This practical problem led to interesting observations. Baumé pointed out that the influence of temperature increased with the degree of alcohol and ventured a general rule: An increase of 5 degrees on the thermometer required an increase of 1 degree on the areometer. Bories noted that the variation of density was also a function of the degrees of temperature. That is why, to build an instrument that could be used for all degrees of spirituosity and under all temperatures, Bories saw no other way than to perform thousands of repetitive experiments with a sensitive hydrometer for determining the density of various mixtures of alcohol and water (from 9 : 1 to 1 : 9) and for each degree on the thermometer between 0° and 20°C.

The resulting table was incorporated into the stem, which for this purpose was made rectangular, with four faces. Titrating alcohol thus became an easy operation "which requires only the eyes," two elementary actions, immersing a thermometer and a hydrometer into the liquor. Bories bragged that it was accessible to lay persons but nevertheless satisfactory for the physicist.

A third difficulty explains why this painstaking empirical procedure for making the table was necessary. When they tried to determine the intermediary graduations of the scale between distilled water and rectified spirit, the instrument makers realized that they could not simply follow the arithmetical progression from 0 to 10. The density is not proportional to the proportions of the mixture, because the total volume of a mixture of water and spirit is less than the sum of the two separate volumes of constituents. This phenomenon, described as *mutual penetration*, was considered the main cause of error in hydrometry. Though it was recognized as a general fact and related to the mutual attraction between the constituent substances of the mixture, the decrease of volume differed for each mixture and prevented the graduation of scales through a simple calculation.

CONTRASTED SCIENTIFIC STYLES

The aforementioned difficulty prompted some general reflections and further scientific investigation related to hydrometry. Facing this problem, investigators took two different attitudes.

In a lengthy essay on pyrometry and areometry, Swiss physicist Jean André de Luc considered the irregular increase of density and described it as a general phenomenon of all "mixtures," one example of a central problem. De Luc thus emphasized the cognitive function of the hydrometer. He viewed it, not as a means to reach a consensus on a price for alcohol, but as an instrument to explore the complexity of nature. He then developed a broad reflection on metrics. Considering that instruments only measure effects (e.g., of heat on mercury), de Luc assigned them two exploratory functions. They allowed one to discover immediate causes through the measurements of effects and to realize how many causes are concurring for one and the same effect.

They also helped disentangle simultaneous effects of one and the same cause.

Because causes and effects are not isolated but interactive, the variation of effects is not proportional to the variation of causes. In this respect, the hydrometer is paradigmatic, because it shows three different effects that are not proportional to their causes: (1) The level of immersion into the liquid is not exactly proportional to its density because of the variations of the volume of the hydrometer with heat and because of the irregularities in the shape of the tube; (2) the changes of density do not follow the same law as the changes in temperature; and (3) the increase of density is not proportional to the quantitative composition of the liquor. Because so many irregularities were incorporated in its construction, the hydrometer epitomized the complexity of nature that prevented the formulation of general and universal laws.[30] Practically, it meant that the graduation of any hydrometer could be determined only by experiment. The instrument maker must determine experimentally the density for each degree of the thermometer and for each proportion of the mixture (figure 7.3).[31]

> Concerning physical co-effects in general, and I venture to say here, in co-effects of every type, if one cannot fix all the relations degree by degree through certain and immediate observations, one must avoid drawing general rules from relations taken at the extremes. The action of causes, whether moral or physical, is too complicated, whether with regard to the subjects on which they act or with regard to the secondary causes which escape our observations, to assume that the variations of the observables increase in the exact proportion to the evident causes, and consequently that the co-effects of the latter are proportional to one another.

Thus, with all its imperfections, the hydrometer nevertheless deserved attention from physicists precisely because it emphasized the limitation of human knowledge and suggested a probabilistic approach to natural laws: "We are obliged to content ourselves with the probable in so many respects in nature, that to search for the physical rules of probability is perhaps more essential for us than to attach ourselves to these mathematical rules concerning the hypotheses."[32]

Remarkably, de Luc's approach revived the kind of philosophy that accompanied Boyle's uses of hydrometers. In his *Medicina Hydrostatica*, Boyle emphasized that specific gravities, like other qualities,

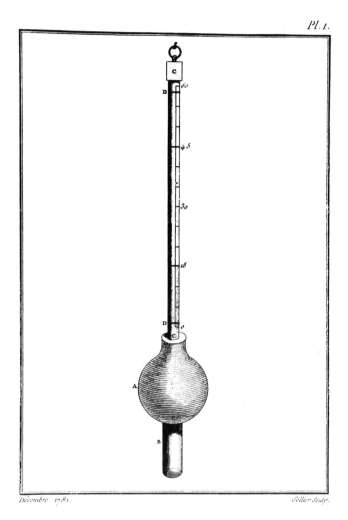

Figure 7.3
Areometer designed by Jean Antoine De Luc to measure the density of spirituous
liquors. A, hollow reservoir; B, mercury ballast. Fixed points marked at "15" and
"45" determined by immersing areometer respectively in two liquids of known
spirituous concentration. From "Suite du Mémoire de M. De Luc, Sur la Pyro-
métrie et l'Aréométrie," *Observations sur la Physique*, 18(1781), Plate I, facing
p. 506.

could never be determined with mathematical precision and that we could reach only the probable.[33]

Confronted with the same difficulty of the uneven increases of volumes, Lavoisier responded in another language. While he was performing many analyses of mineral waters, as a young geologist in the 1760s, he observed that the gravity of water did not increase regularly, according to the proportion of salt added. Like many of his colleagues, Lavoisier first tried to find out whether such irregularities followed a law. Then he realized that this phenomenon could be related to complex variables, such as the shape of molecules of salt and water or the crystallization of water within the salt. He then imagined a "chemical geometry" with analytical formulae permitting the deduction of the state of the fluids in solution. It was a kind of "Newtonian dream" inspired by the hydrometer:[34]

> This was the plan I set out for myself at the beginning, and I have not permanently abandoned it; but I lacked so many of the things necessary to carry it out, there were so many preliminary experiments required to collect a sufficient number of data, that I was deterred by the amount of work.

Although Lavoisier postponed his ambitious program, his memoir on the areometer, nevertheless, emphasized all the benefits that chemistry would derive from the hydrometer. He disparaged those chemists who "content themselves with words," such as elective affinities. The hydrometer would bring clarity into the complex process of combination. It would also bring clarity into the mysterious origin of the "fixed airs" identified by Hales.[35]

> It is mainly in the art of combinations that knowledge of the specific weight of fluids can be most illuminating. That part of chemistry is much less advanced than one thinks. We hardly have established the first elements. We combine acids with alkalis every day; but in what manner is the union of these two entities made? Are the constituent molecules of the acid lodged within the pores of the molecules of alkali, as M. Lemery thought, or are the acid and the alkali perhaps composed of different facets which mesh with one another, or which unite by simple contact in the manner of the Madgeburg hemispheres? How do the acid and the alkali separately seize water? How do they do it after their combination? Does the new salt which is formed occupy only the pores of the water? Is this a simple division of parts, or is there rather a real combination, whether of part to

part, or of one part to many? Finally from whence comes the air that escapes with such vivacity at the moment of combination, and which, manifesting its natural elasticity, occupies at once a vastly greater space than that of the two fluids which it has left? Did that air originally exist in the two mixts? Was it fixed in some manner, as Hales thought, and as most scientists think still, or is it rather a factitious air which is the product of the combination, as M. Eller thought? When queried about such subjects, chemistry responds to us with futile names of relations, analogies, of contacts ... which offer no idea, and which only accustom the mind to be satisfied with words. If it is possible to the human mind to penetrate these mysteries, it is through investigations of the specific weight of fluids that one can hope to arrive there.

The attitude of the young Lavoisier in the face of the difficulty raised by the variation of volume of salt solutions is remarkable mainly for three reasons. Unlike Bories or Baumé, he did not content himself with a description of the phenomenon and a series of experiments to circumvent the difficulty. Like de Luc, he soon realized that the hydrometer raised deep theoretical issues. Unlike de Luc, however, he confined the problem to chemistry instead of dealing with it in terms of natural philosophy. In his view, the hydrometer was a subject for chemical investigation. The most remarkable is that, instead of being stopped by the problem he could not solve, Lavoisier decided to use the hydrometer as an instrument for chemical investigation. So impressive is the conclusion of Lavoisier's memoir on hydrometry that it prompted the following comment by Marco Beretta:[36]

> In emphasizing the central role of the notion of specific gravity, Lavoisier introduced an entirely new approach to the subject of chemical combination. Given a method by which all quantities of matter could be detected, isolated and identified, the old notion of mixt which was the central concept of Stahlian chemistry lost its function and meaning.

In connecting this 1768 conclusion and Lavoisier's "crucial" investigations carried out in 1772 and 1773, Beretta has invited Lavoisier's scholars to reassess the respective role of the balance and of the hydrometer in the process of the Chemical Revolution. One should be cautious, however, before attributing a revolutionary role to hydrometry, for various reasons.

First, Lavoisier never achieved his promising program of investigation of the nature of affinities through a systematic investigation of salt solutions. Rather, it was carried out by such contemporary chemists as Wentzel, Bergman, or Richard Kirwan, who stayed apart from the theoretical shift achieved by Lavoisier. What should be noticed is that in his study of specific gravities and the attraction of various salt substances, Kirwan used only a hydrostatic balance and discouraged the use of a hydrometer as "subject to an infinity of errors".[37]

Lavoisier certainly made extensive use of the hydrometer for chemical investigations. In his detailed study of the 1773 laboratory notebook, Frederic L. Holmes has pointed out that Lavoisier used the hydrometer to decide between Meyer's and Black's alternative theories of acids. Like the balance, the hydrometer was expected to deliver a clear answer "to decide all at one stroke, and in a manner that would leave nothing uncertain, whether lime gives something to a caustic alkali in the fashion of Meyer, or whether, on the contrary, it removes something according to the English hypothesis".[38] The result—a decrease in density—persuaded him in favor of Black's hypothesis, but later, when reported in the *Opuscules Physiques et Chimiques*, the experiment no longer allowed such firm conclusion.[39] Although favorable to Black's theory, the experiment could still be explained in terms of Meyer's theory. The hydrometer did not allow the model crucial experiments that Lavoisier was able to set up with his balance. Even for exploratory purposes, Lavoisier used it only occasionally. A series of experiments performed to identify and characterize the fixed air released from a calcareous earth reacting with acid are reported in the *Opuscules*. Lavoisier determined the density of fixed air with his hydrometer through a comparison of the density of a sample of lime water with the density of another sample of the same lime water with fixed air added, and he repeated similar measurements with distilled water and other fluids.[40] For determining the density of other airs, he later preferred the balance-sheet method, using an air pump and a balance.[41] Although the ratios of weights and volumes proved essential in his investigation of "airs," Lavoisier did not encourage the use of the hydrometer. Thus, clearly Lavoisier did not consider the hydrometer as a reliable instrument of investigation. He sought to avoid all considerations of volumes and favored the action of weighing. The balance successfully displaced the hydrometer, both for exploratory and demonstrative functions.

Surprisingly, Lavoisier's interest in hydrometry was not revived even when he became a commissioner of the Gunpowder and Saltpeter Production in 1775, although the introduction of the hydrometer was a priority both in the making of saltpeter and in its refinery. Patrice Bret has revealed the manuscripts of the *Dictionnaire de l'Artillerie* of the *Encyclopédie Méthodique*, partly written by Lavoisier and his colleagues (for the entries concerning gunpowder production). The single entry devoted to instruments is "aréomètre," but it was not Lavoisier who took charge of it. The article was written by A. Clouet. It recommended a glass hydrometer with centesimal graduations taken from an instrument designed in provincial workshops by workers, whereas the construction of the standards was assigned to Mossy, the instrument maker of the Académie royale des sciences de Paris.[42] The standardized hydrometers of the Régie proved crucial to improve the production of gunpowder, because a better control of the concentration of leaching waters allowed not only increased yields but a reduction of the wood consumed for evaporation.[43] So promising was the use of hydrometers for national defense that in early January 1794, while Lavoisier was imprisoned, Guyton de Morveau ordered 1000 hydrometers for the National Laboratory of Weapons to be used for training citizens in the art of gunpowder in the *Cours révolutionnaire de Salpêtre*.[44]

The hydrometer that the young Lavoisier sought to improve and to promote as an investigative tool was no longer viewed as a scientific instrument in the last decades of the century. More precisely, the hydrometer was still perceived as indispensable for determining the specific gravities of fluids, but it was given a minor status in the laboratory. In the third section of his *Traité élémentaire de chimie*, Lavoisier first recommended the hydrostatic balance for determining the specific gravity of liquids. The hydrometer was recommended only for the analysis of mineral waters, and Lavoisier added that the graduated stems were "convenient" though not accurate.[45] Lavoisier obviously had transferred his early enthusiasm for measurement from hydrometers to gasometers.

IRRETRIEVABLE MULTIPLICITY

The great number of hydrometers available at the end of the eighteenth century certainly revealed a pressing social demand. However, it can be considered also as a symptom suggesting none of these instruments was the best one. The terms *diversity* and *imperfection* were often coupled in the abundant literature on hydrometers. For instance, Richard Kirwan commented on the hydrometers in a letter to Guyton-de-Morveau:[46]

The pèse-liqueur of Baumé is explained very well in his *Eléments de Pharmacie*. It is the simplest one, but it is not yet a perfect instrument, and others are put in fashion that are worth no more, which is still more troublesome, because, as you say, people no longer understand one another.

Although in 1785 Kirwan considered imperfection as a temporary state that could be overcome thanks to a thorough project of standardization of weights and measures, in 1800 Jean Hassenfratz had a more radical view of the imperfection of hydrometers. A former assistant of Lavoisier, Hassenfratz referred the amazing number of hydrometers used for titrating alcohols to the conflicting interests at stake in the use of the hydrometer:[47]

Because the interest of the merchant is to pay the least possible duty, and the interest of the treasury is to collect the largest duty, the former has sought to deceive, and the latter to discover the fraud. Consequently a great number of different areometers have been made.

Does that mean that the hydrometer was too imperfect to allow any consensus to form? Efforts were made to standardize the graduation of the scales of hydrometers in various countries, especially in the United Kingdom. At the request of the government, the Royal Society sponsored an investigation of "the best method for proportioning the excise on spirituous liquors".[48] Charles Blagden, who was the reporter, identified the mutual penetration of water and spirit as the key problem and set up a series of experiments "to ascertain the quantity and law of condensation".[49] Blagden first recommended the determination of the proportion of the mixture by weight, rather than by volume, because it was the only accurate method of fixing proportions and also because if the proportions were to be taken by volumes, a different volume must be prepared at every degree of temperature. For determining the specific gravity of the mixtures, Blagden chose the method of filling a vessel of a known weight and ascertaining the increase of weight it acquires, then computing the specific gravity by comparison with the same amount of pure water at the same temperature. This method, consisting of weighing solids instead of liquids, was considered much more reliable than direct weighing of liquids in the hydrometer. In fact, the hydrometer, presented as "the readiest way of ascertaining specific gravities, and doubtless the most convenient for public business",[50] was confined to the realm of convenience but unfitted for the purposes of standardization.

This view was presumably shared by Haüy and Lavoisier when they set out to determine the standard unit of mass for the metric system. The kilogram or "grave"—conventionally defined as the weight of 1 cubic decimeter of distilled water at the temperature of melting ice—presupposed an accurate weighing of a liquid. In fact, the uncertainties of the hydrometer were avoided through weighing distilled water in a metallic box with a hydrostatic balance.[51]

In spite of the number of hydrometers built in the late eighteenth century and of the number of papers devoted to the subject, areometry did not become an independent domain of study. A series of memoirs entitled "De l'Aréométrie," by Hassenfratz (1797–1800), represents the swan song of a short-lived discipline. In their very fragmented presentation, the four memoirs reflect a disunity that Hassenfratz emphasized by choosing different names for different uses. The hydrometer used for determining the proportion of a salt in a solution of water he named the *salinograde*, thus avoiding the suffix *meter* that he wanted to reserve for the measure of lengths. The hydrometer for determining the proportion of alcohol in water was named the *alcograde*. In addition to different instruments for every fluid under test, different salinogrades should assay every kind of salt.[52]

To cope with such a multiplicity, Hassenfratz tried to shape a research program, apparently connected with the preparation of salt substances for the laboratory classes at the École polytechnique. From a rapid estimation of the number of acids, bases, and metals and their different degrees of oxidation, Hassenfratz inferred that the number of possible salts should be near 1500. As only 40 of them were important or useful, 40 different salinogrades were needed. Hassenfratz effectively built only one, for nitrous salts. To avoid the painstaking and time-consuming series of experiments required for graduating the scales of each hydrometer, Hassenfratz imagined a graphical method for determining the intermediary degrees of the scale by calculation.[53] In a way, Hassenfratz thus achieved the ambitious program of a systematic study of salts outlined by Lavoisier in 1768, but he did it in renouncing the analytical deductive approach suggested by Lavoisier in favor of a more pragmatic graphical method.

The alcograde was described as an instrument able to give only rough approximations: "Up to the present, it is the most suitable instrument to give approximate results." It was not even suited for the purpose of its most general uses because measuring densities, Hassen-fratz noted, is not measuring the quality and the value of spirits, which could be appreciated only through degustation.[54]

In the context of the creation of the metric system in France and of the aspiration to uniform and universal standards, the diversity of hydrometers in use in 1800 and the heterogeneity of the quantity named *a degree* in each different model was perceived as a scandal. Decroizilles, the inventor of the burette or berthollimeter and of the alcalimeter, attempted to apply the logic of the metric system to areom-etry. He thus designed a complex method to remedy the heteroge-neity of hydrometers. His "aréométrotype" was intended to provide the standard reference to graduate commercial hydrometers. It was a tech-nique to determine the relative specific gravity of liquids through weighing them in a glass bottle. In other words, clearly most admitted that accurate weighing of liquors could not be made directly with a pèse-liqueur but had to be mediated through weighing the solid con-tainers of liquids:[55]

> There is scarcely any instrument more precise than a good balance at its point of perfect equilibrium, whereas the line of the level of immersion of a pèse liqueur is difficult to determine, because it varies according to the liquid with which that part above the liquid is moistened.

The terminology associated with the hydrometers themselves was entirely revised in keeping with the revolutionary mood, with its characteristic passion for neologisms. The notion of specific gravity had to be replaced by *pesanteur hydro-majeure* for liquids heavier than water or by *légèreté hydro-majeure* for liquids lighter than water. In full accep-tance of the decimalization imposed by the metric system, the degrees had to be replaced by "centièmes," although Decroizilles allowed both standards to be retained on the same scale to facilitate the conversion. Hassenfratz's and Decroizilles's efforts to promote a revolutionary areometry thus resulted mainly in linguistical inventions and a general discredit of direct weighing of liquors (i.e., of the technique of the pèse-liqueur).

In summary, the story of the hydrometer can be characterized from the three different perspectives outlined at the beginning of this chapter. The "quantifying spirit" of the Enlightenment is certainly illustrated by the revival of interest in the hydrometer in the late eigh-teenth century. A number of circumstances ranging from the regulation of the wine trade to an increasing interest in mineral waters to military production of gunpowder prompted an intense production of hydrometers. This small piece of apparatus became crucial for extending the state bureaucracy and was developed, as was Boyle's hydrometer, as an instrument of control and of detection of frauds and adulterations. Hydrometry also played a crucial role in manufacturing standardized products, such as beers.

Far from suggesting a one-way dissemination from academies to factories, this chapter shows that improving hydrometers was a common concern shared by manufacturers, tax collectors, pharmacists, wine traders, and military defense. This case study in Enlightenment metrology thus leads to a more complex and more conflicting view of the so-called quantifying spirit. Social and commercial interests did not fully converge with scientific interests. The divorce between laboratory instruments and commercial instruments, as well as the emergence of a variety of new analytical tests based on chemical reactions (chlorometers, alcalimeters, etc.) that soon displaced the imperfect hydrometer, suggest that the cult of precision did not reign supreme over the whole society, that it had to compromise with constraints such as convenience, simplicity, and reliability.

Concerning the uses of the hydrometer in chemical laboratories, it was certainly an indispensable piece of apparatus but with minor functions. The hydrometer could not supplant a balance, even for weighing fluids, because building and operating a hydrometer most often implied the use of the hydrostatic balance. Because it also required the use of thermometers and tables or rules for adjusting the temperatures, the hydrometer exemplifies the solidarity between instruments in a laboratory that operated as a system.

In this "metric system," however, the hydrometer never competed with the balance for investigative and demonstrative purposes. The rejection of volumetrics in favor of weighing was a typical attitude of late-eighteenth-century chemists, which prompted the divorce between laboratory instruments and industrial or commercial instruments.

The chemists' interest in hydrometry at the end of the eighteenth century did not provide a future for the hydrometer in chemical investigation. The hydrometer could be recommended for studying the composition of mineral waters but, for determining the proportion of the constituents of water, Lavoisier and his contemporaries rather used gasometers and balances. The singular fate of the volumetric hydrometer, sharply contrasting with the triumph of the balance and gravimetry, has to be understood in the context of late-eighteenth-century instrumental practices.

The story of the hydrometer exemplifies the evolution of the rhetoric of the Paris Academy from utility to accuracy, as pointed out by Licoppe.[56] Undoubtedly, the prevalence of accuracy over utility undermined the importance of the hydrometer. More precisely, the hydrometer challenged the axiomatic principles organizing the emerging culture of precision that, according to Licoppe, were repeatability, comparability, and replicability of experiments.[57] The hydrometer met only the first condition to the extent that the same experiments performed by the same experimenter in the same conditions could be expected to lead to the same result. At least, repeatability was a condition for establishing the tables accompanying the hydrometers. Comparability, on the contrary, was the major obstacle, not only because the same experimenter operating with different instruments in the same circumstances could obtain different results but, more basically, because the same operator with the same instruments operating in different circumstances would obtain different results.

The tables allowing comparison of the results for different temperatures were conceived as a remedy to that weakness but could not take into account the variable composition of liquors according to their geographical origin or the year of their production, which truly determined their quality. Replicability (i.e., the possibility of reproducing the same results in different places, under different circumstances, with different instruments by different operators) is even more problematic. The literature about the hydrometer studied in this chapter, makes clear that the results could not be validated apart from the circumstantial details of the experiments and even of the construction of the instrument. Significantly, the various hydrometers and the tables in use were always named after their designers, thus suggesting that the validity of the performance of a hydrometer was referred to the authority of its author rather than to universal authority.[58]

From an epistemological point of view, according to Gaston Bachelard, precision instruments can be described as "materialized theories".[59] Though hydrometers were the "materialization" of Archimedes' law, they better exemplify the gulf between the general and the peculiar, between principles and material techniques. In the evolution of metrics, the hydrometer appears as the epitome of the "à peu près" ("almost") of approximative measurement. The general impression emerging from the literature of this period is that of a radical imperfection. Far from promoting universalism, a view of nature bound by uniform laws, the evolution of hydrometry in the late Enlightenment suggests that local circumstances cannot be overlooked. Lavoisier's failure to design a universal hydrometer that could measure all kinds of liquors, and the rapid split of hydrometry into a variety of instruments made for special purposes (e.g., acidimeters, salinometers, lactometers, saccharometers) in the nineteenth century, suggest a revenge of particularism over universalism.

In a sense, the hydrometer acted as a counterpoint to the balance. Whereas the general concept of a balance as an equilibrium or equality between two quantities allowed a great simplification of nature and society because it made diverse phenomena commensurable, the hydrometer displayed the difficulty of establishing proportions between different phenomena. Direct weighing of liquors, already a familiar and popular practice, thus raised a host of questions about natural laws. Whereas the balance contributed to the rationalization of nature, the hydrometer suggested its complexity and the limits of our rationalization of the real world.

Notes

I acknowledge with gratitude the help of European colleagues who provided me with bibliographical references and useful comments on hydrometers: Paolo Brenni from Florence, Patrice Bret from Paris, Robert F. Bud from London, Agusti Nieto-Galàn from Barcelona, Yves Noel from Caen, and Elisabeth C. Vaupel from Munich.

1. Eusèbe Salverte, "Note sur l'origine de l'aréomètre," *Annales de chimie*, 27(1798): 113–117.

2. R. Boyle, "A New Essay-Instrument Invented and Described by the Honourable R. Boyle Together with the Uses Thereof, *Philosophical Transactions of the Royal Society* 10(1675): 329–348. See also Simon Schaffer, "Metropolitan Measures and Their Margins of Error," in *Eighteenth Century: Theory and Interpretation*, in press.

3. C. C. Gillispie, *Science and Polity in France at the End of the Old Regime* (Princeton: Princeton University Press, 1980).

4. For instance Mathurin V. Brisson, *Traité élémentaire ou principes de physique*, 3rd ed. (Paris: Bossange, 1800).

5. [P. J. Macquer], "Balance Hydrostatique," *Dictionnaire de chimie* (Paris: Lacombe, 1766), 1: 191–193. Notably, Macquer did not however, insert an entry "aréomètre" or "pèse-liqueur" in his *Dictionnaire*, either in 1766 or in 1778.

6. According to a legend related by Herodotus, to reascertain the amount of land taxes after the annual flood of the river Nile diminished the surface of the land cultivated by the farmers, the royal administrators invented proportions and scales; M. Serres, *Les origines de la géométrie* (Paris: Flammarion, 1993), 318–320.

7. Tore Frängsmyr, J. L. Heilbron, and Robin E. Rider, ed., *The Quantifying Spirit in the Eighteenth Century* (Berkeley: University of California Press, 1990).

8. Schaffer, "Metropolitan Measures."

9. Bernadette B. Bensaude-Vincent, "The Balance Between Chemistry and Politics," *The Eighteenth Century: Theory and Interpretation,* in press; Norton Wise, "Meditations: Enlightenment Balancing Acts, on the Technologies of Rationalism," in *World Changes: Thomas Kuhn and the Nature of Science,* ed. Paul Horwich (Cambridge, MA: MIT Press, 1993), pp. 207–256.

10. Schaffer, "Metropolitan Measures."

11. Pierre Bories, "Sur la manière de déterminer les titres ou degrés de spirituosité des eaux-de-vie et esprits de vin," *Prix des Etats du Languedoc Société Royale des Sciences* (Montpellier, 1772), pp. 18–21.

12. Lavoisier, "Recherches sur les moyens, les plus sûrs les plus exactes les plus commodes de déterminer la pesanteur spécifique des fluides soit pour la physique soit pour le commerce," *Oeuvres de Lavoisier* (Paris: Imprimerie Impériale, 1865), 3: 438.

13. (Bories, "Manière de déterminer," p. 20):
J'ai crû devoir distinguer l'Hydromètre de l'Aréomètre, quoique ces deux mots soient synonymes dans l'acception ordinaire. Ils sont l'un & l'autre dérivés du Grec. Le mot Aréomètre signifie proprement mesure des corps rares, poreux, spongieux, mesure de volume; au lieu que celui d'Hydromètre veut dire mesure de l'Eau, ou en général mesure de Liqueur, pèse liqueur.

14. Entry "alcool," *Encyclopédie Méthodique*: Dictionnaire de: *chymie, pharmacie et metallurgie.* (Paris: Panckouche, 1786–1815), 2: 99–106.

15. See A. D. Morrison-Low, "Hydrometer," in *Instruments of Science: an Historical Encyclopedia,* ed. Robert Bud and Deborah Jean Warner (New York: Garland, 1998), pp. 311–313.

16. Patrice Bret, *Lavoisier et l'Encyclopédie methodique, le manuscrit des regisseurs des poudres et salpêtres pour la dictionnaire de l'artillerie (1787)* (Florence: Olschki, 1997), Intro.

17. Lavoisier, "Analyse du gypses, *Oeuvres,* 3: 111–144; "De la nature des eaux d'une partie de la Franche-Comté, de l'Alsace, de la Lorraine, de la Brie et de Valois," *ibid.,* pp. 145–188.

18. Lavoisier, "Recherches sur les moyens," p. 429.

19. Ibid., 432: "Or cette égalité de volume est, toutes choses égales, un avantage bien réel; elle évite dans la pratique beaucoup de réductions et de calculs, et simplifie par là les opérations".

20. Brisson, *Traité élémentaire*, p. 258.

21. Demachy, "Art du distillateur d'eaux fortes," in *Description des arts et métiers*. (Paris: Desaint et Saillant, 1761–1788), chapter 7, p. 98.

22. Christopher Hamlin, *A Science of Impurity: Water Analysis in Nineteenth Century Britain* (Berkeley: University of California Press, 1990), 16–46. Bergmann recommended a combination of three different approaches: investigation of physical properties, chemical tests, and evaporation, which allowed weighing the residue and the analysis of the salts dissolved in alcohol. 16–46.

23. A. Nieto-Galàn, "Ciencia a Catalunya a l'inici del sigle XIX: Theoria i applicacions," (unpubl. Ph.D. dissertation, University of Barcelona, 1994), I am grateful to Agusti Nieto-Galàn for the helpful information and bibliography that he provided on this crucial aspect of hydrometry.

24. Polycarpe Poncelet, "Mémoire dans lequel on détermine les différents titres ou degrés de spirituosité des eaux de vie ou esprits de vin, par le moyen lc plus sûr et en même temps le plus simple et le plus facilement applicable aux usages du commerce," Prix des Etats du Lanquedoc Société Royale des Sciences (Montpellier, 1772).

25. For instance, the "trois-cinq," usually obtained after eight distillations, means that three parts of the liquor added to two parts of pure water would provide five parts of "Preuve de Hollande." The "trois-six," the result of 12 distillations, is an alcohol that is changed into "Preuve de Hollande" after addition of an equal weight of water. See Bories, 61–65; "Manière de détermines," Louis Sébastien Lenormand, *Vocabulaire de l'art du distillateur*, 2d. ed.

26. See Poncelet, *Mémoire*, Bories, *Manière*, Demachy, "Art du distillateur," J. A. C. Chaptal, "De la distillation du vin," *Mémoires de l'Institut national* [Supplement], 2(1809): 34–45.

27. (Demachy, "Art du distillateur," 95–96):

L'incertitude de tous ces moyens n'échappait pas au physicien qui ne peut être trop scrupuleux, au Fermier qui voudrait que tout fût esprit de vin, et au marchand qui veut bien savoir comment faire illusion à l'Acheteur, mais qui ne veut pas que le fabricant lui en impose. Le pèse-liqueurs s'offrait à tous trois.

28. Some of them are displayed at the University of Oxford Museum of the History of Science; see C. R. Hill, Catalogue I, (Oxford, 1971), 47.

29. Lavoisier, *Oeuvres*, t. III, 451–455.

30. J. A. Deluc, "Sur la pyromètric et l'aréomètrie et sur les mesures physiques en général," *Observations sur la Physique*, 18(1781): 494.

31. Ibid., 497:

Quant aux co-effets physiques en général, & j'ose le dire ici, dans les co-effets de tout genre, si l'on ne peut pas fixer tous les rapports degré par degré par des observations immédiates et sûres, il faut éviter de tirer des règles générales de rapports pris dans les extrêmes. L'action des causes, tant morales que physiques, est trop compliquée, soit par la variété des sujets sur lesquels elles agissent, soit par les causes secondaires qui échappent à nos observations, pour que les modifications observables croissent en proportions exactes des causes évidentes; & par conséquent, pour que les co-effets de celles-ci soient proportionnels entre eux.

32. Ibid., 498:

Nous sommes obligés de nous contenter du probable à tant d'égards dans la nature que chercher les règles physiques de probabilité nous est peut-être plus essentiel que de nous attacher à ces règles mathématiques sur des hypothèses.

33. Schaffer, "Metropolitan Measures."

34. Lavoisier, 1774, *Oeuvres*, T. III, 168–169:

Tel était le plan que je m'étais proposé d'abord et que je n'ai pas abandonné pour toujours; mais il me manquait tant de choses pour l'exécuter, il fallait tant d'expériences préliminaires pour pouvoir rassembler un nombre suffisant de données, que j'ai été effrayé du travail.

35. Lavoisier, 1768, 449–450:

C'est principalement dans l'art des combinaisons que la connaissance de la pesanteur spécifique des fluides peut porter le plus de lumière. Cette partie de la chimie est beaucoup moins avancée qu'on ne pense; à peine en avons-nous les premiers éléments. Nous combinons toujours un acide avec un alcali; mais de quelle manière se fait l'union de ces deux êtres? Les molécules contituantes de l'acide se logent-elles entre les pores de celles de l'alcali, comme le pensait M. de Lémery, ou bien l'acide et l'alcali sont-ils composés de différentes facettes dont l'une peut s'engrener dans l'autre ou s'unir par le simple contact à la façon des hémisphères de Magdebourg? Comment l'acide et l'alcali tiennent-ils séparément à l'eau? Comment y tiennent-ils après leur combinaison? Le nouveau sel qui s'est formé occupe-t-il seulement les pores de l'eau? Est-ce une simple division des parties, ou bien y-a-t-il une combinaison réelle, soit de partie à partie, soit d'une partie à plusieurs? Enfin, d'où vient cet air qui s'échappe avec tant de vivacité dans le moment de la combinaison, et qui, jouissant de son élasticité naturelle, occupe sur le champ un espace énormément plus grand que celui des deux fluides dont il est sorti? Cet air existait-il primitivement dans les deux mixtes? Y était-il en quelque façon fixé comme le pensait M. Hales et comme le pensent encore la plupart des physiciens, ou bien est-ce un air pour ainsi dire factice et qui soit le produit de la combinaison, comme le pensait M. Eller? La chimie, consultée sur ces différents objets, nous répondra par de vains noms de rapports, d'analogues, de frottements … qui ne présentent aucune idée, et qui n'ont d'autre effet que d'accoutumer l'esprit à se payer de mots. S'il est possible à l'esprit humain de pénétrer ces mystères, c'est par des recherches sur la pesanteur spécifique des fluides qu'il peut espérer y parvenir.

36. Beretta, *New Course*, 39.

37. Richard Kirwan, "Experiments and Observations on the Specific Gravities and Attractive Powers of Various Saline Substances," *Philosophical Transactions*, 71(1781): 6–41; 72(1782): 179–236; 73(1783): 15–84. French translation: *Observations sur la physique*, 24(1784): 134–156, 188–189, 356–368; 25(1784): 13–28; 27(1785): 447–457; 28(1786): 94–109.

38. F. L. Holmes, *Antoine Lavoisier—the Next Crucial Year* (Princeton: Princeton University Press, 1997), p. 73.

39. Lavoisier *Oeuvres*, T.I, 1862, 579–581.

40. Lavoisier, *Oeuvres* T.I, 1862, 569–572.

41. The method is described in a letter of Lavoisier to Brisson, September 21, 1784, Archives de l'Académie des Sciences, manuscrits retrouvés en 1993. Lavoisier carefully weighed a vessel from which air had been pumped. He then introduced the air to be tested into the vessel and weighed it again. The increase of weight indicated the weight of the air. In repeating the same experiment with various airs, one could compare their specific weights. To know the absolute specific gravity of the air, it was necessary to determine the volume of the vessel.

42. Patrice Bret, *Lavoisier et l'Encyclopédie* (op. cit. fn. 16), pp. 81–84.

43. *Ibid.*, Intro. See also Patrice Bret, "Lavoisier et l'apport de la chimie académique à l'industrie des poudres et salpêtres," *Archives internationales d'histoire des sciences*, 46(1996): 57–74.

44. Contract between Guyton de Morveau and Betally, signal February 3, 1794. Private papers of Guyton de Morveau, Quoted in Patrice Bret, "Laboratoires et instruments de la recherche—sur les poudres et explosifs et sur la guerre chimique en France (1789–1939)," Communicated to the European Science Foundation workshop on Chemistry, Laboratories and Instruments, Lisbon, November 26–27, 1996.

45. Lavoisier, *Traité élémentaire de chimie* (Paris: Cuchet, 1789), pp. 336–340.

46. Kirwan to Guyton de Morveau, January 6, 1785, in E. Grison, M. Goupil, P. Bret, eds., *A Scientific Correspondence During the Chemical Revolution: Louis-Bernard Guyton de Morveau and Richard Kirwan, 1782–1802* (Berkeley: Office for History of Science and Technology, 1994), 93:

Le pèse liqueur de Baumé est assez bien expliqué dans ses *Eléments de pharmacie*, c'est celui qu'on a trouvé le plus simple, mais ce n'est pas encore un instrument parfait, et puis on en met d'autres en vogue qui ne valent pas mieux, ce qui est encore plus fâcheux, car comme vous le dites on ne s'entend plus.

47. J. M. Hassenfratz, "De l'aréometrie," *Annales de chimie*, 33(1800): 7. VIII, 7: Comme l'intérêt des marchands est de payer le moins de droit possible et que l'intérêt du fisc est de percevoir les plus grands droits, les premiers ont cherché à tromper, les seconds à découvrir la fraude; en conséquence il a été formé un grand nombre d'aréomètres différents.

48. In a competition held in 1802 by the Board of Excise to determine the best hydrometer, 19 instruments were submitted. Eventually, the Acts of Parliament adopted Batholomew Sikes's hydrometer in 1816 and 1818. In the United States, an instrument patented in 1790 by John Dicas of Liverpool was adopted as the standard by the United States government (see A. Morrison Lowe, "Hydrometer").

49. C. Blagden, "Report on the Best Method of Proportioning the Excise upon Spirituous Liquors," *Philosophical Transactions of the Royal Society* 80(1790): 675.

50. Ibid., 684. The instrument adopted by an Act of Parliament for use by the Board of Excise was a hydrometer designed by tax collector Bartholomew Sikes. The Sikes hydrometer, made of gilded brass, used both a graduated stem and weights to keep the stem floating at a definite level. It was supplied with a thermometer and a slide rule to adjust the reading to the standard temperature of 62°F.

51. A. Birembault, 1959 "Les deux déterminations de l'unité de masse du système métrique," *Revue d'histoire des sciences* 12(1959): 25–54; L. Marquet, "Historique de la naissance du kilogramme," *Bulletin du Bureau National de Métrologie*, 20(1989): 9–18; Y. Noel, "Lavoisier et les poids et mesures," in *Il y a deux cents ans Lavoisier*. Actes du Colloque organisé à l'occasion du bicentaire de la mort d'Antoine-Laurent Lavoisier 1794 (Paris: Tcc et Doc. Lavoisier, 1995), pp. 169–179.

52. J. H. Hassenfratz, "De l'aréometrie," *Annales de chimie*, 26(1798): 3–18; 27(1798): 118–140; 31(1799): 125–140; 33(1800): 3–49.

53. Hassenfratz, *Ibid.*, 26(1798): 118.

54. Hassenfratz, *Ibid.*, 33(1800): 39.

55. Decroizilles, "Notices sur l'aréométrie," *Annales de chimie*, 58(1806): 244: Il n'y a guère d'instrument plus précis qu'une bonne balance dans son point d'équilibre parfait, tandis que la ligne de niveau d'immersion d'un pèse-liqueur est difficile à déterminer, parce qu'elle varie, d'ailleurs, en raison de la propreté de l'instrument et de la quantité de liquide dont il reste mouillé dans sa partie surnageante.

56. Christian Licoppe, *La Formation de pratique scientifique. Le discours de l'expérience en France et en Angleterae (1630–1820)* (Paris: Editions la Découverte, 1996).

57. Ibid., 251–257.

58. Jan Golinski has emphasized the contrast between the experimental practices in England, giving priority to the potential appropriation of the instruments and their uses by a wide public, with those of the French academic culture, where the

precision of instruments and the accuracy of the measurements prevailed. The hydrometer being a typical instrument adaptable for a wide variety of uses, one could expect a better future for the hydrometers in England. A thorough study of the hydrometric cultures in England could provide a test for these national styles. See Jan Golinski, *Science as Public Culture: Chemistry and Enlightenment in Britain, 1760–1820* (Cambridge: Cambridge University Press, 1992), esp. pp. 129–152.

59. Gaston Bachelard, *Essai sur la connaissance approchée* (Paris: Librairie Philosophique, 1927), p. 70.

8

"Fit Instruments": Thermometers in Eighteenth-Century Chemistry

Jan Golinski

The story of the thermometer in its first century and a half has mostly been told in terms of the gradual achievement of standard scales of measurement. In his authoritative study of the development of the instrument, W. E. Knowles Middleton surveyed the scene from the mid–seventeenth century: "The history of thermometry for the succeeding century and more," he wrote, "is largely a record of attempts to make thermometers universally comparable."[1] Accordingly, the emphasis in Middleton's account was placed on efforts to standardize designs, on the definition of scales, and on procedures for calibration. The thermometer was traced from its first appearance as an untested and confusing device, with workings not clearly understood, to its widespread acceptance as a taken-for-granted tool that could yield reliable and conformable measurements.

This history offers itself for reinterpretation in the light of a significant body of more recent work on the construction of instruments in scientific practice.[2] Bruno Latour has indeed referred to the thermometer in connection with stories of the "black-boxing" of scientific instruments—their gradual acceptance as well-understood pieces of apparatus that can henceforth be used unproblematically.[3] In studies of parallel cases, such as the air pump and the telescope, historians have explored how new devices came to be accepted as reliable means for eliciting natural phenomena in the seventeenth century.[4] The thermometer emerged in the same period; it also, for a while, was subject to debate as to the kind of phenomena it could be used to display. In each case, as the working of the apparatus came to be more generally understood, a particular class of phenomena emerged as separable from the "noise" characteristic of the instrument. The apparatus could then be viewed as a transparent medium for the elucidation of a natural effect.

Recent historical studies have particularly emphasized the social dimension of this process. Instruments have been shown to have achieved their taken-for-granted status through the formation of a consensus as to their proper use and through the disciplining of their users. Such practices as standardization of manufacture, the creation of uniform scales of measurement, and the routinization of methods of calibration have come to seem of critical importance in the social process of stabilizing and extending instrumental use. Latour's description of these practices and his suggestion that the history of the thermometer exemplifies them are promising, as I hope to show in what follows.

The thermometer will, indeed, prove to have been an uncertain apparatus for much of the eighteenth century, its behavior being as much in question as the phenomena it was supposed to reveal. Argument continued as to whether it provided an accurate measure of heat and as to what conditions were necessary to increase its accuracy. The achievement of agreed methods of standardization and calibration was an important aspect of the eventual stabilization of the instrument, as a consensus emerged about how it functioned and how it could be used for best results. The thermometer was recognized as working reliably as an instrument to display temperature; as Latour notes, one sign of this was its incorporation as a component of more complex ensembles of equipment.

Consideration of the specific role of the thermometer *in chemistry* enables us to fill in more of the details of this story and, in some respects, to complicate the narrative for, although a common project in the eighteenth century was to standardize the results of thermometric measurement, considerable disagreement ensued as to its chemical significance. During this period, the thermometer was an important device in at least some chemical laboratories, but its use and interpretation depended on fairly localized contexts of practices and assumptions. Even when the goal of standardization of measurements was shared, beliefs about the chemical uses of the apparatus varied significantly. Very different—sometimes competing—characterizations of the discipline of chemistry were connected with alternative modes of deploying the thermometer. The device was more central to some definitions of the field than to others. We can best illuminate the issues at stake here, within the compass of a short chapter, by focusing on three pivotal figures: Herman Boerhaave, Joseph Black, and Antoine Laurent Lavoisier. We shall see that all invested time and energy in the creation of standard scales of temperature and in calibrating thermometers, but we shall also see that they had rather different views of the chemical uses of the instrument. In each case, the thermometer was deployed in a specific setting of practices and beliefs, linked with a particular view of the scope and limits of chemistry.

One dimension of the connection was theoretical. Conceptions of the nature of heat were integral to the development of chemistry in this period, as a line of distinguished historians, including Hélène Metzger, Henry Guerlac, Douglas McKie, Arthur Donovan, and Robert Fox, has shown.[5] The question was not so much what heat *was*—material fluid or corpuscular motion—as whether it was viewed as an external instrument of chemical change or as a chemical constituent of certain bodies. Did heat impinge on chemistry as a physical agent that remained apart from chemical combination, or was it itself a chemical entity, capable of binding as a component of chemical compounds? Framing the issues in this way shows us how conceptions of the scope and character of chemistry were bound up with the question of heat.

Thinking about the thermometer in this context reminds us that these issues were not purely conceptual. Theoretical notions of the nature and role of heat in chemistry were sustained by material and social practices. The manufacture and calibration of instruments posed a series of challenges on the material level, and social resources were also mobilized to ensure their standardization, replication, and reliability. Boerhaave, Black, and Lavoisier all built relationships with the instrument makers who manufactured their apparatus as they grappled with the problems of calibration and replication of thermometers.

The functioning of the thermometer as an instrument depended on a willingness to prefer its indications over the bodily sensation of heat or cold, even in some instances of a plain contradiction. The device thus played its part in the history of the displacement of authority from the senses to their instrumental surrogates, a process that occurred repeatedly in chemistry in this period. In a variety of fields of chemical investigation, instrumental and sensory means of assessing qualities and changes were being compared with one another, sometimes arousing debate about their respective merits.

In pneumatics, for example, various tests were proposed to discriminate between different gases, and considerable discussion concerned the relation of experimental criteria to those of smell or breathability. In mineralogy also, the traditional skills of appraising minerals by appearance, texture, smell, and taste gave way to methods of chemical analysis. In neither case, of course, were sensory discriminations entirely dispensed with: Joseph Priestley's tests for the identity

of gases relied on visual observations of color changes and other transformations, and similar ocular observations were required in mineral
analysis.

Nonetheless, an important displacement of sensory attention—a
new regime of sensory discipline—was required by the introduction of
new instruments. As Lissa Roberts has recently argued, new apparatus
thus played a critical part in a profound reconfiguration of the sensory
and bodily deportment of chemists, which was characteristic of the late
eighteenth and early nineteenth centuries.[6] The thermometer was certainly involved in this transformation, although it had first provoked
debate about the reliability of an instrumental proxy for sensory experience more than a century earlier.

Before we discuss Boerhaave, we can get a feel for the issues surrounding the thermometer and the senses by examining an incident
from its early history in the seventeenth century. The problem for the
first advocates of the device was well expressed by Francesco Sagredo,
friend of Galileo and maker of air thermometers: One had to have faith
in the reliability of the apparatus, "although our feelings seem to indicate the contrary."[7] One simply had to be willing not only to delegate
to an instrument the perception of heat and cold, qualities that had
always been defined in intrinsically sensory terms, but to trust the
apparatus even when the senses were flatly contradicted.

This was what Robert Boyle designated a "paradox," in his *New
Experiments and Observations Touching Cold* (1665). Boyle acknowledged
that to propose an alternative means of measuring cold, other than by
sensation, would appear "a work of needless curiosity, or superfluous
diligence" to most philosophers, who had always identified cold as a
tactile quality best apprehended by the senses. He proposed to make
the "paradox" plausible (but no more than that) by levering it on the
back of a corpuscularian hypothesis concerning the perception of heat
or cold. He stipulated that "we are wont to judge a body to be cold . . .
[when] we feel its particles less vehemently agitated than those of our
fingers, or other parts of the organ of touching."[8] This hypothesis made
the perception of heat or cold a subjective judgment, superimposed on
objective motions of particles in the organ of sensation and in the body
that was examined. A possibility, therefore, was that the perception of
heat or cold could be mistaken; one could be misled, for example, by
preexisting motions in the particles of the organ. Hence, the same
bucket of water could appear either warm or cool to the touch,
depending on whether the hand had previously been in cold water or

hot. "[W]e ... impute that to objects, whereof the cause is in our-selves," claimed Boyle.[9]

Once this gap had been opened up between subjective percep-tions and objective motions of particles, Boyle could introduce the thermometer as a putatively more direct measure of the latter, although he proposed no hypothesis to account for *its* response to temperature changes. "Fit instruments," such as Boyle's sealed air or water ther-mometers, could be more reliable than the senses; they could show, for example, that people were easily deceived into thinking that damp air was colder than it was. The crucial case concerned the effect of wind. Boyle proposed that blowing air appeared colder than it really was, for two reasons. First, it drove away the "warm steams of the body" that normally shielded the skin from the ambient cold. Second, the wind could "pierce deeper than the calm air is wont to do into the pores of the skin, where, by comparison to the more inward and hotter parts of the sensory, it must needs appear less agitated and consequently colder."[10]

This slightly problematic explanation, in which blowing air pos-sesses a motion that enables it to penetrate the skin but not the kind of motion that constitutes heat, was used by Boyle to argue against an alternative theory of cold set forth by Thomas Hobbes. Hobbes had claimed that cold simply *was* a certain motion of the particles of the air; his theory, like Boyle's, distinguished sensations from the corpus-cular motions that caused them. Boyle, however, was able to deploy the thermometer to argue that it gave an incorrect account of the motions of the corpuscles as these were revealed in their effects on nonhuman instruments. Air blown through bellows onto the hand seemed cold, Boyle recorded, but it did not depress the level of liquid in a thermometer—or not much, anyway.

By the time he wrote this, Boyle had already applied the same theory of heat as corpuscular motion to chemistry in his *Sceptical Chy-mist* (1661). In that work, he discussed heat (or "fire") as the agent of analysis of material bodies. As is well-known, he expressed skepticism as to the continuous identity of any supposedly elementary chemical substances through the transformations wrought by fire. Although he subsequently sustained an interest in thermometry, he does not appear to have applied the instrument to chemistry. A contemporary, how-ever, did so. Nicaise Le Febvre included a diagram of a "Weatherglass, Thermometer, or Engin to judge of the equality or degrees of heat" in his chemical textbook (originally published in 1660). The device was an S-shaped, sealed tube with bulbs at both ends and containing a quantity of tinted water whose motions would show the expansion and contraction of the enclosed air. It was proposed for use by the chemist who "will proceed with more nicety in observing the exact degrees of heat" in the furnace.[11]

The thermometer was first brought systematically into chemistry in the 1720s by Herman Boerhaave, who reiterated Boyle's arguments in defense of the reliability of the instrument. Boerhaave also repeated Boyle's experiment of blowing air through bellows onto the thermometer bulb, showing that the level of liquid in the tube was not depressed. Expansion or contraction of a liquid, whether mercury or alcohol, in the sealed glass tube of the thermometer was proposed by Boerhaave as an invariable sign of heat or of what he called "elemental fire"[12]:

> [T]his expansion of bodies by heat, if it be performed in a glass hermetically sealed, cannot arise from any other physical cause hitherto known, but from fire alone; so that the thing desired is now found, *viz.* a true, certain, inseparable, peculiar, characteristic of fire; and this alone we shall use in the sequel for discovering the nature of fire; taking it for granted, that in all phænomena where this rarefaction is observed, there fire is discovering itself to us in proportion.

For Boerhaave, then, the thermometer was a sure and unique indicator of the quantity of elemental fire present in a body. This entity was to be distinguished from the "vulgar fire" fueled by combustible matter and manifested in flame. Boerhaave's elemental fire, as Hélène Metzger and Rosaleen Love have detailed, was a universally distributed, subtle matter, with an inherent power of expansion and an ability to cause motion in the particles of other bodies.[13] It could be "fixed" in, or could "adhere" to, normal ponderable matter, but it did not itself have weight, nor did it participate in normal chemical combination. Its presence in bodies was signaled more reliably by the thermometer than by the senses.

In its role as the prime instrument of chemical change, this elemental fire had a central place in Boerhaave's didactic exposition of chemistry. In his celebrated lectures, given at the University of Leiden for most of the first three decades of the eighteenth century, he presented the subject by way of an elaborate unpacking of the terms of its definition. Chemistry was defined as, "An art, whereby sensible bodies contained in vessels, . . . are so changed by means of certain instruments, and especially fire, that their several powers and virtues are thereby discover'd; with a view to the uses of medicine, natural philosophy, and other arts and occasions of life."[14] Unlike Boyle, then, Boerhaave saw fire as an agent of change *revealing* the chemical components of bodies and their properties rather than disrupting the continuous existence of chemical substances. For Boerhaave, the thermometer was thus a crucial chemical instrument, key to the mastery of fire. It was, so to speak, a second-order instrument, the artifact that gave humans the control of the cosmic instrument of fire that was the basis of chemistry's claim to the status of an art.

This conviction was reflected in Boerhaave's close interest in the work of instrument maker Daniel Gabriel Fahrenheit. Fahrenheit wrote at least 18 letters to Boerhaave from Amsterdam, beginning in 1718 and increasing in frequency as Boerhaave was preparing his chemical lectures for publication in the late 1720s.[15] In his text, Boerhaave was the first chemist to advocate use of the mercury thermometer, endorsing Fahrenheit's instruments as particularly sensitive and reliable. He specified the appropriate degrees of heat for certain chemical operations by reference to points on the Fahrenheit scale, recorded the use of thermometers to discern the heating and cooling effects of certain combinations, and described how measurements of boiling points could be used for chemical identification and analysis.[16]

Fahrenheit informed Boerhaave of a series of ingenious and precise experimental investigations that went a long way toward satisfying the need for a standard method for calibrating thermometer scales. Since Boyle in the 1660s, researchers had regularly complained about the absence of a standard scale that would allow for comparisons between readings made with different instruments. The problems, however, were not easily overcome. All the different thermometric liquids (water, alcohol, linseed oil, mercury) created problems of limited range or expansibility. Comparisons between them suggested that at least some expanded in a nonuniform manner. Fahrenheit shared with some of his contemporaries (including Carlo Renaldini, René-Antoine Réaumur, and Anders Celsius) the conviction that calibration should be effected by means of one or more fixed points that could be set uniformly on different instruments, but disagreement continued about how many of these should be used and which phenomena were appropriate choices.[17]

In an article in the *Philosophical Transactions* in 1724, Fahrenheit suggested three fixed points: blood heat (96°), the "commencement of freezing" of water (32°), and the temperature of a mixture of ice, water, and salt (0°). Use of the boiling point of water was problematic because of its apparently wide variation in different atmospheric conditions (a variation that Fahrenheit overestimated, perhaps because of impurities in his water samples).[18] He also became aware that variations in blood heat made that a somewhat unreliable calibration point. For his own use, Fahrenheit adopted a method of replicating a standard thermometer by measuring precisely the amount of mercury introduced into a tube of specified dimensions. The method relied on his determination of the expansion coefficient of mercury, which he measured with alcohol thermometers across a limited range and then extrapolated to the entire range required. For those who could not obtain his instruments, Fahrenheit noted to Boerhaave in a letter of 1729, he would recommend use of the fixed points of melting ice and boiling water, though the latter would have to be adjusted for atmospheric pressure in a way he did not specify.[19]

From Boerhaave's point of view, perhaps the most intriguing of Fahrenheit's investigations was that in which a mercury thermometer and an alcohol thermometer were compared in different conditions. Fahrenheit supplied both instruments to Boerhaave in 1717, "in order to demonstrate ... the precise agreement between the two." Over the course of a few months, however, Boerhaave noted discrepancies, which he brought to the instrument maker's attention. Fahrenheit responded with a careful investigation of the problem, from which he concluded that the disparities were due to the different expansion properties of the two types of glass used in the tubes of the instruments. Slight differences in the composition of glass, even from the same factory in Amsterdam, could cause significant differences in the liquid levels in the tubes, "for, this glass factory having frequently changed hands, the composition of the glass has also frequently changed, and therewith its expansion, as I have several times discovered to my great frustration, harm and sorrow."[20] For Boerhaave, the significance of this finding was not the frustration that it caused the instrument maker but the confirmation it offered to the claim that expansion of any particular body is directly proportional to the heat it has absorbed.[21]

Boerhaave maintained his conviction that a perfect thermometer —one which measured the expansion of a single body—would offer a direct representation of the amount of heat in that body. He anticipated that some readers would find this implausible[22]:

Here some will be apt to cry out, that I am dealing in chimæras, and alledging things, not only vain and false, but contrary to common sense and experience, which manifestly shews that iron, in winter, is colder than feathers, and quicksilver than alcohol. But let it be considered, I am not here speaking of fire appearing to the senses by its degree of heat and cold, but of fire discover'd by that characteristic above settled, of rarefying bodies.

He went on to indicate an explanation for this kind of contradiction between thermometer readings and the experiences of the senses. Bodies that are maintained in the same conditions must contain equal quantities of elemental fire, but the *speed* with which that fire was absorbed and could subsequently be released varied with the density of the material. Denser bodies were slower to heat and to cool than rarer ones. Hence, presumably, iron is warmed by the hand more slowly than are feathers and thus seems more cold to the touch.

Boerhaave had thus been led from his trust in the thermometer as an instrumental measure of heat content to a focus on the dynamics of heat transfer phenomena. The issue of the speed of heat exchange between bodies, whether some materials take longer than others to heat and to cool, was, he said, a "question, which I have much considered."[23] This concern had a precedent among experimental researchers using the thermometer. The members of the Academia del Cimento, in the late 1650s and 1660s, had also been interested in the timing of thermal effects. Rather than trying to establish a standard temperature scale, they had primarily aimed to monitor the movements of the liquid level in the thermometer tube; they had therefore chosen alcohol as their thermometric fluid because it responded quickly to changes.[24] Boerhaave's recommendation to chemists to observe the time dependence of thermal phenomena was echoed in the words of his pupil Cromwell Mortimer, who published "A discourse concerning the usefulness of thermometers in chemical experiments," in the *Philosophical Transactions* in 1746 and 1747. Mortimer advised the chemist, "to observe his Clock with as much Exactness as the Astronomer doth in his Observations; and . . . let him have his Laboratory furnish'd with various Sorts of Thermometers, proportion'd to the Degree of Heat he intends to make use of."[25]

In the late 1750s and early 1760s, Joseph Black was to undermine the foundations of the Boerhaavian perspective at a number of points. In place of an emphasis on the dynamics of heat transfer, he gave priority to the different *capacities* of bodies to absorb heat. The existence of these different heat capacities and of "latent" heat revealed in transitions to a different physical state demonstrated that the thermometer was not an infallible index of the heat content of a body. Additionally, the body's density was not a reliable indicator of its heat capacity, which could be known only by experiment. Finally, the crucial experimental phenomena for Black were those in which thermal equilibrium had been reached between bodies with initial differences in temperature; he ignored the question of how this state had been achieved.

For all that these conceptual changes were substantial, threads of continuity also connected Black's achievements back to Boerhaave's. One link was through continuing work on the standardization of thermometer scales. In an article in the *Philosophical Transactions* in 1722 and 1723, Brook Taylor had already brought into question the supposition that thermometric liquids expanded uniformly with increased heating. Boerhaave did not recognize this as an empirical problem; he defined heat by reference to expansibility, as it was registered by the thermometer, assuming that the relationship was linear. Taylor, on the other hand, was able to subject the expansion of linseed oil to empirical examination because he produced an independent way of measuring quantities of heat. He pioneered the "method of mixtures," later put to telling use by Black.

Mixing hot and cold water in different proportions, Taylor wrote that he found the rise in level of the oil in the thermometer tube, "or the Expansion of the Oil, was accurately proportional to the Quantity of hot Water in the Mixture, that is, to the Degree of Heat."[26] Although Boerhaave does not seem to have noticed this finding, it produced nothing to disturb him, provided he had been willing to see his stipulation of uniform expansion subjected to empirical test. He could even have appreciated Taylor's method, as he agreed with its assumption that substances contained a quantity of heat in proportion to their bulk. What was distinct about Taylor's approach, however, was its focus on thermal equilibrium and its disregard of the dynamics of heat exchange that had preceded it. Boerhaave's claim that bodies transferred heat with a speed inversely proportional to their density was thus circumvented.

Taylor's kind of investigation was taken further by Jean André de Luc and Adair Crawford later in the eighteenth century.[27] Black himself did research on the expansion properties of mercury, and one can easily see how such thermometric inquiries might have contributed to the recognition of specific heat capacities. Comparisons of different thermometric liquids, as they expanded in response to heat increments, might well have suggested that each had a specific capacity for absorbing heat. Another investigator, whose work in this field in the late 1730s impressed Black considerably, was Scottish physician George Martine. Martine (unlike Taylor) did explore the dynamics of heat transfer; he set up samples of heated mercury, water, and oil and measured their temperatures at regular intervals as they cooled. Boerhaave's supposition of an inverse proportionality between rate of heat transfer and density was empirically falsified, according to Martine, who declared: "I know of no stronger instance than this of the weakness, or, if I may venture to say so, of the presumptuousness of the human understanding, in pronouncing too hastily concerning the nature of things from some general preconceived theories."[28]

Martine's prime interest was in thermometric calibration and the creation of a universal standard scale. He gave a thorough review of the history of the problem, beginning in the late seventeenth century, when comparison of measurements between scattered observers had been "morally impossible" because "every workman made ... [thermometers] according to his own way and fancy."[29] Martine described the arguments that surrounded the designation of putative fixed points on a standard scale. He dismissed some evidence that water did *not* always freeze at the same temperature in all locations as due to "the carelessness of observers, or the errors of the workmen." Variations of the boiling point were said to be insignificant within the normal range of atmospheric pressure.[30] Martine's suggestion was therefore that these two serve as the fixed points, with the lower set at 32° and the upper at 212°. The levels of the mercury under these two conditions should be engraved on the scale, and the interval between the marks should be divided into 180°. In this way, Martine took Fahrenheit's scale and linked it (as Fahrenheit had not) to a publicly described method of calibration, which was widely replicable. This was essentially the method that became general thereafter. A committee of the Royal Society, under the chairmanship of Henry Cavendish, set further specifications in 1777 for the techniques to be used to measure the fixed points, but Martine's basic proposal was upheld.[31]

Black was one of the earliest admirers of Martine's work. Students in his lectures at Edinburgh were shown and made to copy Martine's table of comparison of 15 different thermometer scales. Black greeted this conversion table as the creation of a "universal language" for translating between the measurements of different observers.[32] Martine's standard method of calibration and his conversion table were vital resources for Black in his research on latent and specific heats, as vital as the thermometers themselves with which he was supplied by Glasgow instrument makers James Watt and Alexander Wilson.[33]

Ironically, however, Black's discoveries ushered in a crucial displacement of the thermometer from its Boerhaavian role in chemical practice. His novel doctrine of heat stipulated that the instrument was not in fact capable of detecting all the heat present in a body. John Robison, editor of the posthumous edition of Black's lectures, stated this well. "The experience of more than a century had made us consider the thermometer as a sure and an accurate indicator of heat ... We had learned to distrust all others," Robison recalled. Yet, Black's disclosure that a change in the physical state of a body would release or absorb a significant quantity of heat indicated that it could be present although "concealed from our sense of heat, and from the thermometer."[34]

Notwithstanding this result, Black's discoveries were connected in their origin with thermometric research. According to Robison, at least, they arose fairly directly from a novel approach to the traditional problem of the expansion of thermometric liquids. In 1760, Black reported to a philosophical society in Glasgow the results of his experiments using the method of mixtures to investigate the uniformity of expansion of mercury. He identified a slight increase in the metal's expansibility at higher temperatures but concluded that this did not pose significant problems for its use in thermometers. Black's interpretation of this effect, however, linked it with realms wider than the confines of thermometry. He saw it as lending support to his notion that heat was composed of a subtle, self-repellent matter, capable of weakening the mutual attraction of particles of normal matter: "[t]herefore when a body is expanded by heat, and the distance of its particles is a little increased, its attraction of cohesion being thus weakened, a further addition of heat to it can easily be imagined to produce more effect in expanding it further."[35]

According to Robison, this kind of account rendered plausible the supposition that heat could be contained in bodies in a way that was not revealed by the thermometer. Heat was envisioned as a subtle material fluid, the role of which was not (as for Boerhaave) the communication of motion but the inducement of expansion by a weakening of cohesive attraction. This expansion had, of course, also been crucial for Boerhaave but, in Black's formulation, it was not necessarily detected by the thermometer in an unmediated way. It did, however, have a direct connection with the body's physical state, as an increase of heat would ultimately cause the transition from a solid to a liquid to a gaseous state. Black (at least on Robison's reading) had generalized his understanding of the nature of heat, working outward from the specific realm of the thermometric expansion of mercury to forge a connection between the heat contained in a body and its physical state or "fluidity." This was to widen considerably the implications of the studies of expansibility undertaken by Taylor, Renaldini, and de Luc. As Robison put it, theirs "were the thoughts of a philosopher interested only in an instrument of research.... Dr. Black's surmises about the thermo-metrical scale were those of a chemist, studying the nature of fluidity."[36]

Some doubt must remain as to the accuracy of Robison's reconstruction of the path of Black's investigations. His edition of the lectures was prepared from his mentor's distressingly fragmentary manuscript remains, with the expressed aim, "to support the Claims of [D]r Black to his various chemical discoveries."[37] Robison was more concerned with securing Black's posthumous reputation in the wake of Lavoisier's revolution in chemical theory than with producing a historically accurate document. His text continues to thwart the ambitions of historians who seek to read through it to discover what Black actually said. It is nonetheless clear that the relationship between fluidity and the "latent" heat present in a body (but hidden from the thermometer) was indeed a specifically chemical problem, within the terms of Black's understanding of the scope of the discipline. His exposition of chemistry particularly stressed the agency of heat, "the chief material principle of activity in nature," which he discussed at considerable length. Chemistry was defined by Black as the science of the operations of heat, an emphasis already present in Boerhaave. However, insofar as Black saw heat as a "material principle" whereas Boerhaave had seen it as an external "instrument," his approach rendered heat a direct participant in chemical operations.[38]

Although Black imported thermometric techniques into chemistry from other physical sciences, to describe his view of heat as "physical" would be inaccurate. Rather, as chemistry was understood in terms of heat, so heat was understood by Black in terms of chemistry. As Robison clarified, Black perceived the latent presence of heat in bodies in terms of its chemical combination with ponderable matter:[39]

He now saw good chemical reasons for considering this concealed heat as *united* with the substance of the body, in a way very much resembling many chemical combinations. A warm body carried its warmth about with it. It shared its warmth with other bodies, just as a brine shares its salt with more water. In liquefaction and vaporization it attached to itself a determinate quantity, and no more than the proper quantity, just as water does of salt, or as salt does of water in forming a crystal. The appearance, and many properties and modes of activity, are changed by this union, just as in other chemical combinations. He was therefore led to consider heat as chemically combined, in all its modes of existence, whether in expanding, in melting, or in boiling off a fluid.

The important point here seems to be not so much Black's theory of the *nature* of heat as his account of its *function*. Heat, in Black's view, acts by entering into chemical combination with ponderable matter. Therefore, to class it as an "instrument" of chemical change, as Boerhaave had, is inappropriate. Black reserved that designation for the material implements of the laboratory. Heat is itself a chemical entity; it exerts its effects through chemical combination with other substances. Its mode of agency is comparable to that of other matter, though of course its specific effects are unique, as are those of other chemical substances.

This vision seems critical to Black's approach to studying heat experimentally. As has been mentioned, he focused attention on the state of equilibrium, claiming with justification that it had not previously been well understood. Boerhaave's supposition that bodies acquired heat at different speeds to reach this state was dismissed as "entirely the work of imagination, without any foundation in fact."[40] The important thing was the equilibrium, not how it was achieved; this was the stable state of combination, such as the compounds that chemists were accustomed to studying. Further, as in chemistry generally the properties of substances were studied as they underwent combination and separation, so heat was studied in its absorption and release by bodies. Different substances proved to have fairly different specific capacities for heat; Black concluded that this could not be explained simply in terms of their densities or indeed on the basis of any other general principle. Specific heat capacities were analogous to the specific affinities that bodies possessed for combination with other bodies: affinities that chemists could measure, compare, and tabulate but could not hope to explain by reduction to some more fundamental level of physical ontology.

In line with this approach to the study of heat, Black disqualified the thermometer from serving as a direct instrumental indicator of its quantity. To apply the device to two bodies in thermal equilibrium and to conclude from their identical readings that the two had the same heat content was a fundamental error, in Black's view: "It is confounding the quantity of heat in different bodies with its general strength or intensity, though it is plain that these are two different things, and should always be distinguished."[41] The thermometer measures the strength or intensity of heat (its temperature) but not its quantity, which can be known only through the traditional chemical methods of combination and separation.

The distinction can be experienced phenomenologically when we touch a piece of iron and a piece of wood that have been covered for a while with snow. Black gave an explanation of this experience very different from Boerhaave's: the speed of heat transfer does not make the iron feel colder than the wood, at least *not solely* that speed. Rather, in Black's view, the iron, "will draw heat faster, *and will continue for a longer time* to draw heat from the hands, before it assumes the temperature of that hand. It is therefore plain, that the iron has a greater *capacity and attraction* for heat, or the matter of heat, than the wood has."[42] By explicitly mentioning "attraction" in conjunction with heat capacity, Black seems to have indicated how he conceived of heat transfer between bodies as analogous to chemical combinations governed by affinities. Like the dissolving power of water for certain salts, for example, the capacity of bodies to contain heat was thought to be related to the degree of attraction between them.

The influence of Black's "chemical theory of heat" on his Scottish contemporaries was considerable, as Arthur Donovan, among others, has emphasized.[43] His student and (indirect) successor in the Glasgow chemistry chair, William Irvine, proposed an ingenious combination of the theories of latent and specific heats. Irvine's idea was that bodies release or absorb heat when they change their physical state, because of a dramatic fall or rise in their specific heat capacities. A gas will have a higher heat capacity than will the corresponding liquid, which will have a higher capacity than that of the corresponding solid. Adair Crawford subsequently built on this an attempt to account for the phenomenon of animal heat released in respiration.

Although Black never endorsed these theories, they were, in at least two critical respects, founded on his own approach. First, both distinguished the heat apparent to the senses *or the thermometer* from the total heat present in a body. Irvine and Crawford nominated the latter "absolute heat" (as opposed to "sensible heat") and proposed that it could be calculated as the product of the body's heat capacity and its temperature above absolute zero. Second, they continued the attempt to link heat exchanges with chemical combinations governed by affinities. Crawford proposed that heat and phlogiston, when combined with a body, were each capable of diminishing the attraction of that body for the other principle. Hence, air that becomes "phlogisticated" in the course of respiration loses part of its capacity for heat, which is consequently released from it in the form of "animal heat."[44]

Also well recognized is that Black's theory paved the way for Lavoisier's account of *caloric*, the material fluid of heat that combines with ponderable substances as they assume the liquid or gaseous state. Lavoisier's caloric theory played a pivotal role in his accounts of combustion, calcination, and respiration. Notably, Lavoisier was already thinking in the early 1770s of elemental fire as capable of existing in two states: "free," when it would be apparent to the senses or to the thermometer, or "combined," united to normal matter by a process of chemical combination. The extent to which he was influenced by Black and his followers in the formulation of these ideas remains unclear. Possibly Lavoisier's path to the caloric theory was largely independent of the Scottish work. Black himself, however, recognized a lineage, asserting that the caloric theory "is founded on the doctrine of latent heat, and is, indeed, an extension of it."[45]

Given this convergence of attention on the chemical role of heat, it is not surprising that Lavoisier also was interested in problems of thermometry. In a document from the mid-1760s, which Marco Beretta has plausibly argued presents Lavoisier's scheme for an elementary course of lectures on chemistry, the author devoted considerable attention to thermometers.[46] He discussed the relative merits of various instruments and scales, concluding with an endorsement of the Réaumur scale but urging the use of mercury, rather than Réaumur's favored alcohol-water mixture, as the thermometric liquid. One decade later, Lavoisier continued to pursue the issue of the accuracy of Réaumur thermometers, comparing a number of them placed in melting ice in the cellars of the Paris Observatory.[47] Further, in the 1780s, he again used the very small variations in temperature in the observatory cellars to test the accuracy of mercury thermometers calibrated by different methods.[48]

Lavoisier's interest in thermometry was symptomatic of his ambition to bring methods of precision measurement to bear in chemistry. Early in his career, the thermometer took its place alongside the hydrometer, the pyrometer, and the barometer in his experimental practice. The distinctive practices of accurate measurement made possible by this ensemble of apparatus allied Lavoisier with some other experimenters, such as Henry Cavendish, but it distanced him from some of his contemporary chemists. The thermometer itself was a sign of this divergence of approaches. In the generation before Lavoisier, chemist Gabriel-François Rouelle had asserted—in what could almost have been a riposte to Cromwell Mortimer—that the genuine chemist would have "his thermometer at the tip of his fingers and his clock in his head." Even in the 1780s, Swedish chemist Torbern Bergman admitted that he knew of a colleague who regarded thermometers as "physical subtleties, superfluous and unnecessary in a laboratory."[49]

Although thermometers played an important role in Lavoisier's culture of precision, they could not directly deliver a measure of the heat that was concealed by chemical combination with ponderable matter. In the early 1780s, Lavoisier worked with Pierre Simon de Laplace to design a new apparatus that could serve this purpose.[50] The machine, later named the *calorimeter*, effected a crucial displacement in the role of the thermometer in chemical practice. Henceforth, the thermometer became a component part of a more complex assembly of equipment. Its working was taken for granted as a condition of its functioning as part of the new device; in other words, it was "black-boxed."

In the calorimeter, the heat released by reactions was gauged by measuring the amount of water produced by melting ice. The device converted the heat of reaction into the latent heat of fusion. Not surprisingly, therefore, Black recognized the principle underlying what he called "a most ingenious apparatus."[51] Though its principle was clear to them, the Scots had not, in fact, anticipated the calorimeter, preferring to gauge heat exchanges by the rise or fall of temperature in a medium such as water. The invention of Lavoisier and Laplace removed the need for a thermometer to measure the temperature change in such a situation. Instead, heat that was thermometrically undetectable was measured by its effect of changing physical state. The apparatus thus shifted the burden of precision measurement from the thermometer to the balance, which was used to weigh the melted ice.

This move was significant in Lavoisier's campaign to introduce accurate measurement into chemical research. Notwithstanding the attention he had devoted to thermometric calibration, the balance remained his favored instrument for precision measurement. As Bernadette Bensaude-Vincent has discussed in connection with the hydrometer, Lavoisier's search for methods of exact measurement came to rest, in many respects, in determinations of *weight* (see chapter 7). He invested considerable expertise and capital in commissioning the manufacture of balances of almost unrivalled accuracy.[52] He turned them to telling account in giving precise weights of substances involved in chemical reactions, most dramatically in the controversy over the analysis and synthesis of water, in the mid-1780s, when the exact equivalence of the weights of reactants and products was made to carry the burden of the proof that water was a compound.[53] The balance, applied to the products of chemical reactions, permitted an accurate measure of a total quantity that could then be set against the weights of the reactants. The use of the balance in connection with the calorimeter allowed for the same process of accountancy to be applied to exchanges of heat in chemical reactions.[54]

The thermometer had a new role in this apparatus, but it could not be discarded entirely. For the calorimeter to work properly, the ice melted by the heat of reaction had to be previously on the point of melting, so that no extra heat would be consumed in raising its temperature. This could only be determined by thermometry. In addition, the device employed an extra layer of ice for insulation, to protect the measured ice from being melted by ambient heat. Lavoisier and Laplace reported that the insulation was most effective when the temperature of the surroundings was a few degrees above freezing, suggesting that a thermometer would also be required to gauge that. The calorimeter, though in some respects displacing the thermometer in its role as measuring instrument, continued to require for calibration the use of the older device.[55]

Getting the new apparatus to work could indeed be a troublesome matter. In a paper published in *Philosophical Transactions* in 1784, Josiah Wedgwood initiated what was to be a long-running critique of the calorimeter by British chemists over the following few decades. Wedgwood pointed to two major problems. First, the crushed ice retained water through capillary attraction, so the water failed to flow into the collecting container to be weighed. Second, melted ice tended to refreeze in contact with unmelted ice, so the processes of thawing

and freezing were going on simultaneously.[56] Two possible explanations for this latter phenomenon were offered by Wedgwood, by
Thomas Beddoes (who commented on it in his translation of Torbern
Bergman's 1785 *Dissertation on Elective Attractions*), and by Richard
Kirwan, participating in the discussions of the Coffee House Philosophical Society in April 1785. One possibility was that water vapor
tended to freeze at a temperature higher than that of the liquid (say at
33°F), so that it might freeze in contact with ice and yet release enough
heat to melt some of the ice. The alternative was that melting ice had a
positive attraction for the heat of surrounding bodies, so that it would
rob them of their heat, even at the cost of freezing surrounding water.
The latter possibility was formulated within the framework in which
heat exchanges were viewed as governed by relative affinities, the same
framework that was applied to the understanding of chemical reactions.
It was invoked by Kirwan and others to raise objections to Lavoisier's
theories of combustion and calcination. They pointed out that the new
accounts of heat exchanges in these reactions ought to be reconciled
with a consistent interpretation of the relative affinities of different
substances for the matter of heat.[57] In the view of several of his British
critics, Lavoisier had failed to do this.

Lavoisier, however, sidestepped these objections as he attempted
to respond to doubts about the accuracy of the calorimeter. In his
textbook, *Traité élémentaire de chimie* (1789), in which the calorimeter
was first given its name, the apparatus was described as a straightforward device for measuring caloric. It was accorded a prominent place
among the author's descriptions of the instruments used in chemical
operations, whereas the thermometer was not described at length at
all. Lavoisier addressed some of the practical difficulties that his critics
had raised, reasserting that with careful use, the apparatus could yield
accurate measurements of quantities of heat exchanged in chemical
reactions. He declined, however, to enter into the subject of the affinities between caloric and ponderable chemical substances, declaring in
justification that "our conceptions of the nature of these combinations
are not hitherto sufficiently accurate." Whereas Black had talked of the
"capacity and attraction" of bodies for heat, Lavoisier mentioned only
"the capacity of bodies for containing caloric," thereby evading the
issue of relative attractions.[58]

The calorimeter, then, was upheld in the *Traité* as an instrument
by which the quantity of heat released by a body could be determined.
It was contrasted with the thermometer, which Lavoisier relegated to a

subsidiary role. Like the organs of sensation, he claimed, the thermometer could detect only the *motion* of heat passing between bodies. Quantitatively, it registered only the portion of that heat absorbed by the instrument itself. Just as one sensed heat or cold only when a temperature difference existed between the hand and the object felt, so that a transfer of heat occurred between the two, so "The change ... which takes place upon the thermometer, only announces a change of place of the caloric in those bodies, of which the thermometer forms one part; it only indicates the portion of caloric received, without being a measure of the whole quantity disengaged, displaced, or absorbed."[59] As a putative measure of heat, the thermometer was demoted to the level of the sensory organs; because it participated itself in the changes to which it responded, its status as an autonomous indicator of those changes was undermined.

The thermometer was disqualified as a means of measuring heat, though its status as an instrumental index of temperature was, if anything, enhanced by Lavoisier's use of it. He declined to analyze its mode of working, but he routinely used it in this capacity. For controlling the conditions in which the calorimeter was deployed and for correcting the measures of gas volumes in the gasometer, the thermometer was indispensable. Lavoisier gave instructions for reducing gas volumes to those that would be occupied at a standard temperature of 10° Réaumur, which he designated a convenient operating temperature "most easily approached to at all seasons." To facilitate these calculations, his English translator, Robert Kerr, included a conversion table for reckoning from degrees Réaumur to the Fahrenheit scale more familiar in Britain. In relation to this kind of function, the thermometer was accepted as an unproblematic instrument. Lavoisier did not feel obliged to describe it or give instructions as to how it should be made.[60]

From the vantage point of Lavoisier's textbook, then, we can see how the thermometer had assumed an instrumental function that was so routine that it required no specific discussion. The index of the device's unproblematic acceptance is that it had been built into larger ensembles of equipment, such as the calorimeter or the gasometer, in which its regular working was anticipated as a condition of its use. However, this black-boxing was achieved only by delimiting the range of phenomena that the instrument was taken to reveal. Calibration was coextensive with the formation of a consensus as to what the instrument registered: hence the agreement that the thermometer measured temperature, although Black (with his talk of "intensity") and Lavoisier (who spoke of "motion") disagreed about how temperature should be understood in relation to heat. Temperature was defined operationally—as what the thermometer measured—more clearly than it was defined theoretically.

By the end of the eighteenth century, the thermometer had been thoroughly uncoupled from direct sensory apprehensions of heat. The displacement had been, in fact, a double one: First, in the late seventeenth century, authority for perceiving heat had been shifted from the senses to the thermometer; then, with the work of Black, the thermometer itself was demoted to a register of temperature, incapable of detecting the heat that was latent within a body. This innovation prepared the way for Lavoisier and Laplace's introduction of the calorimeter, proposed as a new and direct instrumental measure of heat. In relation to this apparatus, and in other respects, Lavoisier treated the thermometer as a device that could be taken for granted and used to regulate the conditions for chemical experiments.

The course of development we have surveyed was bound up with a succession of different characterizations of chemistry as a discipline. For Boerhaave, the thermometer was the true and unique sign of the presence of elemental fire, which was itself regarded as the primary "instrument" of chemical change. In Black's view, the thermometer could not be taken as a direct measure of the presence of heat, but it did register its intensity in its chemical combinations with ponderable matter. For Lavoisier, the instrument was a tool for the introduction of methods of accurate measurement into chemistry, but it discharged this function only in conjunction with other apparatus. Processes of heat transfer were to be gauged by the calorimeter, which produced a measurable quantity of melted ice at the end of the reaction, but told the experimenter nothing about the specific affinities between chemical substances and caloric. The role of the thermometer in this apparatus was to control the physical circumstances in which chemical events unfolded.

Lavoisier's experimental practice established a model for generations of subsequent chemists. The thermometer remained an indispensable tool in chemical laboratories; lists of the apparatus in stock at Oxford and Harvard universities in the 1820s record numerous different types of thermometers.[61] However, the device was used by chemists to establish and monitor the physical conditions for experiments or to determine the physical characteristics of substances. It was no longer viewed as a direct indicator of heat, and heat was no longer regarded as a chemical entity. In this sense, we might say that the thermometer had ceased to have an intrinsically *chemical* significance.

NOTES

1. W. E. Knowles Middleton, *A History of the Thermometer and Its Use in Meteorology* (Baltimore: Johns Hopkins University Press, 1966), 39. See also Henry C. Bolton, *The Evolution of the Thermometer, 1592–1743* (Easton, PA: Chemical Publishing, 1900).

2. See, for example, the studies collected in Robert Bud and Susan E. Cozzens, eds., *Invisible Connections: Instruments, Institutions, and Science* (Bellingham, WA: SPIE Optical Engineering Press, 1992); Albert Van Helden and Thomas L. Hankins, eds., *Instruments* (*Osiris*, new series, 9) (Chicago: University of Chicago Press, 1994); and David Gooding, Trevor Pinch, and Simon Schaffer, eds., *The Uses of Experiment: Studies of the Natural Sciences* (Cambridge: Cambridge University Press, 1989).

3. Bruno Latour, *Science in Action: How to Follow Scientists and Engineers Through Society* (Cambridge, MA: Harvard University Press, 1987), 68.

4. Steven Shapin and Simon Schaffer, *Leviathan and the Air-Pump: Hobbes, Boyle and the Experimental Life* (Princeton, NJ: Princeton University Press, 1985); Albert Van Helden, "Telescopes and Authority from Galileo to Cassini," in Van Helden and Hankins, eds., *Instruments*, 9–29.

5. Hélène Metzger, *Newton, Stahl, Boerhaave et la doctrine chimique* (Paris: Albert Blanchard, 1930); Henry Guerlac, "Chemistry as a Branch of Physics: Laplace's Collaborations with Lavoisier," *Historical Studies in the Physical Sciences* 7(1976): 193–276; Douglas McKie, *Antoine Lavoisier: Scientist, Economist, Social Reformer* (London: Constable, 1952); D. McKie and Niels H. de V. Heathcote, *The Discovery of Specific and Latent Heats* (London: Edward Arnold, 1935); Arthur L. Donovan, *Philosophical Chemistry in the Scottish Enlightenment: The Doctrines and Discoveries of William Cullen and Joseph Black* (Edinburgh: Edinburgh University Press, 1975); and Robert Fox, *The Caloric Theory of Gases from Lavoisier to Regnault* (Oxford: Clarendon Press, 1971).

6. Lissa Roberts, "The Death of the Sensuous Chemist: The 'New' Chemistry and the Transformation of Sensuous Technology," *Studies in History and Philosophy of Science* 26(1995): 503–529.

7. Quoted in Martin K. Barnett, "The Development of Thermometry and the Temperature Concept," *Osiris* [1st series] 12(1956): 277.

8. Robert Boyle, "New Experiments and Observations Touching Cold," in Thomas Birch, ed., *The Works of the Honourable Robert Boyle*, vol. 2 (London: J. and F. Rivington, 1772), 481.

9. Ibid., 482.

10. Ibid., 482.

11. Nicaise Le Febvre, *A Compleat Body of Chymistry*, translated by P.D.C. Esq. (London: O. Pulleyn, 1670), 87, diagram facing p. 82.

12. Herman Boerhaave, *A New Method of Chemistry*, 2nd ed., vol. 1 translated by Peter Shaw, (London: T. Longman, 1741), 214.

13. Metzger, *Newton, Stahl, Boerhaave*, 209–228; and Rosaleen Love, "Herman Boerhaave and the Element-Instrument Concept of Fire," *Annals of Science* 31(1974): 547–559.

14. Herman Boerhaave, *A New Method of Chemistry*, translated by Peter Shaw and Ephraim Chambers (London: Osborn and Longman, 1727), 51.

15. Pieter van der Star, ed., *Fahrenheit's Letters to Leibniz and Boerhaave* (Leiden: Museum Boerhaave, 1983); John Powers, "Some Like It Hot: Herman Boerhaave, Daniel Fahrenheit, and the Multiple Meanings of Thermometry," presented at the Department of History of Science, Harvard University, 29 May 1996. I am grateful to John Powers for sharing with me his work on Fahrenheit and Boerhaave.

16. Star, *Fahrenheit's Letters*, 42–43; Jon Eklund, *The Incomplete Chymist: Being an Essay on the Eighteenth-Century Chemist in His Laboratory, with a Dictionary of Obsolete Chemical Terms of the Period* (Smithsonian Studies in History and Technology, no. 33). (Washington: Smithsonian Institution Press, 1975), 7; and James R. Partington, *A History of Chemistry*, vol. 2 (London: Macmillan, 1961–70), 747.

17. Barnett, "Development of Thermometry"; Middleton, *Thermometer*, 65–105.

18. D. G. Fahrenheit, "Experimenta circa gradum caloris liquorum nonnulorum ebulientium instituta," *Philosophical Transactions of the Royal Society of London* [henceforth *Philosophical Transactions*] 33(1724): 1–7; Barrett, "Development of Thermometry"; and Star, *Fahrenheit's Letters*, 40.

19. Star, *Fahrenheit's Letters*, 18–31 (calibration methods), 163 (letter from Fahrenheit to Boerhaave, 17 April 1729).

20. Star, *Fahrenheit's Letters*, 147, 149.

21. Boerhaave, *New Method of Chemistry*, vol. 1, 216.

22. Ibid., vol. 1, 246.

23. Ibid., vol. 1, 252.

24. Barnett, "Development of Thermometry"; and Middleton, *Thermometer*, 32–37.

25. Cromwell Mortimer, "A Discourse Concerning the Usefulness of Thermometers in Chemical Experiments," *Philosophical Transactions* 44(1746–47): 672–675, 673.

26. Brook Taylor, "An Account of an Experiment, Made to Ascertain the Proportion of the Expansion of the Liquor in the Thermometer, with Regard to Degrees of Heat," *Philosophical Transactions* 32(1722–23): 291.

27. Middleton, *Thermometer*, 124–126; and Adair Crawford, *Experiments and Observations on Animal Heat and the Inflammation of Combustible Bodies* (London: J. Murray, 1779), 15–39.

28. George Martine, *Essays Medical and Philosophical* (London: A. Millar, 1740), 175–214; and Martine, *Essays and Observations on the Construction and Graduation of Thermometers and on the Heating and Cooling of Bodies*, 3rd ed. (Edinburgh: Alexander Donaldson, 1780), 73.

29. Martine, *Essays Medical and Philosophical*, 179–181.

30. Ibid., 192.

31. [Henry Cavendish, et al.], "Report of the Committee Appointed by the Royal Society to Consider the Best Method of Adjusting the Fixed Points of Thermometers," *Philosophical Transactions* 67(1777): 816–857.

32. Thomas Cochrane, *Notes from Doctor Black's Lectures on Chemistry 1767/8*, Douglas McKie, ed., (Wilmslow, Cheshire: Imperial Chemical Industries Limited, Pharmaceuticals Division, 1966), 7; and Joseph Black, *Lectures on the Elements of Chemistry*, John Robison, ed., vol. 1 (Philadelphia: Matthew Carey, 1807), 65.

33. R. G. W. Anderson, *The Playfair Collection and the Teaching of Chemistry at Edinburgh, 1713–1858* (Edinburgh: Royal Scottish Museum, 1978), 21.

34. Note by Robison, in Black, *Lectures on Chemistry*, vol. 1, 343.

35. Ibid., 52–58.

36. Ibid., Robison's preface, xxxi.

37. Robison to James Black, 21 April 1802, in Douglas McKie and David Kennedy, "On Some Letters of Joseph Black and Others," *Annals of Science* 16(1960): 129–170, 165. Regarding the background to Robison's edition, see John R. R. Christie, "Joseph Black and John Robison," in A. D. C. Simpson, ed., *Joseph Black 1728–1799: A Commemorative Symposium* (Edinburgh: Royal Scottish Museum, 1982), 47–52.

38. Black, *Lectures on Chemistry*, vol. 1, 11.

39. Ibid., note by Robison, 344.

40. Ibid., vol. 1, 82.

41. Ibid., vol. 1, 74.

42. Ibid., vol. 1, 75 (emphasis added).

43. Donovan, *Philosophical Chemistry*, 250–277; and Donovan, "James Hutton, Joseph Black and the Chemical Theory of Heat," *Ambix* 25(1978): 176–190.

44. Crawford, *Experiments and Observations*; and Fox, *Caloric Theory*, 22–28.

45. R. J. Morris, "Lavoisier on Fire and Air: The Memoir of July 1772," *Isis* 60(1969): 374–380; Morris, "Lavoisier and the Caloric Theory," *British Journal for the History of Science* 6(1972): 1–38; and Black, *Lectures on Chemistry*, vol. 1, 230.

46. Marco Beretta, *A New Course in Chemistry: Lavoisier's First Chemical Paper* (Florence: Leo S. Olschki, 1994), 60–69. This is an edition of Dossier 380 from the archives of the Académie des Sciences, Paris. It probably dates from 1764 to 1766 and is in an unknown hand with Lavoisier's corrections. Beretta suggests that it was dictated by Lavoisier to an amanuensis and then was reviewed by him. The lectures were never given publicly.

47. Middleton, *Thermometer*, 119.

48. A. L. Lavoisier, "Thermomètre des caves de l'Observatoire. Précautions pour construire et graduer ce thermomètre," in *Oeuvres de Lavoisier*, vol. 3 (Paris: Imprimerie Impériale/Imprimerie Nationale, 1862–93), 421–426.

49. Rouelle, quoted in Roberts, "Sensuous Chemist," 509; Torbern Bergman, *Physical and Chemical Essays*, vol. 1, translated by Edmund Cullen (London: J. Murray, 1784), *xxx*.

50. A. L. Lavoisier and P. S. de Laplace, *Memoir on Heat, Read to the Royal Academy of Sciences, 28 June 1783*, translated by Henry Guerlac (New York: Neale Watson Academic Publications, 1982); and Lissa Roberts, "A Word and the World: The Significance of Naming the Calorimeter." *Isis* 82(1991): 199–222.

51. Black, *Lectures on Chemistry*, vol. 1, 168.

52. Maurice Daumas, *Lavoisier: Théoricien et expérimentateur* (Paris: Presses Universitaires de France, 1955).

53. Jan Golinski, "'The Nicety of Experiment': Precision of Measurement and Precision of Reasoning in Late Eighteenth-Century Chemistry," in M. Norton Wise, ed., *The Values of Precision*, (Princeton, NJ: Princeton University Press, 1995), 72–91.

54. M. Norton Wise, "Mediations: Enlightenment Balancing Acts, or the Technologies of Rationalism," in Paul Horwich, ed., *World Changes: Thomas Kuhn and the Nature of Science*, (Cambridge, MA: MIT Press, 1993), 207–256; and Bernadette Bensaude-Vincent, "The Balance: Between Chemistry and Politics," *The Eighteenth Century: Theory and Interpretation* 33(1992): 217–237.

55. Lavoisier and Laplace, *Memoir on Heat*, 9–14.

56. Josiah Wedgwood, "An Attempt to Compare and Connect the Thermometer for Strong Fire, Described in Volume LXXII of the *Philosophical Transactions*, with the Common Mercurial Ones," *Philosophical Transactions* 74(1784): 358–384, 371–384; and T. H. Lodwig and W. A. Smeaton, "The Ice Calorimeter of Lavoisier and Laplace and Some of Its Critics," *Annals of Science* 31(1974): 1–18.

57. Wedgwood, "Thermometer," 380; note by Thomas Beddoes in Torbern Bergman, *A Dissertation on Elective Attractions*, translated by Beddoes (London: J. Murray, 1785), 350–351; and *Minutes of the Coffee House Philosophical Society, 1780–1787* (Oxford: Museum of the History of Science), Gunther MS 4, minutes of 1 April, 1785.

58. Antoine Lavoisier, *Elements of Chemistry*, translated by Robert Kerr (Edinburgh: William Creech, 1790), 19, 182, 343–356.

59. Ibid., 20–21.

60. Ibid., 313–314, 335–339, 348–349, 484.

61. Inventory of Charles Daubeny (1823), in C. R. Hill, *Catalogue I: Chemical Apparatus* (Oxford: Museum of the History of Science, 1971), 69–75; and catalogue of equipment by Samuel Webber (1821), in I. Bernard Cohen, *Some Early Tools of American Science: An Account of the Early Scientific Instruments and Mineralogical and Biological Collections in Harvard University* (Cambridge, MA: Harvard University Press, 1960), 172–176.

9

PLATINUM AND GROUND GLASS: SOME INNOVATIONS IN
CHEMICAL APPARATUS BY GUYTON DE MORVEAU AND
OTHERS
William A. Smeaton

When Arthur Young, the English agronomist, visited chemist Louis
Bernard Guyton de Morveau (1737–1816)[1] in Dijon in 1789, he found
a laboratory containing "such a variety and extent of apparatus, as I
have seen nowhere else," and he regretted that Guyton had not published an account of it.[2] Young had done some chemical research, but
his knowledge of laboratories was probably limited, and much of the
apparatus that he saw in Guyton's house may have been similar to that
used by other chemists. However, Guyton had an inventive mind and,
from various manuscript and printed sources, we can learn about some
original apparatus that he devised. We discuss certain applications of
two materials that became available in the middle of the eighteenth
century. One is platinum, a newly discovered metal remarkable for its
very high melting point and its resistance to attack by most reagents,
and the other is ground glass, an ancient substance treated in a new
way. In each case, we consider Guyton's work in the context of that of
some of his contemporaries.

In 1764, at the age of 27, Guyton was elected to the Académie
des Sciences, Arts et Belles-Lettres de Dijon as a literary man, but he
soon became interested in chemistry and installed a laboratory in the
house that he bought in 1768. During a visit to Paris in that year, he
met P. J. Macquer, from whose books he had studied chemistry, and
began a friendship and correspondence that lasted until Macquer's
death in 1784. His early letters show his interest in mineral analysis;
when, in 1773, he received from the Comte de Buffon a piece of
native platinum, a metal only recently introduced into Europe from
South America, he made a careful study of its properties. When
reporting his results to Buffon, a fellow-member of the Dijon Academy
(though he rarely visited Dijon), he expressed the hope that a use
would be found for the metal and also for an alloy of platinum and steel
that he had prepared.[3]

Nothing further was heard of the alloy, but Guyton described an
important use for platinum in 1785, in notes to his translation of T. O.

Bergman's chemical works, especially the essay on the blowpipe.[4] This was the mouth blowpipe through which the operator directed the flame of a candle onto the specimen being examined. Mineralogists used it to heat a small amount of a mineral, either alone or with a flux, on a metallic spoon, or a piece of charcoal that might react with it and give rise to a color change or the formation of a metal. Bergman's blowpipe was made of silver and contained a little platinum to make it harder and less easily oxidized. Guyton concentrated on the nozzle, the part that was most fragile and susceptible to oxidation, and made one of a very hard alloy of silver with two-sevenths of platinum, which he had cast only with some difficulty. For substances that reacted with charcoal, Bergman used a small silver or gold spoon, and Guyton improved this with an alloy of equal parts of gold and platinum, which withstood the heat of a flame excited by oxygen and was easily cleaned with nitric acid. An unnamed colleague in the Dijon Academy had made a spoon of "pure" platinum by melting it into a button and grinding it with emery on the lathe, but we now know that the platinum must have contained other metals. However, the nozzle and spoon served their purpose; later, when pure platinum was easily obtained, we find Michael Faraday recommending the provision of several interchangeable platinum nozzles with different apertures and of platinum foil, instead of a spoon, for mounting specimens.[5] The small amount of platinum required was not expensive.

Normally, mineralogists analyzed a specimen in the laboratory at the end of their journey, but Guyton realized that it would be advantageous to do this soon after collecting it, for the result might suggest further investigations at the site of its discovery. He had a personal interest in the subject, having reported to the Dijon Academy the results of many mineral analyses, some of which had been published. His translation of Bergman's essay on the blowpipe first appeared in 1781, but without his notes,[6] and several years earlier he had designed a portable chemical outfit (nécessaire chymique), which contained chemicals and apparatus for analysis in two wooden boxes, each measuring $7 \times 4 \times 1.5$ inches, small enough to be carried in a coat pocket. A number of them must have been made by 1783, for he then said that some owners had them bound in leather, whereas others preferred to have the wood polished.[7]

The first box contained compartments for six bottles, each 1 inch in diameter and perhaps 2.75 inches high, containing substances that had to be handled with great care (figure 9.1).[8] They were, in modern

Figure 9.1
The nécessaire chymique (1783). Nouveaux Mémoires de l'Académie de Dijon, 1783 (semester 1). (See note 7.)

language, nitric acid and solutions of silver and mercurous nitrates, barium chloride, oxalic acid, and ammonia. This is not the occasion for a discussion of their applications in analysis, but an interesting note is Guyton's comment that the ammonia was also useful for treating snake bites, evidently an occupational hazard for mineralogists. Each bottle had a stopper, accurately ground with emery to fit its neck and with the top half removed on a grindstone to leave a flat surface, which was held firmly in place by a thick screw made of service wood (*bois de cormier*), noted for its fine grain and hardness, that passed through a threaded hole in the top of the box. When all the screws were tightened, the box was under considerable strain, so the central partition was made thicker than the others to strengthen it, a refinement that must have been introduced as a result of trial and error. As we shall see, ground-glass stoppers had been in use since the 1750s, but the retaining screws seem to have been devised by Guyton, probably in collaboration with a Dijon craftsman.

The second box, which was not illustrated or described in detail, contained a blowpipe and spoon with three fluxes recommended by Bergman: sodium carbonate, sodium borate, and sodium ammonium hydrogen phosphate ("microcosmic salt"), together with a steel blade for testing the hardness of minerals, a magnetized needle for detecting iron, tweezers for handling small specimens, and five reagents that did not have to be handled as carefully as those in the first box. These were crystalline ferrous sulfate and alum, potassium ferrocyanide solution, lime water, and alcoholic tincture of gall nuts. Understanding why Guyton considered these three solutions to require less careful handling is difficult, for any of them would damage the other equipment in the box, and he must have taken some steps to avoid leakage. Test papers colored with litmus, turmeric, and Brazil wood; filter papers; and a bent glass tube for stirring solutions and collecting gases completed the outfit that, with domestic glass vessels found everywhere, enabled traveling mineralogists to perform all the usual tests.

The blowpipe was first used for mineral analysis in Sweden in the 1740s, and small boxes for carrying it and the necessary chemicals were introduced there by A. F. Cronstedt, according to G. von Engeström, whose "Description and use of a mineralogical pocket laboratory, and especially the use of the blowpipe in mineralogy" was published in 1770 as an appendix to his English translation of Cronstedt's *An Essay Towards a System of Mineralogy.*[9]

A second edition of von Engeström's "Description..." appeared in 1772 and it was translated into Swedish (in 1773) and German (in 1774), but Guyton seems to have seen none of these and probably devised his outfit independently. This can be inferred from a comparison of the contents of the two pocket laboratories, particularly two very small retorts provided only by von Engeström. These were of a type used in Bergman's laboratory in Uppsala, but Guyton knew nothing about them until they were described in a letter from C. A. H. G. de Virly, a mineralogist from Dijon who studied chemistry under Bergman from December 1781 to June 1782.[10] He arrived in Uppsala with an introduction from Guyton, who began writing to Bergman, the first foreign chemist with whom he corresponded, in 1779.

Guyton referred to de Virly's letter in his description of an ingenious portable retort stand and distillation apparatus, published with his account of the *nécessaire chymique.* The retort stand, apparently made to his own design, was 10 inches high, and the retorts, only 1 or 2 inches in diameter and based on de Virly's description, could be heated in various positions by means of an alcohol burner with wicks of three sizes (figure 9.2). Guyton had adapted the household alcohol burner that kept food warm at the table; he may have been the first chemist to utilize this source of heat in the laboratory. This apparatus, far smaller than that generally used, was intended for use in ordinary laboratories and not just by the traveler, but it does not seem to have been widely adopted. One reason may be Guyton's observation that it was difficult to make retorts thin enough to avoid cracking when heated.

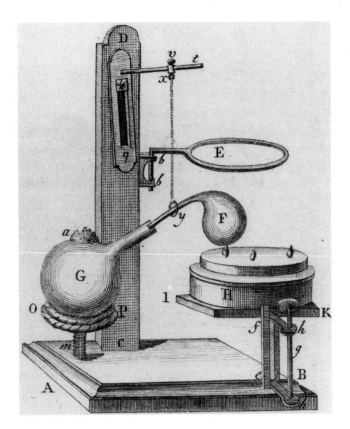

Figure 9.2
The portable retort stand (1783). Nouveaux Mémoires de l'Académie de Dijon,
1783 (semester 1). (See note 7.)

J. H. de Magellan's new English edition of Cronstedt's book,
published in 1788, included von Engeström's account of the blowpipe
with only a few modifications, but the pocket laboratory that he
described was more elaborate than the original, and one of the two
boxes would have fitted only into a very large pocket. A portable laboratory of a different kind, not intended to be carried in the pocket,
was designed by J. F. A. Göttling, professor of chemistry at Jena and
announced in the edition for 1789 of his *Almanach oder Taschenbuch für
Scheidekünstler und Apotheker*, published in 1788. He sold it for 3 golden
louis (approximately £3 sterling) if ordered before 31 December 1788
and charged an extra half-louis thereafter, an early example of a selling
technique that is familiar in the twentieth century. No extra charge was
levied for carriage as far as Leipzig, some 70 km from Jena; beyond that

point, it was distributed by booksellers. Leipzig was an important center for the book trade; interestingly, this use was being made of the services developed for distributing books throughout Europe. In 1790, Göttling gave a full account of his portable laboratory and its uses in a book that was translated into English in 1791 as *Description of a Portable Chest of Chemistry*.[11] This influenced both Richard Reece, who began to sell his chemical and medical chests circa 1800, and William Henry, who made and sold chemical chests of various sizes in Manchester from 1801 onward.[12]

Since 1775, Guyton had known that platinum could be fused with a flux containing white arsenic (arsenious oxide), but some arsenic remained, and it was unsuitable for the blowpipe nozzle and spoon. In 1779, F. K. Achard showed that the arsenic was expelled on strong heating and, in 1784, he made a crucible by casting the arsenical platinum into a mold before heating.[13] Guyton found that Achard's flux of white arsenic and potash (potassium carbonate) left the platinum in a friable state; in 1785, he was critical of Achard's method when writing to Richard Kirwan, his regular correspondent in London. However, he soon obtained a good result by adding common salt and powdered charcoal and, at a meeting of the Dijon Academy, he showed three crucibles that he had made.[14] One was sound, even-walled, and free from blisters and he found it far more useful than the iron crucibles recommended by Bergman for mineral analysis. This was probably the first practicable platinum crucible.

Neither Achard nor Guyton seems to have made more crucibles, but their manufacture was taken up by Marc Étienne Janety, the Paris goldsmith who was already making plate and jewelry from platinum prepared by the arsenic process, which yielded a metal suitable for hot forging rather than for casting.[15] While in Paris from February to August 1787, Guyton met Janety and advised him to remove final traces of other metals by quenching the platinum in molten saltpeter (potassium nitrate). This rendered it more malleable and enabled Janety to produce larger vessels, including several crucibles for Guyton, one of them 30 lines (68 mm) in diameter, which Guyton described as "very fine" in a letter to his friend Pierre Louis Baudot, a Dijonnais then living in Paris.[16] Platinum was both smuggled and legally imported from the Spanish colony of New Granada (now Colombia), and platinum apparatus soon ceased to be a rarity. In 1812, Guyton heated diamond to a high temperature in a platinum cylinder when studying its combustion,[17] and he devised a pyrometer for measuring very high temperatures by recording the thermal expansion of a platinum rod mounted on a ceramic base. An early version was made circa 1810 by instrument maker Étienne Lenoir (or by his son, Paul Étienne), but it did not come into general use, and the full description of the final version that Guyton wrote before his death in 1816 was never published.[18] Platinum was always an expensive metal but, after use in apparatus made for a particular purpose, it could be recycled, so old platinum apparatus has rarely survived.

When writing to Baudot on 10 January 1788, Guyton asked him to visit "Meignie, ingénieur en instruments de mathématiques," who lived in the same quarter of Paris, to discuss the price of a Volta eudiometer that he had made for the Dijon Academy. It was intended for use in his lecture course to demonstrate the formation of water from hydrogen and oxygen, as he later told Martinus van Marum, his correspondent in Haarlem.[19] The instrument, invented in 1777 by Alessandro Volta and one of many eudiometers in use in the 1780s (see chapter 5), consisted of a vertical graduated glass tube in which a mixture of air and hydrogen (or another inflammable gas) was ignited by an electric spark, permitting measurement of the change in volume. The gases were introduced through a wide metal funnel attached to the lower end of the tube and dipping into water in a pneumatic trough. The funnel also acted as a support for the apparatus; later versions included metal taps at the ends of the tube.[20] "Meignie" was asking for 190 livres, but Fortin, another Paris instrument maker, had charged only 96 livres for a similar vessel in copper. Guyton thought that 124 livres would be appropriate, as the new one was made of iron (which, unlike copper, did not form an amalgam) so that it could be used in mercury instead of water; this would add to the cost, presumably because iron was more difficult to work into the required shape. It must have been about this time that Guyton had the idea of using platinum wires instead of silver (which formed an amalgam) when passing electric sparks through gases over mercury.[21]

The instrument maker in question was Pierre Bernard Mégnié (1751–1807), the son of a master locksmith in Dijon. He had been known to members of the Dijon Academy since 12 November 1773, when they considered and suggested improvements to his memoir about the pantograph. On 4 May 1775, a revised version was submitted on his behalf by Guyton, who had evidently met him in Paris and, on 24 May, the Academy heard a favorable report from two members and elected him as a corresponding member, a rare honor for a craftsman.[22]

On both occasions, Guyton spoke in praise of Mégnié, describing him as a protégé of astronomer J. F. Lalande, from whom he had learned much about astronomy, and as a worker who was skilled with his hands and had a good knowledge of mathematics and mechanics. He added that Mégnié's election might help his career, as he was entering a competition to be appointed "engineer for the construction of mathematical instruments" attached to the Paris Academy of Sciences. He was not given the title, but he received a half-share of the cash prize offered for the instrument, a quarter-circle 3 feet in diameter, which he made for the competition.[23] Mégnié visited Dijon about this time and, on 17 June 1775, he showed the Academy a balance that he had made for Guyton "during his sojourn in this town." It incorporated several original features and was intended to give weights down to "the smallest parts of a grain" for chemical analysis or the determination of specific gravities.[24] In 1789, Guyton owned a balance, apparently a different one, as Young stated that it was made in Paris, which turned with a twentieth of a grain (approximately 2.6 mg) when loaded with 3,000 grains (approximately 160 gm). This was comparable to one of A. L. Lavoisier's balances, which weighed 18 or 20 oz. (some 550 or 600 gm) with a precision of a tenth of a grain (5.3 mg).[25] The maker of Guyton's balance could have been either Mégnié or Fortin, both of whom provided precision balances for Lavoisier.

Young also commented on Guyton's "new and most ingenious inventions for facilitating enquiries in the new philosophy of air." Guyton had learned about Joseph Priestley's and Lavoisier's experiments on gases before visiting Paris for several weeks between February and April 1775. His original purpose was to discuss legal matters concerning the purchase of a house by the Dijon Academy but, as a correspondent of the Paris Academy of Sciences since 1772, attached to Macquer, he had the right to attend meetings. Possibly it was there that he became acquainted with Lavoisier and Louis Alexandre, duc de la Rochefoucauld d'Enville, an enthusiastic amateur chemist.

On 18 April 1775, soon after returning to Dijon, Guyton told Macquer that he had had apparatus for experiments on gases sent to him and wished to show his compatriots all that Lavoisier had so obligingly taught him. He had already begun to decant air from one vessel to another "very passably," to the great astonishment of his colleagues.[26] At the Dijon Academy on 18 May 1775, he demonstrated experiments with "different kinds of air," procedures intended to convince the public of the reality of Priestley's discoveries[27] and, in a letter to la Rochefoucauld on 25 May 1775, he referred to "the seven kinds of air."[28]

The minutes of the Academy mention fixed, nitrous, acid, alkaline, and phlogisticated airs (carbon dioxide, nitric oxide, hydrogen chloride, ammonia, and nitrogen), but he seems also to have demonstrated inflammable air (hydrogen), and the seventh may have been common air.

In the same letter to la Rochefoucauld, Guyton described a striking experiment in which acid air (hydrogen chloride) and alkaline air (ammonia) were collected over mercury in two glass tubes of equal diameter separated by an iron tap to which they were both cemented. When the tap was opened, the gases mixed, the volume decreased, and the ammoniacal salt (ammonium chloride) was deposited on the inside of the glass. Guyton had designed the apparatus, and it had been perfectly made by Mégnié who, as we have seen, was then in Dijon. The iron was not affected, but Guyton recognized that it would eventually be corroded by the acid air. He intended to try to overcome this problem by having a similar apparatus made entirely of glass, but this is never mentioned in any of his subsequent letters or publications.

An airtight tap would have had to be made of closely fitting ground glass, a very difficult task for a glassworker; as late as 1827, Faraday described only metal taps for gases, generally of brass, which had to be inspected and cleaned regularly.[29] The introduction of burettes with glass taps for use in volumetric analysis was also long delayed. In 1846, E. O. Henry had a burette with a glass tube and a copper tap, but even after glass taps were eventually brought into use, they were not at first popular on account of the difficulty of avoiding leakage. Though Friedrich Mohr described them in 1855, he preferred a stopcock made of a rubber tube with a metal clip.[30]

Although they did not find favor with chemists until the late nineteenth century, glass taps were made and used for at least one purpose as early as the 1780s. Priestley's recently invented method for dissolving carbon dioxide in water to produce carbonated water was criticized in 1775 by J. M. Nooth, who claimed that the bladder used by Priestley as a container for the gas gave the water an unpleasant flavor. He therefore designed an apparatus made entirely of glass,[31] which was soon improved by William Parker, a London glass manufacturer, possibly in collaboration with Magellan, who published a description of it in 1777.[32] The carbonated water was drawn off from the central part of the apparatus by removing a ground-glass stopper, a

Figure 9.3
Nooth's apparatus, improved by Parker and Magellan (left) and Blades (right). John
Elliot, the Principal Mineral Waters of Great Britain and Ireland, London, 1781.
(For the full title, see note 34.)

procedure that must have been inconvenient; however, in 1783, in the
third edition of his book, Magellan stated that the stopper had been
replaced by a "glass-cock."[33]

Many readers would assume that the improvement was made by
Parker or Magellan but, in 1781, it was stated by John Elliot, in the
appendix to his account of mineral waters,[34] that it was the work of
"Mr Blades, of Ludgate Hill," who is more precisely identified in a
London directory as John Blades,[35] a glass manufacturer. The frontispiece of Elliot's book shows both versions of Nooth's apparatus (figure
9.3), and the advantage of the tap over the stopper (k in the earlier
version) is obvious. An example of the apparatus with a ground-glass
tap is located in the George III collection in the Science Museum,

London,[36] and another is clearly visible in an illustration, first published in 1816, of the chemical lecture theater at Guy's Hospital, London.[37]

Glass taps must have been made in large numbers, for the Nooth apparatus became popular for domestic use; why then were they not adopted by makers of chemical apparatus? The answer to that question seems to be given by Elliot, who explained that Nooth's apparatus must stand on a wooden tray to catch any water that spilled. Any leakage, however slight, would be unacceptable to chemists. We must, however, note that the inventory of apparatus destroyed when Priestley's house was burnt down in 1791 includes "Landriani's Eudometer [*sic*] with Glass Cocks sent from Italy."[38] What is not known is whether glass taps were present in all of Landriani's eudiometers or were fitted to only a few special examples. However, all other eudiometers at that time (and later) were made with metal taps, so we can infer that glass was found to be unsatisfactory, probably on account of leakage. Solid ground-glass stoppers did, however, come into the chemical laboratory in the 1770s.

When conserving a gas collected in a jar in the pneumatic trough, Priestley kept the jar inverted in a shallow dish of water or mercury, and this simple procedure, described and illustrated in 1774 in his *Experiments and Observations on Different Kinds of Air*, was adopted by Lavoisier and other French chemists. However, when the French translation appeared in 1777, it included the text of a paper that was read to the Paris Academy of Sciences on 15 May by the Duc de Chaulnes to describe various improvements to the pneumatic trough and related apparatus.[39] He dispensed with the shallow dishes and kept his gases in jars sealed, while still below the surface of the trough, by plates of sheet metal covered with greased leather, and he stressed the need to make the ends of the jars smooth by grinding with emery. He may have been inspired by the moist leather covering on the plate of the air pump devised in 1740 by J. A. Nollet,[40] which remained in use at least until 1801.[41]

A further development was due to J. A. Sigaud de la Fond and published by him in 1779.[42] He was aware of Chaulnes's use of leather-covered sheet metal, but he preferred to store gases that had been collected over water in bottles with tightly fitting corks. For acidic gases collected over mercury because they dissolved in water, he used small phials, 3 inches (81 mm) long and 1 inch (27 mm) in diameter, with ground-glass stoppers. He considered that the jars used by Priestley were too wide to allow the gas to be decanted below the surface of the water or mercury in the pneumatic trough without loss of gas or introduction of common air. Corks had been used for closing wine bottles since the late seventeenth century, but glass stoppers came into use only around 1750, originally, it seems, for wine decanters.[43] In 1827, Faraday collected gases in bottles with greased, well-fitting ground-glass stoppers and stored them for months (or even years) inverted, with the necks immersed in water. The glass normally was ground with emery powder, an impure form of the mineral corundum (aluminium oxide), but fine sand sometimes was used.[44]

When examining the effect of heat on nitric acid, Priestley heated it in a "glass phial with a ground stopper and tube," having found that the evolved fumes affected cork.[45] This must have been one of the first times that a ground-glass joint was used on a hollow tube rather than a solid stopper, and the procedure does not seem to have been frequently adopted. In 1795, Guyton described a eudiometer of his invention in which oxygen was absorbed by solid sulfide of potash (probably potassium polysulfide) in a small retort containing the gas mixture being analyzed, which was heated by a candle.[46] The amount of oxygen absorbed was shown by the rise in the level of water in a graduated vertical glass tube leading from the neck of the retort. To prevent loss of gas, the tube was formed in two parts that the operator joined together below the water surface, and it was made airtight by means of a ground-glass joint (at C on figure 9.4). The eudiometer was tested successfully at the École Polytechnique, where Guyton was one of the chemistry professors; in 1801, it was described and illustrated by P. Jacotot, professor of chemistry at the École Centrale in Dijon and Guyton's former confrère in the Dijon Academy.[47] It was also described in 1802 by E. J. B. Bouillon-Lagrange, who had worked in the laboratory of the École Polytechnique.[48] However, it was not mentioned in 1805 in an account of eudiometers by Guyton's fellow professor, A. F. Fourcroy, so it can have been used only rarely.[49] Neither Jacotot nor Bouillon-Lagrange mentioned the essential ground-glass joint, so possibly they described the apparatus without having used or even seen it—not, perhaps, an uncommon practice among authors of textbooks.

At the end of his lecture course at the École Polytechnique in 1797, Guyton demonstrated a set of retort stands and a distillation apparatus, derived from the small-scale apparatus that he described in 1783 but now using retorts of a conventional size. However, he said that the earlier equipment, which could be packed in a small box, still

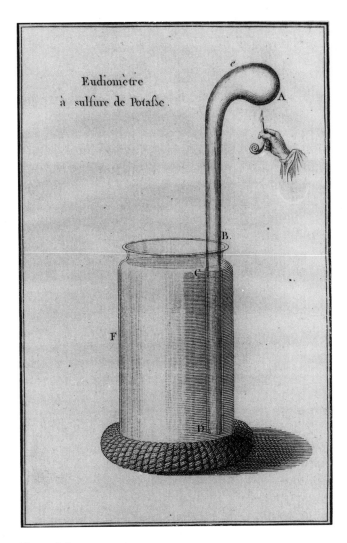

Figure 9.4
Guyton's eudiometer (1795). *Journal de l'École Polytechnique*, 2 (1795). (See note 46.) The legend reads "Sulfide of Potash Eudiometer."

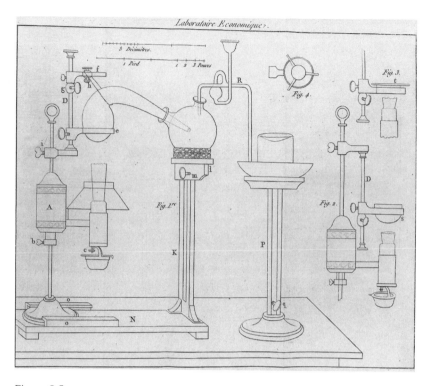

Figure 9.5
Guyton's "economical laboratory" (1797). Annales de Chimie, 24 (1797). (See note 51.) There is a scale of 3 decimetres and another of 2 feet, with inches shown.

found favor with travelers and that several sets had been made by the Dumotiez brothers ("les citoyens Dumotier"), who were well-known Paris instrument makers.[50] Guyton's account of the new apparatus was published later in 1797, with an illustration (figure 9.5).[51] The source of heat was now an oil lamp with a tubular wick and glass chimney, invented by Aimé Argand and usually called the *Argand lamp*.[52] The lamp and a retort were mounted on one stand, a receiver on another, and a pneumatic trough, if required, on a third. This could be set up even when the lamp was in normal use for illumination, thus ensuring that the heat issuing from the top was not wasted but was used to produce distilled water.

Guyton explained at some length the value of distilled water in practical chemistry; surprisingly, he stated that many chemists still used rain water or water from springs or rivers, all of which he found to contain dissolved salts. During a normal winter evening, Guyton could prepare up to 3 dl of distilled water with no trouble or expense with his "Economical Laboratory," as he called it. In 1799, an English translation of Guyton's paper, with the illustration, appeared in the journal edited by William Nicholson,[53] and the part of the illustration depicting the distillation apparatus was included in the frontispiece of *The Chemical Pocket-Book* by James Parkinson, who added a short description.[54] Parkinson did not, however, include the modification shown on the right of the illustration, in which the chimney of the lamp was shortened and the retort stand was modified to support a platinum or silver evaporating basin or a platinum crucible on a triangle of iron wire. These were suitable for evaporation, fusion, and other processes requiring a high temperature. The apparatus seems to have come into general use, for in 1816 "Guyton's lamp furnace" was included in the catalog of apparatus sold by F. C. Accum.[55]

Industrial and other applications of chemistry were always regarded as important by Guyton, so for several months in 1783 and 1784, he worked enthusiastically on the large-scale production of hydrogen and other gases for filling the newly invented balloons. He made two flights in the balloon that he and two colleagues designed for the Dijon Academy. The balloon was filled with hydrogen generated from zinc or iron and sulfuric acid at a rate of up to 845 cubic feet (some 29 m^3) per hour in apparatus that he developed. Ten years later, in 1793 and 1794, he was also involved in the production of hydrogen from steam and red-hot iron for military balloons. Both phases of Guyton's interest in the large-scale production of gases have been discussed in detail by P. J. Austerfield,[56] who considers his achievements in the context of the similar work that was being done elsewhere in France and in other countries. J. Langins has given a separate account,[57] which complements Austerfield's, of Guyton's activity in 1793 and 1794, so this important aspect of Guyton's work on gas apparatus is not pursued in this chapter.

Guyton's interest in using acid gases to disinfect the air in unhealthy places, such as hospitals and prisons, led him to an entirely new application of ground glass. In 1773, he first used muriatic (hydrochloric) acid gas generated by heating common salt with sulfuric acid, believing that it neutralized ammonia, which was emitted from decaying animal matter and waste products and carried into the air the noxious effluvia then thought to be responsible for infection. This caused the effluvia to be brought down with the solid salt that was formed, the experiment that he demonstrated in 1775. After accepting Lavoisier's oxygen theory of acids, he abandoned his original explanation and now believed that the effluvia were oxidized by the acid; irrespective of the theory, the method gave good results. It was recommended by the Royal Society of Medicine in Paris and was widely adopted in France, especially in military hospitals during the wars in the 1790s. Acid fumigation also became popular in Britain, where James Carmichael Smyth disinfected naval establishments with nitric acid vapor and William Cruickshank used oxymuriatic acid (chlorine) in the military hospital at Woolwich.[58]

In all these processes, the gas was generated by heating in open vessels, and the volume depended only on the quantities of reagents used; however, from circa 1800, Guyton developed a method of controlling the release of the gas. Before moving from Dijon to Paris in 1791, he had begun to experiment with chlorine evolved from an unheated mixture of hydrochloric and nitric acids (which, in fact, yields chlorine and nitrosyl chloride). On returning several years later to sell his house, he cleared his laboratory and found a bottle containing the mixture that still emitted chlorine when the glass stopper was removed.

The gas was more conveniently prepared, again without heating, from a mixture of black manganese (manganese dioxide), common salt (sodium chloride), and sulfuric acid. Consequently, with the collaboration of one of the Dumotiez brothers, Guyton designed an apparatus that allowed the reaction to be performed safely by people with no laboratory experience. In 1805, he described and illustrated the apparatus in the third edition of his treatise on disinfection.[59]

The reaction mixture was prepared in a bottle kept in a wooden case, with a ground-glass stopper held in place by a wooden screw, a method similar to that used in his *nécessaire chymique*. When the screw was loosened, the stopper could be released, emitting as much gas as was needed. However, the stopper had to be screwed down tightly; after loosening the wooden screw, it was sometimes difficult to remove the stopper without breaking the neck of the bottle, a problem that was not mentioned in the account of the *nécessaire chymique*. Dumotiez therefore replaced the stopper by a glass plate, again held down by a wooden screw, but the pressure of the screw often broke the glass. Eventually, Guyton and Dumotiez produced a satisfactory apparatus (figure 9.6).

Both the neck of the bottle, of approximately 100-ml capacity, and the glass plate covering it were made of exceptionally thick glass so

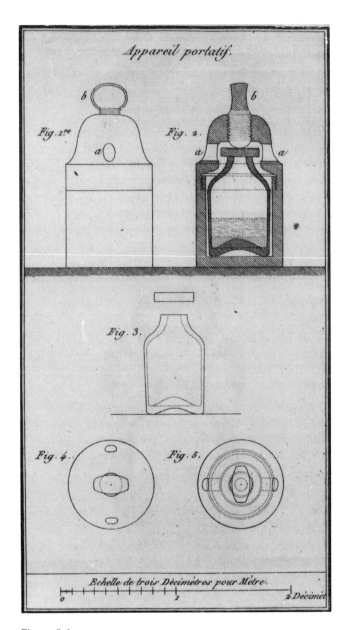

Figure 9.6
The portable disinfection apparatus (1805). L. B. Guyton, Traité des moyens de désinfecter l'air, 3rd ed., 1805. (See note 59.) The legend reads "Scale of three decimetres to a metre."

that they could withstand the pressure of the screw holding the plate down. The entire apparatus was enclosed in a wooden case made in two parts screwed together, so that it could be opened to remove the bottle for refilling. When the top screw was loosened, the pressure of the gas raised the plate slightly, and it escaped through holes in the lid (see figure 9.6). These holes were placed in positions enabling the plate to be raised manually, presumably by inserting a pointed object, if the gas pressure was too low. The device involved no risk of breakage, so it could be safely carried by persons entering a locality where they might be exposed to infection. Guyton once used it at a funeral in hot weather, when an unpleasant odor came from the coffin, and he believed he was better protected from infection than were the priests burning aromatic herbs.

In his book, Guyton also described a modified version with a different method of releasing the chlorine (not discussed here) and a larger "permanent" apparatus, which was intended mainly for use in hospital wards. The chlorine was generated in a cylindrical glass vessel of 600- to 700-ml capacity, not enclosed in wood as was the portable apparatus but held in a wooden press (figure 9.7). It was covered by a glass plate (labeled I in the figure) attached to the lower surface of a wooden disc (H) that was raised and lowered by the screw (F). The gas escaped when the glass plate was raised and, as in the case of the portable apparatus, an essential feature was for the disc to be ground flat but not polished, and the edge of the cylinder had to be similarly prepared "with the same exactitude as those of the receivers of an air pump."[60] Examining surviving air pumps made by Dumotiez to see whether they satisfy this requirement has not been possible, but we know that such glass surfaces were in general use by 1819 in the type of air pump described in the *Encyclopédie Méthodique*,[61] and it has been suggested that the designer of the large apparatus was familiar with air pumps.[62] It certainly has an important feature in common with air pumps made circa 1710 by Francis Hauksbee Sr.[63] and in 1761 by George Adams.[64] Each has a screw press to hold the receiver in place, similar in general appearance to that in the disinfection apparatus. To see whether this was used in air pumps from the Dumotiez workshop would be desirable. Both the permanent and portable versions were sold by Dumotiez and by Boullay, a Paris pharmacist, but the detailed descriptions and drawings (clearly the work of a professional draftsman) were intended by Guyton to ensure that the apparatus could be copied anywhere by a craftsman with the necessary skill. An example of the

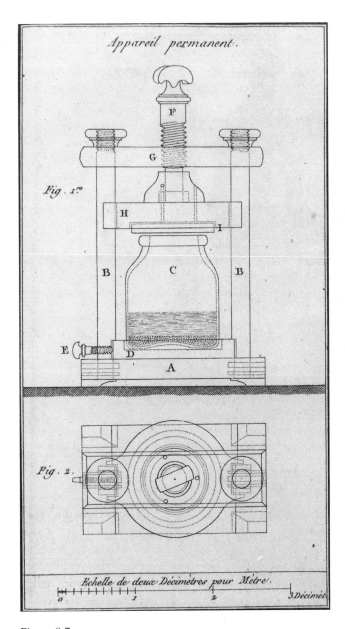

Figure 9.7
The permanent disinfection apparatus (1805). L. B. Guyton, Traité des moyens de désinfecter l'air, 3rd ed., 1805. (See note 59.) The legend reads "Scale of two decimetres to a metre."

Figure 9.8
Guyton's large apparatus, made by Boullay. Height, 34 cm; base, 24 × 14 cm.
Private collection of W. A. Smeaton.

permanent apparatus in the *Museo di Storia della Scienze* in Florence
conforms to the original specification but has a more elegant shape.
One in the *Musée d'Histoire de la Médecine* in Paris has been modified by
the insertion of a copper or bronze tap to allow the chlorine to escape,
eliminating the need to raise the cover by means of the screw but surely
leading to corrosion of the tap by chlorine.[65] A third, in the private
collection of W. A. Smeaton, is in its original condition, with Boullay's
label (figure 9.8). Two examples of the portable apparatus have been
seen, both with Dumotiez's label: The one in the Science Museum in
London still contains manganese dioxide;[66] the other, in the Schweizer
Pharmazie-historisches Museum in Basel, has not been closely examined. Guyton's method was approved by several official organizations

and was used in army hospitals. The Spanish government recommended it and, in 1816, Accum described it as having the support of the Royal College of Physicians of Edinburgh when he offered both versions for sale.[67]

A description translated from part of Guyton's book with the illustrations was published in Italy in 1832,[68] but in France chlorine gas was being replaced as a disinfectant by a solution of "chloride of oxide of sodium" (sodium hypochlorite) or of bleaching powder (containing calcium hypochlorite), a process introduced in the early 1820s by A. G. Labarraque.[69] Some opposition came from supporters of chlorine disinfection, and much debate ensued among chemists about the mode of action of hypochlorites, an interesting topic beyond the scope of this chapter. Guyton's apparatus probably was rarely used by 1856, when John Taylor was granted a British patent for "an improved vessel for containing chemicals for the generation of disinfecting gases."[70] This was made of glass or earthenware and, as in Guyton's apparatus, the upper part could be raised by means of a screw to permit gas to escape. However, the seal between the two parts was not made of ground glass but of vulcanized rubber, a substance that had recently become available.

Rubber, which originated in South America, was studied by chemists in Europe after circa 1750, but it found few uses before 1821, when Thomas Hancock patented a method of working it into blocks that could be cut into sheets.[71] In 1827, Faraday described various applications of Hancock's rubber, including the use of rubber washers instead of leather or cardboard for rendering taps airtight. However, he pointed out that if a washer was drawn up too tightly, the rubber was squeezed out and reusing the washer was impossible, even though the joint remained airtight while it was in place.[72] This limitation would have made Hancock's rubber unsuitable for use in Taylor's apparatus where, as in Guyton's, a tight seal was essential.

However, this problem did not arise with rubber vulcanized by heating with sulfur. Vulcanization, a process discovered and patented by Charles Goodyear in 1841, resulted in a product that kept its shape when tightly compressed, and manufacturers soon found many uses for it. Taylor's apparatus, with its vulcanized rubber seal, would have served its purpose well, but we do not know whether it was manufactured on a large scale. However, it is a convenient topic with which to end this chapter for, like platinum and ground glass, vulcanized rubber was a new material that proved to be invaluable to the chemist.

NOTES

1. Until 1789, he was known as "Guyton de Morveau," but, during and after the French Revolution, he adopted at various times the names "Guyton-Morveau" and "Guyton," which was his family name. In this chapter, he always is called *Guyton*.

2. All references to Young in this chapter are taken from Arthur Young, *Travels in the Kingdom of France in the Years 1787, 1788, and 1789*, vol. 1 (Bury St Edmunds: J. Rackham, 1792), 148–154.

3. L. B. Guyton, "Lettre . . . à M. le Comte de Buffon," *Observations sur la Physique* 6(1775): 193–203. This is discussed by W. A. Smeaton, "L. B. Guyton de Morveau: Early Platinum Apparatus," *Platinum Metals Review* 10(1966): 193–203; and Donald McDonald and Leslie B. Hunt, *A History of Platinum and Its Allied Metals* (London: Johnson Matthey, 1982), 59–66.

4. Torbern Olof Bergman, *Opuscules Chymiques et Physiques*, vol. 2 (Dijon: Frantin, 1785), 91, 460 (notes by Guyton).

5. Michael Faraday, *Chemical Manipulation* (London: W. Phillips, 1827), 111, 118. A facsimile edition is included in "The Royal Institution Library of Science" (New York: Wiley, 1974).

6. T. O. Bergman, "Mémoire sur le Chalameau à Souder, & son Usage dans l'Analyse," *Observations sur la Physique* 18(1781): 207–235.

7. Guyton, "Description et Usage du Nécessaire Chymique, et de l'Appareil d'Expériences sur le Réchaut à l'Esprit-de-Vin," *Nouveaux Mémoires de l'Académie de Dijon*, 1783 (semester 1), 159–176.

8. Before the introduction of the metric system, local variations were seen in weights and measures in France, but scientists generally used Paris units: 1 inch equalled 2.71 cm; 1 line equalled 2.3 mm; 1 ounce equalled 30.57 gm; 1 grain equalled 53.1 mg.

9. Gustav von Engeström, "Description and Use of a Mineralogical Pocket Laboratory; and Especially of the Use of the Blowpipe in Mineralogy," in Axel Fredrik Cronstedt, *An Essay Towards a System of Mineralogy* (London: Dilly, 1770). For details of the various editions see W. A. Smeaton, "The Portable Chemical Laboratories of Guyton de Morveau, Cronstedt and Göttling," *Ambix* 13(1966): 84–91.

10. See letters from Charles André Hector Grossart de Virly to Bergman, 10 December 1781, 22 June and 4 July 1782, Göte Carlid and Johan Nordström, eds., *Torbern Bergman's Foreign Correspondence*, vol. 1 (Stockholm: Almqvist & Wiksell, 1965), 397–398.

11. See W. A. Smeaton, "The Portable Chemical Laboratories."

12. John Keith Crellin, "Portable Chemical Chests (of R. Reece, F. Joyce and R. B. Ede)," *Ambix* 14(1967): 60; and E. Leonard Scott, "William Henry's Portable Chemical Chests," *Ambix* 14(1967): 61–62. For later developments to circa

1850, see Brian Gee, "Amusement Chests and Portable Laboratories," in Frank A. J. L. James, ed., *The Development of the Laboratory* (Basingstoke: Macmillan, 1989), 37–59.

13. D. McDonald and L. B. Hunt, *A History of Platinum*, 75–77.

14. Guyton to Kirwan, 20 September 1785, Emmanuel Grison, Michelle Goupil and Patrice Bret, eds., *A Scientific Correspondence during the Chemical Revolution: Guyton de Morveau and Kirwan, 1782–1802* (Berkeley: University of California, 1994), 132; and L. B. Guyton, "Mémoire sur la Fabrication des Ustensiles de Platine," *Nouveaux Mémoires de l'Académie de Dijon*, 1785 (semester 2), 106–112.

15. D. McDonald and L. B. Hunt, *A History of Platinum*, 78–81.

16. Guyton to Baudot, 20 October 1787, Bibliothèque Municipale de Dijon, MS 1181, f. 165.

17. L. B. Guyton, "Nouvelles Expériences sur la Combustion du Diamant," *Annales de Chimie* 84(1812): 20–33.

18. John A. Chaldecott, "The Platinum Pyrometers of Guyton de Morveau," *Annals of Science* 28(1972): 347–368.

19. Guyton to Baudot, 10 January 1788, note 16, ff. 166–167, and Guyton to van Marum, 26 February 1789, E. Lefebure and J. G. de Bruin, eds., *Martinus van Marum. Life and Work*, vol. 6 (Leyden: Noordhoff International, 1976), 139.

20. For details of the Volta eudiometer, with illustrations, see Abraham Wolf, *A History of Science, Technology and Philosophy in the Eighteenth Century*, 2nd ed. (London: Allen & Unwin, 1952), 355–357; and James Riddick Partington, *A History of Chemistry*, vol. 3, (London: Macmillan, 1962), 326–327, 814–815.

21. L. B. Guyton, *Encyclopédie Méthodique, Chymie*, vol. 1, part ii, (Paris: Panckoucke, 1789), 738.

22. Académie de Dijon, Procès-verbaux, registre 7, f. 25 (verso), 12 November 1773; registre 7, ff. 276(verso)–277(recto), 4 May 1775; registre 8, ff. 2(verso)–3(verso), 24 May 1775.

23. For an account of Mégnié, see Maurice Daumas, *Les Instruments Scientifiques au XVIIe et XVIIIe Siècles* (Paris: Presses Universitaires de France, 1953), 360–363. Daumas points out that some confusion has been caused by the presence in Paris of another instrument maker named Mégnié, probably an older relation of Pierre Bernard Mégnié.

24. Académie de Dijon, Procès-verbaux, registre 8, f. 16(recto), 17 June 1775.

25. Daumas, *Lavoisier, Théoricien et Expérimentateur* (Paris: Presses Universitaires de France, 1955), 135.

26. Guyton to Macquer, 18 April 1775, Bibliothèque Nationale, Paris, MS 12306, letter 9.

27. Académie de Dijon, Procès-verbaux, registre 7, f. 293(recto), 18 May 1775.

28. Guyton to la Rochefoucauld, 25 May 1775, Bibliothèque de Besançon, MS 1441, ff. 456–457.

29. Faraday, *Chemical Manipulation*, 356–357.

30. Ferenc Szabadvary, *History of Analytical Chemistry* (Oxford: Pergamon, 1966), 236–237, 246–247.

31. John Mervin Nooth, "The Description of an Apparatus for Impregnating Water with Fixed Air . . . ," *Philosophical Transactions of the Royal Society* 65(1775): 59–66.

32. Jean Hyacinthe de Magellan, *A Description of a Glass Apparatus for Making in a Few Minutes, and at a Very Small Expence* [sic], *the Best Mineral Waters* (London: W. Parker, 1777).

33. J. H. de Magellan, *A Description . . . Mineral Waters*, 3rd ed. (London: printed for the author, 1783), 7–8.

34. John Elliot, *An Account of the Nature and Medicinal Virtues of the Principal Mineral Waters of Great Britain and Ireland and Those Most in Repute on the Continent* (London: J. Johnson, 1781), 59.

35. He is included as "John Blades, Glass Manufactury, 5, Ludgate Hill" in *Kent's Directory for the Year 1783* (London: Richard and Henry Causten, 1783), 21.

36. Science Museum, London, inventory number 1927–1198. An illustration is found in Alan Q. Morton, *Science in the Eighteenth Century. The King George III Collection* (London: Science Museum, 1993), 60.

37. Illustration of the "Chemical Theatre, Guy's Hospital," reproduced from William Babington, *A Syllabus of a Course of Chemical Lectures Read at Guy's Hospital* (London: W. Phillips, 1816), in Samuel Wilks and G. T. Bettany, *A Biographical History of Guy's Hospital* (London: Ward, Lock, Bowden & Co, 1892), facing p. 473. I thank Dr Robin Spring for this reference.

38. Douglas McKie, "Priestley's Laboratory and Library and Other of His Effects." *Notes and Records of the Royal Society* 12(1956): 114–137, especially 122.

39. Marie Joseph Louis d'Albert d'Ailly, Duc de Chaulnes, "Description d'un Appareil Nouveau. . . ," in Joseph Priestley, *Expériences et Observations sur Différentes Espèces d'Air*, vol. 3, (Paris: Nyon, 1777), 313–352, with three plates.

40. Daumas, *Les Instruments Scientifiques*, 286.

41. It is described and recommended by Pierre Jacotot, *Cours de Physique Expérimentale et de Chimie*, vol. 1, (Paris: Richard, Caille and Ravier, 1801), 41–42.

42. Joseph Aignon Sigaud de la Fond, *Essai sur Différentes Espèces d'Air* (Paris: Nyon, 1779), 28–30, 309.

43. William Bowyer Honey, *Glass. A Handbook for the Study of Glass Vessels of all Periods and Countries* (London: Ministry of Education, 1946), 110.

44. Faraday, *Chemical Manipulation*, 355.

45. Joseph Priestley, *Experiments and Observations on Different Kinds of Air ... Abridged ... with Many Additions*, vol. 3, (London: J. Johnson, 1790), 70.

46. L. B. Guyton, "Description et Usage d'un Eudiomètre à Sulfure de Potasse." *Journal de l'École Polytechnique*, 2(1795): 166–168.

47. P. Jacotot, *Cours de Physique Experimentale*, vol. 2, 147.

48. Edmé Jean Baptiste Bouillon-Lagrange, *Manuel d'un Cours de Chimie*, 3rd ed. (Paris: Bernard, 1802), vol. 1, 215, and vol. 3, plate 9. The description was repeated unchanged in the fourth (1808) and fifth (1812) editions.

49. A. F. Fourcroy, *Encyclopédie Méthodique, Chimie*, vol. 4, (Paris: Agasse, 1805), 276–279.

50. For an account of the brothers Louis Joseph and Pierre François Dumotiez, see Daumas, *Les Instruments Scientifiques*, 378–379.

51. L. B. Guyton, "Nouveaux Moyens de Fournir, Presque sans Frais, le Feu et l'Eau pour les Expériences Chimiques," *Annales de Chimie* 24(1797): 310–326.

52. See Michael Schrøder, *The Argand Burner: Its Origin and Development in France and England, 1780–1800* (Odense: Odense University Press, 1969).

53. L. B. Guyton, "New Methods of Affording, at Inconsiderable Expence [sic], the Heat and Water Required for Performing Experiments in Chemistry," *Journal of Natural Philosophy, Chemistry and the Arts* 2(1799): 209–215, plate facing p. 240.

54. James Parkinson, *The Chemical Pocket-Book*, 2nd ed. (London: H. D. Symons, 1801). The frontispiece, which includes "The Chemical Characters of Hassenfratz & Adet" and "The Economical Laboratory of Guyton," is described on pp. *iii–vi*. The frontispiece and description are also in the third (1803) and fourth (1807) editions. The first edition (1800) has not been examined. The first American edition (James Humphrey, Philadelphia, 1802), based on the second London edition, has not been examined. It includes an appendix by James Woodhouse, who describes and illustrates his own "Economical Apparatus" and criticizes Guyton's for being expensive and too complicated. The relevant passage is quoted (with no page reference) by Edgar F. Smith, *James Woodhouse, a Pioneer in Chemistry* (Philadelphia: John C. Winston, 1918, 207–213; facsimile reprint by New York, Arno Press: 1980).

55. Frederick Christian Accum, *Descriptive Catalogue of the Apparatus and Instruments Employed in Experimental and Operative Chemistry ... Manufactured and Sold by Accum and Garden, Operative Chemists, Compton Street, Soho, London*. The title page of this separately paginated 60-page catalog is printed on the verso side of the last page (p. 263) of Accum's *A Practical Essay on Chemical Reagents or Tests* (London: J. Callow, 1816). "Guiton's [sic] lamp-furnace" is included on p. 5 of the catalog,

priced at £2 12s 6d to £5 5s (£2.625 to £5.25). An illustration ("fig. 2, plate ix") is mentioned, but no plates are included in the two copies examined in the British Library, London, and in the Cambridge University Library.

56. Peter J. Austerfield, "The Development of Large-Scale Production and Utilisation of Lighter-Than-Air Gases in France, Britain and the Low Countries from 1783 to 1821, with Special Reference to Aeronautics and the Coal-Gas Industry," Ph.D. Thesis, University of London, 1981. For a useful short account, see P. J. Austerfield, "From Hot Air to Hydrogen: Filling and Flying the Early Gas Balloons," *Endeavour* 14(1990): 194–200.

57. Janis Langins, "Hydrogen Production for Ballooning During the French Revolution: An Early Example of Chemical Process Development," *Annals of Science* 40(1983): 531–558.

58. For a general account of disinfection by gases, in Swedish with English summary, see Lars Oberg, "De Mineralsura Rökningarna," *Lychnos* (1965–66), 153–192.

59. L. B. Guyton, *Traité des Moyens de Désinfecter l'Air*, 3rd ed., (Paris: Bernard, 1805), 386–395.

60. Ibid., 388.

61. Gaspard Monge, et al., *Encyclopédie Méthodique, Physique*, vol. 3 (Paris: Agasse, 1819), 768.

62. Private communication from Robert G. W. Anderson to William A. Smeaton, 24 May 1996.

63. Daumas, *Les Instruments Scientifiques*, plate 53.

64. A. Q. Morton, *Science in the Eighteenth Century*, illustration on p. 37 (Science Museum, London, inventory number 1927–1624/1).

65. "Flacon à Fumigations de Guyton de Morveau." *Revue d'Histoire de la Pharmacie* 24(1977): plate facing p. 286.

66. Science Museum, London, Wellcome Collection: accession number A 99551, registered number 1650/1938. Dumotiez's label shows that his workshop was in the Rue du Jardinet, behind the École de Médecine, Paris.

67. F. C. Accum, *Descriptive Catalogue*. On page 20 of the catalog, "Portable disinfecting, or preservative vials, with directions for use" are advertised at a price of 10s6d to 12s6d (£0.525–£0.625), and "The same apparatus, for large wards of hospitals, prisons etc." for £1 1s to £1 10s (£1.05 to £1.50). The description given by Accum clarifies that they were made to Guyton's design, though his name is not mentioned.

68. L. B. Guyton, *Trattado de' Mezzi di Disinfettare l'Aria*, (Torino: Mancio and Speirini, 1832).

69. Antoine Germain Labarraque, *De l'Emploi des Chlorures d'Oxide de Sodium et de Chaux* (Paris: Huzard, 1825). He had been advocating his process for several years before 1825.

70. John Taylor, British Patent 1478 (1856).

71. S. Pickles, "Production and Utilisation of Rubber," in Charles Singer et al., eds., *A History of Technology*, vol. 5 (Oxford: Clarendon Press, 1958), 732–775.

72. Faraday, *Chemical Manipulation*, 358.

BIBLIOGRAPHY

Accum, Frederick Christian. *A Practical Essay on Chemical Reagents or Tests* (London: J. Callow, 1816).

Chaldecott, John A. "The Platinum Pyrometers of Guyton de Morveau." *Annals of Science* 28(1972): 347–368.

Daumas, Maurice. *Les Instruments Scientifiques au XVIIe et XVIIIe Siècles* (Paris: Presses Universitaires de France, 1953).

Daumas, Maurice. *Lavoisier, Théoricien et Expérimentateur* (Paris: Presses Universitaires de France, 1955).

Elliot, John. *An Account of the Nature and Medicinal Virtues of the Principal Mineral Waters of Great Britain and Ireland and those most in repute on the Continent* (London: J. Johnson, 1781).

Gee, Brian. "Amusement Chests and Portable Laboratories," in Frank A. J. L. James, ed., *The Development of the Laboratory* (Basingstoke: Macmillan, 1989), 37–59.

Guyton de Morveau, Louis Bernard. "Description et Usage d'un Eudiomètre à Sulfure de Potasse." *Journal de l'École Polytechnique* 2(1795): 166–168.

Guyton de Morveau, Louis Bernard. "Nouveaux Moyens de Fournir, Presque sans Frais, le Feu et l'Eau pour les Expériences Chimiques." *Annales de Chimie* 24(1797): 310–326.

Guyton de Morveau, Louis Bernard. *Traité des Moyens de Désinfecter l'Air*, 3rd ed. (Paris: Bernard, 1805).

Honey, William Bowyer. *Glass. A Handbook for the Study of Glass Vessels of All Periods and Countries*, (London: Ministry of Education, 1946).

McDonald, Donald, Hunt, Leslie B. *A History of Platinum and Its Allied Metals* (London: Johnson Matthey, 1982).

Smeaton, William Arthur. "L. B. Guyton de Morveau: Early Platinum Apparatus." *Platinum Metals Review* 10(1966): 193–203.

Smeaton, William Arthur. "The Portable Chemical Laboratories of Guyton de Morveau, Cronstedt and Göttling." *Ambix* 13(1966): 84–91.

III

THE NINETEENTH AND EARLY TWENTIETH CENTURIES

At the end of the eighteenth century, the chemical revolution provided chemistry with such basic principles as the oxygen theory of combustion, a pragmatic definition of the chemical element, and an understanding of gases as one of the three states of matter. Experimentally, chemistry could now handle gases as well as solids and fluids, and Lavoisier's "balance sheet" method provided a new standard for chemical experimentation and reasoning. During the first decade of the nineteenth century, John Dalton's atomic theory added both a conceptual foundation and an experimental program: the determination of atomic weights and the combining proportions of the atoms composing chemical compounds. On these foundations, chemistry grew, during the first half of the nineteenth century, into the first large-scale modern scientific discipline. It was practiced in laboratories of increasing size and complexity and in which, following Liebig's example at the University of Giessen, students were trained systematically in the craft of experimental investigation.

The scale and complexity of chemistry in this era preclude a systematic survey here of its proliferating experimental methods and apparatus. The first four chapters of part III focus on examples that are representative of nineteenth-century chemical practice. Each chapter describes in some detail either the development of a particular apparatus or the application of contemporary methods to a particular experimental problem. Each is also attentive to a context within which the methods and equipment were devised or used. The contexts vary from the style of an individual scientist to that of a national culture, and from the demands of a fundamental theory to those of a practical military problem. Together, they illustrate the diverse ways in which the roles of instruments and experiments can be situated within the multiple dimensions of scientific activity.

Melvyn Usselman treats in a fresh way the familiar events in the origin and acceptance of Dalton's atomic theory by focusing on the experimental data critical to its support. Putting the long-debated question of how Dalton came to his ideas about atoms in the background, Usselman points out that convincing evidence for Dalton's theory required the identification of compounds consisting of the same two substances combined with each other in different proportions and standing in exact integral ratios to each other. He assesses the quality of the experimental results that Dalton himself and several other chemists obtained. Here are found no strikingly novel instruments or methods but investigators who apply their craft skills and their ingenuity in the design of experiments adequate to the task.

The determination of accurate atomic weights, in which J. J. Berzelius played a leading part, rendered possible during the second and third decades of the century the establishment of the elementary formulas of compounds of increasing complexity. The application of these methods to organic compounds proved to be so technically demanding that chemists engaged in extended searches for better analytical methods and for improved apparatus in which to carry them out. Alan Rocke recounts the advent of one of the most famous of these innovations, the combustion apparatus designed by Justus Liebig, which reduced the elementary analysis of organic compounds to a simple, routine procedure. Besides describing the method itself and its advantages over previous methods, Rocke shows how the new method changed the way chemists practiced organic chemistry and even how it fit into the rivalry and contrasting style of contemporary French and German chemistry.

By midcentury, organic chemistry had become the dominant subfield of chemistry, and its experimental techniques had become highly refined. Colin Russell describes several technical innovations introduced by one of the masters of these techniques, Edward Frankland. Russell's primary aim is to portray the great experimental skill and ingenuity of Frankland, but his essay, which can also stand as evidence for the mature state of organic chemistry, is illustrative of the many special experimental problems that investigators in this field encountered and the many modifications of basic methods that they invented to handle these problems.

Chemistry emerged as a basic science out of such practical activities as metallurgy and pharmacy (and such impractical activities as the search for alchemical gold), and it has always retained close connections with practical applications. Seymour Mauskopf illustrates these ongoing connections in the nineteenth century by describing experiments designed to measure the force of gunpowder. He shows also how the nature of the problem brought together chemical and physical traditions of research, and contrasts laboratory style research on the problem with military-style field research.

The nineteenth century marks in some respects a zenith in the evolution of chemistry as an autonomous science. Although it shared some fundamental and practical problems with physics, the methods and concepts that stood at the core of experimental practice had originated within chemistry. The situation changed dramatically at the end of the century with the emergence of the "physical" atom. As physicists developed theories about the structure of the atom and methods with which to investigate atomic structure, chemical principles of composition began to appear derivable from a more fundamental science, and experimental techniques originating in physics seemed applicable to chemical problems.

The final chapter in this volume illustrates experimental practice in chemistry under these changed circumstances. Mary Jo Nye describes the career of Michael Polanyi, including his role in the adaptation to chemical problems of methods of x-ray crystallography that had originated in physics. Nye is concerned also, however, with the vicissitudes that influence whether an experimental practice leads an individual investigator to landmark discoveries and personal distinction or only to useful secondary contributions. The frustrations experienced by Polanyi are those that all scientists risk when they set out to make a mark in a very competitive world. Nye's chapter reminds us that our concentration in this volume on the role of experiments and instruments in the history of chemistry should not cause us to lose sight of the very human condition of those who deploy such means in the quest for new knowledge.

10

MULTIPLE COMBINING PROPORTIONS:
THE EXPERIMENTAL EVIDENCE
Melvyn C. Usselman

WEIGHT RELATIONSHIPS IN CHEMISTRY

In the hands of scientists, instruments provide a means to an end. By extending the powers of the senses and quantifying phenomena that provide sensory input, instruments yield data that generate a synergistic impulse to the human creativity that has crafted our scientific view of the world. Simple theories are sufficient for understanding simple databases, but the breadth and depth of natural phenomena measured instrumentally with accuracy and precision require explanatory edifices of equivalent complexity. Thus, although science analysts correctly emphasize the underdetermination of theory by experiment, they do not always delineate the severe constraints that experiment, especially instrumentally dependent experiment, places on explanatory possibilities.

An experimental result, once it has become accepted by the scientific community, can never by itself point the way to a single explanatory construct, but it may exclude a large number of alternatives. The efficacy of an experimental result in reducing the appeal of competing theoretical explanations is largely a sociocultural process, one that is greatly enhanced if the obtaining of the result is acknowledged to be more than a simple process of self-validation. Thus, for an instrument's output to be scientifically valuable, its users must ensure that the instrument's operation is replicable. Experimental design, often critically dependent on apparatus and instruments, is a key feature in the winnowing of hypotheses and is vitally important in chemistry.

Instruments and apparatus can vary widely in complexity and sophistication, but their efficiency is judged by the reliability of their output. As a perceptual aid, an instrument can serve a useful purpose with a modest level of precision, but it must be reliable and accurate. Thus, in many cases, very simple instruments have contributed to dramatic results because their quantitative output is relatively uncontroversial. The analytical balance is an important example. Modern chemistry is built on a consistent measurement of mass relationships, and the balance remains chemistry's most important instrument. An early problem in chemistry, whose resolution relied on the accuracy and precision of balances, was that of multiple combining proportions in the early years of the nineteenth century.

With the recognition of gases as chemical substances in the eighteenth century, chemical experimenters were able to achieve a better mass balance for chemical transformations. In the hands of a master experimentalist, such as Cavendish, the combining weights of hydrogen and oxygen in the production of water were accurately established, and others were later to establish definite combining proportions as a defining characteristic of most compound chemical substances. It was Lavoisier, of course, who foresaw most clearly the fundamental importance of mass relationships in chemistry. Some historians mark the beginning of modern chemistry with Lavoisier's famous statement that we know as the law of conservation of mass (emphasis added):[1]

> We may lay it down as an incontestable axiom, that, in all the operations of art and nature, nothing is created; an equal quantity of matter exists both before and after the experiment; the quality and quantity of the elements remain precisely the same; and nothing takes place beyond changes and modifications in the combination of these elements. *Upon this principle the whole art of performing chemical experiments depends*: We must always suppose an exact equality between the elements of the body examined and those of the products of its analysis.

Lavoisier used the law as a guide to his search for, and collection of, reaction products. When a reasonably good match was seen between the weights of starting materials and products, Lavoisier felt satisfied that no material components of a chemical transformation had been overlooked and that the finding obviated the need to invoke the participation of imperceptible massless chemical principles (such as phlogiston).

Lavoisier's *Traité* was also notable for its experimentally based "Table of Simple Substances," a list of 33 substances believed to be the fundamental elements of bodies.[2] To warrant inclusion in the table, the basic substances had to fit Lavoisier's criterion:[3]

> ... if, by the term *elements*, we mean to express those simple and indivisible atoms of which matter is composed, it is extremely

probable we know nothing at all about them; but, if we apply the term *elements*, or *principles of bodies*, to express our idea of the last point which analysis is capable of reaching, we must admit as elements, all the substances into which we are capable, by any means, to reduce bodies by decomposition. Not that we are entitled to affirm, that these substances we consider as simple may not be compounded of two, or even of a greater number of principles; but, since these principles cannot be separated, or rather since we have not hitherto discovered the means of separating them, they act with regard to us as simple substances, and we ought never to suppose them compounded until experiment and observation has proved them to be so.

Thus, by the close of the eighteenth century, chemists had a list of operationally defined, generally accepted chemical elements and a mass-based criterion for analyzing and interpreting chemical change. The chemical balance had become the defining instrument of chemical philosophy. Although Lavoisier appreciated that weight measurements could give no direct evidence of "those simple and indivisible atoms of which matter is composed," others were to make the suggestion that combining weights, especially multiple combining weights, provided convincing evidence for chemical atoms. The experiments that are especially relevant to this development were to come from John Dalton, Thomas Thomson, William Hyde Wollaston, and Jacques Bérard.

JOHN DALTON AND INTEGRAL MULTIPLE COMBINING PROPORTIONS

Despite Lavoisier's revolutionary contributions to chemistry, John Dalton is probably the historical figure best known to modern chemists. This is largely because the logical reconstruction of historical ideas is such an effective pedagogical tool that Dalton's atomic theory is presented as the foundation of chemical knowledge in the opening pages of most introductory chemistry textbooks. The genesis of Dalton's atomic theory has been well studied,[4] and we now know that his early tables of relative atomic weights emerged from his meteorological interests and investigations of gas solubilities in water. Although he was not able to correlate a substance's solubility in water with its calculated atomic weight, Dalton began to use his relative atomic weights for a far more interesting purpose.

By melding the idea of chemical atoms of fixed atomic weights with concurrent researches on the combining proportions of gases, Dalton was able to formulate an atomic hypothesis of compound formation that made sense of known examples of multiple proportions. In addition, his emerging hypothesis made specific predictions about the ratios to be expected between the combining atoms, predictions that were not an essential feature of affinity theory,[5] the prevailing explanation of combining proportions. Dalton stated the mature version of his combining proportions theory in 1810:[6]

> ... [w]e shall find the more complex compounds to consist of 3, 4 or more elementary principles, particularly the salts; but in these cases, it generally happens that one compound atom unites to one simple atom, or one compound to another compound, or perhaps to two compound atoms; rather than 4 or 6 simple elementary atoms uniting in the same instant. Thus the law of chemical synthesis is observed to be simple, and always limited to small numbers of the more simple principles forming the more compound.

Dalton's earliest investigations of combining proportions, among his first chemistry experiments of any kind, were carried out in 1803. Three interesting entries appear in his laboratory notes.[7] The first, dated 21 March, 1803, reports, "Nitrous gas—1.7 or 2.7 [parts] may be combined with [one part of] oxygen, it is presumed."[8] The second, dated 4 August, 1803, reports, "It appears, too, that a very rapid mixture of equal parts com. air and nitrous gas, gives 112 or 120 residuum. Consequently that oxygen joins to nitrous gas sometimes 1.7 to 1, and at other times 3.4 to 1."[9] The third entry appears in a part of the notebook recorded between 10 October and 13 November 1803 (a few weeks after Dalton had compiled his first table of atomic weights): "It appears that 100 com. air + 36 nit. gas give 79 or 80 azot, and that 100 com. air to 72 nit. gas, in a broad vessel, and suddenly mixed, also give 79 or 80 azot."[10]

These results reveal that Dalton had been able to combine nitrous gas with oxygen under two different sets of conditions, with differing consumptions of nitrous gas by the same volume of oxygen. In the first result, the ratio of nitrous gas consumed under one set of conditions to the amount consumed under the second was 2.7 : 1.7 or 1.6 : 1, in the second, it was 3.4 : 1.7 or 2 : 1. This last result was published in a more informative way in 1805:[11]

> 2. If 100 measures of common air be put to 36 of pure nitrous gas in a tube 3/10 ths of an inch wide and 5 inches long, after a few minutes the whole will be reduced to 79 or 80 measures, and exhibit no signs of either oxygenous or nitrous gas.

3. If 100 measures of common air be admitted to 72 of nitrous gas in a wide vessel over water, such as to form a thin stratum of air, and an immediate momentary agitation be used, there will, as before, be found 79 or 80 measures of pure azotic gas for a residuum.

Several points are interesting in this experimental result. Most importantly, the integral ratio of 2 : 1 observed for the combining proportions convinced Dalton that nitrous gas did combine with oxygen in whole-number ratios. In his words,[12]

> [t]hese facts clearly point out the theory of the process: the elements of oxygen may combine with a certain portion of nitrous gas, or with twice that portion but no intermediate quantity. In the former case *nitric* acid is the result; in the latter *nitrous* acid: but as both these may be formed at the same time, one part of the oxygen going to *one* of nitrous gas, and another to *two*, the quantity of nitrous gas absorbed should be variable; from 36 to 72 percent for common air. This is the principle cause of the diversity which has so much appeared in the results of chemists on this subject.

Here Dalton recognizes his combining ratios as the two extremes of an observational continuum. Oxygen can combine with a certain quantity of nitrous gas to form the more highly oxygenated nitric acid and with a doubled quantity to form the less oxygenated nitrous acid (the presence of water always being assumed). With intermediate quantities of nitrous gas, oxygen reacts to form mixtures of the two acids. Although several of his contemporaries were considering similar issues emanating from a variety of analytical results,[13] Dalton's 1805 publication is generally cited as the first unambiguous example of multiple combining proportions.[14]

A second point of interest is the accuracy of the results. Although simple apparatus could be used (some sort of pneumatic trough and a few glass vessels), only one reaction vessel is sufficiently described as "a tube 3/10 ths of an inch wide and 5 inches long." The other is vaguely specified as a "wide vessel."[15] As the results depended on differences in the size and shape of the reaction vessels, Dalton's experimental procedure would have been difficult to repeat. In addition, the notebook entries clarify that Dalton's procedure yielded the published results only after a sequence of repetitions with, presumably, an evolution of technique. We know of no confirmation of Dalton's results in the contemporary literature, but experimental difficulties are suggested. In an 1808 paper to be discussed later, Wollaston wrote,[16]

However, since those who are desirous of ascertaining the justness of this observation [Dalton's integral combining proportions] by experiment, may be deterred by the difficulties that we meet with in attempting to determine with precision the constitution of gaseous bodies, for the explanation of which Mr. Dalton's theory was first conceived, and since some persons may imagine that the results of former experiments on such bodies do not accord sufficiently to authorize the adoption of a new hypothesis, it may be worth while to describe a few experiments, each of which may be performed with the utmost facility, and each of which affords the most direct proof of the proportional redundance or deficiency of acid in the several salts employed.

Also, in his influential 1809 publication, Gay-Lussac reported different combining proportions for the nitrous gas–oxygen reactions. Nitric acid, he claimed, was produced by the combination of 200 parts nitrous gas with 100 parts of oxygen, whereas nitrous acid gas resulted from the combination of 300 parts nitrous gas with 100 of oxygen.[17] Shortly after reading the paper, Thomas Thomson alerted Dalton:[18]

He [Gay-Lussac] says your experiments on the nitrous gas eudiometer are inaccurate. Oxygen either combines with 2^{ce} its bulk of nitrous gas or with 3^{ce} its bulk. . . . I thought it right to let you know them, that you might repeat the experiments, and either constate or refute them in your next volume.

Thus, although Dalton's results were of crucial importance to the comprehension of chemical combination, the experiments that generated them were remarkably ill-suited to their purpose. The reason Dalton chose the nitrous gas–oxygen reactions constitutes a third point of interest. As mentioned, Dalton was primarily interested in studying the gases of the atmosphere when he began chemical experiments in 1802. In his second paper on the topic, he states his purpose:[19]

The objects of the present essay are,

1. To determine the weight of each simple atmosphere, abstractedly; or, in other words, what part of the weight of the whole compound atmosphere is due to azote; what to oxygen, &c.&c.

2. To determine the relative weights of the different gases in a given volume of atmospheric air, such as it is at the earth's surface.

3. To investigate the proportions of the gases to each other, such as they ought to be found at different elevations above the earth's surface.

Whereas in his first paper on atmospheric gases Dalton had reported literature values for the proportions of gases in the atmosphere,[20] in the second paper, he cites his own experimental results. The first gases studied were oxygen and azote, the principal components of the atmosphere. Dalton begins by listing the five methods in general use for determining the amount of oxygen in atmospheric air, the first of which was the mixture of nitrous gas and air over water. Of this process, Dalton wrote,[21]

> As the first of the processes above-mentioned has been much discredited by late authors, and as it appears from my experience to be not only the most elegant and expeditious of all the methods hitherto used, but also as correct as any of them, when properly conducted, I shall, on this occasion, animadvert upon it.

Dalton's animadversion, in the form of the nitrous gas–oxygen experiments reported earlier, followed immediately. Dalton's rationale for his choice of analytical method contains three contributing factors. The first, curiously, was the fact that the method was "much discredited by late authors." Why would a discredited method appeal to Dalton? Perhaps he wished to rehabilitate the process discovered by Joseph Priestley. In a recent book that sheds an interesting light on this issue, Golinski emphasizes the resonances between Priestley's and Dalton's chemical work,[22] and good reason underlies the belief that Dalton admired Priestley's work.

The second feature of the chosen procedure was that Dalton found it to be the "most elegant and expeditious." Expeditious the procedure certainly is, for the complete process, including the measurement of volume changes accompanying the reactions, can be completed in 10 minutes or less. However, the process is elegant only if the word is used to denote simplicity of materials and apparatus. We know that both Dalton and Priestley had a special affection for experiments that required a minimum of costly apparatus. The only reagent required was nitrous gas, which could be obtained in pure form by pouring nitric acid over copper metal.

Last, Dalton had found the procedure to be "as correct as any of [the others], when properly conducted." This probably means that when Dalton used the process to measure the amount of oxygen present in atmospheric air, it produced results that compared favorably with the other four commonly used procedures. Insofar as the methods are compared with respect to their ability to remove oxygen completely from a given quantity of air, Dalton's chosen method can be made to give reliable results.

Why, then, had it become "much discredited"? Probably because the amount of nitrous gas required to remove oxygen from atmospheric air was variable. It was well-known that nitrous gas combined with oxygen over water to produce nitrous acid, or nitric acid, or a mixture of both, depending on the nitrous gas–oxygen ratio. These aqueous reaction products could not be isolated, nor could their oxygen content be measured. Thus, the process did not allow the passage of oxygen from starting materials to products to be followed quantitatively. As an analytical procedure, in which the weight equivalence of starting materials and products was to be sought, the nitrous gas reaction with oxygen had reason to be discredited.

Clearly, from Dalton's presentation, he was not interested in the weight relationships of the reaction, nor was he seeking to verify that atmospheric air contained 21% oxygen. He came to favor the nitrous gas test for oxygen, I believe, because *it was the only procedure that could be made to give evidence of multiple combining proportions*, for a well-known fact was that oxygen combined with azote in different amounts to produce nitrous oxide, nitrous gas, and nitric acid.

Several entries in Dalton's notebooks in 1803 make use of Cavendish's, Lavoisier's, and Davy's analyses of the three azote-oxygen compounds, all of which recognized the fact that they differed only in the proportions of oxygen and azote.[23] Although we now know through the researches of Nash, Thackray, and others that Dalton's measurement of multiple combining proportions was not the immediate cause of his 1803 atomic theory, the integral ratio of the combining proportions took on added meaning when viewed from an atomic perspective. Thus the nitrous gas reaction had evolved from a mere test for oxygen in March 1803 (before Dalton's first table of atomic weights) to an important example of binary combination in whole-number proportions (consistent with atomic theory) by October 1803. By the time Dalton's 1802 paper was published in 1805, the nitrous gas test and the combining proportions it was claimed to reveal had become a centerpiece of Dalton's experimental work.

The relationship of the nitrous gas–oxygen results to atomic theory constitutes a final point of interest in the reaction. By late 1803, just prior to his first lecture to the Royal Institution (22 December, 1803), Dalton had begun comparing analyses of the azotic oxides with theoretical values based on atomic theory. In a notebook entry dated

22 December, 1803, Dalton compared Davy's measured compositions to his own theoretical values.[24]

DAVY'S EXPTS

	Azote	Oxygen
Nitrous oxide	63.3	36.7
Nitrous gas	44.05	55.95
Nitric acid	29.5	70.5

Proportions of compounds according to theory

	Azote	Oxygen
Nitrous oxide	62	38
Nitrous gas	42.1	57.9
Nitric acid	26.7	73.3

The theoretical proportions had been calculated by combining relative atomic weights with assumed atomic combinations of nitrous oxide (two azote + one oxygen), nitrous gas (one azote + one oxygen), and nitric acid (one azote + two oxygen).[25] The compositions of these compounds correlated well with Dalton's evolving ideas on chemical combination. In his 1802 paper, Dalton explained the homogeneity of the atmosphere by preparing what has come to be known as his *first theory of mixed gases*: "When two elastic fluids, denoted by A and B, are mixed together, there is no mutual repulsion among their particles; that is, the particles of A do not repel those of B, as they do one another."[26]

From this novel idea emerged Dalton's rules for determining relative atomic weights and subsequent recognition of the importance of multiple proportions. These were:[27]

> If there are two bodies, A and B, which are disposed to combine, the following is the order in which the combinations may take place, beginning with the most simple: namely,
>
> 1 atom of A + 1 atom of B = 1 atom of C, binary
> 1 atom of A + 2 atoms of B = 1 atom of D, ternary
> 2 atoms of A + 1 atom of B = 1 atom of E, ternary
>
> The following general rules may be adopted as guides in all our investigations respecting chemical synthesis.
>
> 1st. When only one combination of two bodies can be obtained, it must be presumed to be a *binary* one, unless some cause appear to the contrary.

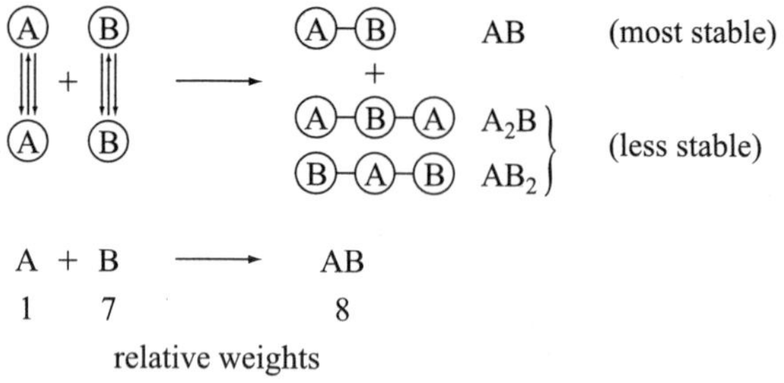

Figure 10.1
Dalton's view of chemical combination.

2nd. When two combinations are observed, they must be presumed to be a *binary* and a *ternary*.

3rd. When three combinations are obtained, we may expect one to be a *binary*, and the other two *ternary*.

The application of these rules is illustrated in figure 10.1.

In his first theory of mixed gases, Dalton had suggested that like atoms repel each other but unlike atoms do not. Thus, if atom A were to combine with atom B, the simple binary compound AB would be the most stable because no interatomic repulsions would be present. The ternary compounds A_2B and AB_2 would be less stable because of the like-atom repulsions but would be more stable than quaternary or higher-order combinations. In a lecture to the Manchester Literary and Philosophical Society in 1830, Dalton reported,[28]

> So far back as the year 1803 I had resolved in my mind the various combinations then known of azote and oxygen, and had determined almost without doubt, that nitrous gas is a binary compound and nitrous oxide a ternary as may be seen in the table of atoms published at the conclusion of my essay on the absorption of gases by water (*Manchester Memoirs*, volume 1, new series).

The azotic oxides, so familiar to Dalton from his atmospheric studies, serendipitously provided strong corroboration for his incipient atomistic theorizing. When chemical combination is viewed in such atomic terms, interesting insights follow. The first, obviously, is that the analysis of binary compounds provided Dalton with data that allowed him to construct a table of relative atomic weights. If, for example, the lightest element, A, combines with a fixed weight of B to form AB, the relative weight of atom B can be determined. More importantly for the theme of this chapter, the ratio of the amount of substance B in AB_2 relative to that in AB is predicted to be *exactly* $2:1$. The atomic view of multiple proportions thus makes a specific prediction about the value of the multiple proportion: It must be an integer, and for the simplest proportion, equal to 2. As Thackray has noted, "It was this *a priori* conviction about the proportions in which particles of different chemicals will combine which guided, sustained—and in some cases even pre-determined—the results of Dalton's chemical inquiries."[29]

Thackray is correct in emphasizing the importance to Dalton's theorizing of multiple proportions, but neither he nor others stress the crucial requirement that the combining ratio be exactly equal to two. I believe that Dalton's recognition of this fact made the nitrous gas test such a key experiment for him. If nitrous gas is formulated as NO, nitrous oxide as N_2O, and nitric acid as NO_2 (as Dalton did), published analyses of the three compounds should approximate the required integral ratios. Using Davy's analyses quoted earlier in this chapter, one gets the analytical formulas as NO (nitrous gas), $N_{2.20}O$ (nitrous oxide), and $NO_{1.88}$ (nitric acid), an approximate but unconvincing fit. Others, such as Berzelius, were later to take corresponding (but more accurate) analyses of metal oxides as evidence for integral proportions, but we must conclude that the ratios available to Dalton by analysis of the azotic oxides did not adequately suit his purposes. It seems probable that he felt that the exact $2:1$ ratio obtained by the two combinations of nitrous gas with oxygen provided a more compelling result, one that he himself had discovered. On this view, we can understand why Dalton took such effort to refine the experimental conditions of the nitrous gas–oxygen reactions until the desired $2:1$ ratio in combining proportions was obtained, and why the successful experiments received such prominence in his 1805 paper.

In conclusion, the nitrous gas–oxygen reactions were central to the development of Dalton's atomic theory, and the reported integral combining proportions appeared to corroborate, for the azotic oxides at least, the predictions made by atomic theory about chemical combination. However, the experiment failed to convince others, and no evidence corroborates that anyone else was able to replicate the results. Even had others been able to replicate Dalton's experiments, doubt remains as to whether his contemporaries would have been willing to accept a law of combining proportions constructed on a single experimental example. In 1808, however, new evidence appeared in the literature to give atomic theory added respectability and reduced the importance of Dalton's experiment.

Thomas Thomson and Salt Chemistry

Development of Dalton's atomic theory in the first quarter of the nineteenth century owes a great deal to Scottish chemist Thomas Thomson, who was the first well-known chemist to adopt the key points of the theory and became its principle popularizer. Thomson recounted his first exposure to Dalton's ideas as follows:[30]

> In the year 1804, on the 26th of August, I spent a day or two at Manchester, and was much with Mr. Dalton. At that time he explained to me his notions reporting the composition of bodies. I wrote down at the time the opinions which he offered, and the following account is taken literally from my journal of that date:
> The ultimate particles of all simple bodies are atoms incapable of further division. These atoms (at least viewed along with their atmospheres of heat) are all spheres, and are each of them possessed of particular weights, which may be denoted by numbers.
> It was this happy idea of representing the atoms and constitution of bodies by symbols that gave Mr. Dalton's opinions so much clearness. I was delighted with the new light which immediately struck my mind, and saw at a glance the immense importance of such a theory, when fully developed.

The last sentence in this account is especially interesting, for it illustrates the strong appeal an atomic theory of chemical combination could have on a prepared mind, as Thomson's undoubtedly was. As we have learned from Mauskopf's report on the subject,[31] Thomson was an acknowledged authority on combining proportions by nature of his discussions on affinity theory in the *Supplement* to the third (1797) edition of the *Encyclopaedia Britannica* and the first two editions (1802, 1804) of his popular textbook, *A System of Chemistry*.

By 1804, many known examples demonstrated that two substances combined with each other in more than one weight proportion to form well-characterized compounds, such as the metallic and non-metallic oxides and acidic and neutral salts. Thomson himself had published in 1804 an analysis of the oxides of lead, in which the weight ratio of oxygen to 100 parts of lead ranged from 10.6 in the yellow protoxide to 13.6 in the red deutoxide to 25 in the brown peroxide

but, like others at the time, saw no significance in the oxygen proportions.[32]

Prior to his encounter with Dalton, Thomson believed that definite proportions could be explained as a consequence of the geometrical shape of the particles involved in combination. The combination
of substance *a* with substance *b* could produce both a minimum proportion of *a* to *b*, in which all the particles of *a* are saturated by those
of *b*, and a maximum proportion, in which all the particles of *b* are
saturated by those of *a*. This model, however, could not readily be
extended to account for instances of a third, or higher, value of combined proportion, such as those found for the lead oxides. However, in
the third (1807) edition of *A System of Chemistry*, Thomson replaced
his geometrical rationale of multiple proportions with Dalton's atomic
interpretation which, as we have seen, easily accommodated a series of
multiple proportions and made specific predictions about the values of
the simplest proportions (i.e., they must be whole numbers).

In the process of promulgating his new views on chemical combination, Thomson acknowledged Dalton as the originator of these
views and went on to present the first published account of Dalton's
atomic theory. However, Thomson was aware that experimental support for the theory was not adequate. As he was later to say,[33]

> Let not the reader suppose that this [Dalton's calculation of relative
> atomic weights] was an easy task. Chemistry at that time did not pos-
> sess a single analysis which could be considered as even approaching
> to accuracy. A vast number of facts had been ascertained, and a fine
> foundation laid for future investigation; but nothing, as far as weight
> and measure were concerned, deserving the least confidence, existed.
> We need not be surprised, then, that Mr. Dalton's first numbers
> were not exact. It required infinite sagacity, and not a little labour, to
> come so near the truth as he did. How could accurate analyses of
> gases be made when there was not a single gas whose specific gravity
> was known, with even an approach to accuracy; the preceding
> investigations of Dalton himself paved the way for accuracy in this
> indispensable department; but still accurate results had not yet been
> obtained.

Although it is likely that Thomson in 1830 exaggerated somewhat the analytical uncertainties to emphasize the impact of his own
work on multiple proportions (which was presented immediately after
the foregoing extract), truly neither Dalton's nitrous gas proportions
nor the known examples of multiple proportions in oxides demanded

theoretical reinterpretation. What Dalton's theory required, at least in the context of multiple combining proportions, was accurate analytical data that yielded integral values for combining proportions, ideally by mainstream experiments that could be replicated with confidence. Such results appeared in two successive papers in the opening 1808 volume of *Philosophical Transactions*, the first of which was Thomson's study of oxalic acid.[34]

Oxalic acid was a well-known, naturally occurring vegetable acid that had attracted chemical interest both for its unusually high oxygen content and for its ability to combine with alkalies in two different proportions. It had been first noted by Scheele, and Aikins's *Dictionary* (1807) had a lengthy entry for the substance. A relevant excerpt reports,[35]

> The oxalic acid unites with all the alkalies, earths and most metallic oxides, forming salts of which a very few only are of importance. This acid like the tartarous, combines with the alkalies in two different proportions, in each of which it forms crystallizable salts. One of these states is with the acid in excess, forming therefore salts known by the general term of *acidula*, the other is when the two ingredients are in perfect saturation. Of these the *Acidulous oxalat of potash*, or according to Dr. Thompson's [sic] useful nomenclature, the Super-oxalat of potash, has been examined with the greatest care.

Thomson's article reported his analyses of the acid and a variety of its salts, a study of its decomposition products, and an interesting atomistic interpretation of the results.[36] Of relevance here is the inclusion of two examples of integral multiple proportions in oxalate salts.

After discussing the physical properties of the oxalate of potash, Thomson reported his first example of integral proportions:[37]

> This salt [the neutral oxalate of potash] combines with an excess of acid, and forms a superoxalate, long known by the name of *salt of sorrel* ... The acid contained in this salt is very nearly double of what is contained in oxalate of potash. Suppose 100 parts of potash; if the weight of acid necessary to convert this quantity into oxalate be x then $2x$ will convert it into superoxalate.

In terms of experimental observation, however, nothing is striking in Thomson's statement. The "acidulous oxalat of potash" (to use Aikins's term) was generally known to contain a quantity of acid in excess of the amount required for "perfect saturation." The statement becomes noteworthy only if one accepts uncritically Thomson's theory-driven claim that a measured proportion of acid in one salt that is "very nearly double" that in another is evidence for an integral ratio of $2x$ to x. Because no experimental technique was reported, no analytical data were given, and only a modest extension of existing knowledge was included, this first example of integral proportions has achieved historical luster only (unless we count also hindsight) by association with Thomson's second example.

An investigation of the strontian salts of oxalic acid led Thomson to the reported discovery of a new compound and to his second instance of integral proportions. Whereas Thomson's analyses of several oxalates were consistent with known values, his preparation of oxalate of strontian yielded a salt with an anomalous composition. Synthesis by a second method yielded a salt of different composition. For 100 parts of acid, one salt contained 151.51 parts of strontian, and the second contained only 75.7, for a ratio of 2.001 : 1. So soon after becoming an advocate of Dalton's theory, Thomson had discovered an instance of integral combining proportions in salt chemistry, one of the most thoroughly studied and best understood chemical topics. Thomson wrote,[38]

> Thus it appears that there are two oxalates of strontian, the first obtained by saturating oxalic acid with strontian water, the second by mixing together oxalate of ammonia and muriate of strontian. It is remarkable that the first contains just double the proportion of base contained in the second.

This result, announced in the pages of *Philosophical Transactions* just a short time after Dalton's atomic theory had been described in Thomson's third (1807) edition of *System of Chemistry*, gave the theory its first predictive success. The result has been included in the Alembic Club reprint volume *Foundations of the Atomic Theory*,[39] and Partington has written that "Thomson gave the first experimental example of the law of multiple proportions."[40] Despite such historical approbation, however, some evidence suggests that Thomson's result is in error.

Shortly after learning of Thomson's discovery, Berthollet encouraged a young colleague at Arcueil, Jacques Étienne Bérard, to verify the result. Bérard worked through Thomson's procedure but could produce only one oxalate of strontian, the composition of which was very different from that of each of Thomson's two salts. In a persuasive passage, Bérard wrote,[41]

Dr. Thomson has certainly been deceived in his calculation respecting the oxalate of strontian. It would follow indeed, from the numbers he gives, that barytes has a greater capacity for saturation than strontian, which is contradictory to all the analyses of salts with base of strontian and barytes hitherto known ...

The little solubility of the oxalate of strontian perhaps misled Dr. Thomson: but it seems to me demonstrated that, in precipitating neutral muriate of strontian by the neutral oxalate of ammonia, a salt with excess of acid cannot be formed, for the residuum remains neutral. In the next place, I do not think, that an acid oxalate of strontian exists; for I have not been able to form it, in employing the same means as for the other oxalates; and, besides, the neutral oxalate of strontian is very little soluble in an excess of its acid. Lastly, since the proportions he gives for the neutral oxalate are not accurate, his simple ratio between the neutral and acid oxalate is done away.

Bérard's critique, published in one of England's most widely read journals, must have been damaging to Thomson's discovery, for beyond claiming an inability to obtain the purported result, Bérard presented sound experimental and theoretical reasons why the result was at odds with established views. Thomson was certainly aware of the problem, for he mentioned it in his 1825 textbook:[42]

Bérard, in a paper on the oxalates, which he published by way of correction of mine, though his results are in general more erroneous than my own, denies the existence of binoxalate of strontian. I have often made the salt, and regularly exhibit it to my students. It differs very much from the oxalate.

A revealing note is that Thomson does not counterbalance Bérard's criticism by reference in *First Principles* to any corroborating evidence. To our knowledge, no one else has ever been able to make two oxalates of strontian.[43] The questionable validity of Thomson's integral proportion has escaped the attention of historians, perhaps because the result supported a law now believed to be correct but more likely because Thomson's paper was made redundant by one that immediately followed it in *Philosophical Transactions* by William Hyde Wollaston.

WILLIAM HYDE WOLLASTON AND INTEGRAL PROPORTIONS

William Hyde Wollaston had abandoned his London medical practice in 1800 to form a partnership with Smithson Tennant for the production and sale of chemical products. Largely because of his skills as a keen and careful analyst, Wollaston was able to produce and market malleable platinum from the crude alluvial ore, in the process discovering the new elements palladium and rhodium.[44] Another, less financially successful initiative was the synthesis and marketing of tartaric acid, oxalic acid, and oxalates, especially superoxalate of potash (salt of sorrel). This work, begun in 1802, led Wollaston to an investigation of the proportions of oxalic acid and potash in their salts to gain better control over the chemical processes required to generate pure product from the wine lees (argol) used as starting material. Events in 1807, however, caused Wollaston to look at his analytical data in a new light.

As a secretary of the Royal Society and as a member of the Society's Committee of Papers (which recommended which of the papers read to the Society were to be printed in *Philosophical Transactions*), Wollaston was at the information hub of British science. He had learned the details of Dalton's theory from Thomson's 1807 *System of Chemistry* and had been impressed by its explanatory power. Thomson claimed Wollaston as a strong advocate of the theory from the outset:[45]

> There were, however, some of our most eminent chemists who were very hostile to the atomic theory. The most conspicuous of these was Sir Humphry Davy. In the autumn of 1807 I had a long conversation with him at the Royal Institution, but could not convince him that there was any truth in the hypothesis. A few days after I dined with him at the Royal Society Club, at the Crown and Anchor, in the Strand. Dr. Wollaston was present at the dinner. After dinner every member of the club left the tavern, except Dr. Wollaston, Mr. Davy, and myself, who stayed behind and had tea. We sat about an hour and a half together, and our whole conversation was about the atomic theory. Dr. Wollaston was a convert as well as myself; and we tried to convince Davy of the inaccuracy of his opinions. . . .

On learning that Thomson was to have his paper read to the Royal Society on 14 January 1808, Wollaston quickly put his paper "On Super-acid and Sub-acid Salts" together for reading on 28 January, 1808. It was published immediately after Thomson's in *Philosophical Transactions* and presented a compelling argument for the generality of integral multiple proportions and Dalton's theory. Wollaston's opening paragraphs are worth citing:[46]

In the paper which has just been read to the Society, Dr. Thomson has remarked, that oxalic acid unites to strontian as well as to potash in two different proportions, and that the quantity of acid combined with each of these bases in their super-oxalates, is just double of that which is saturated by the same quantity of base in their neutral compounds.

As I had observed the same law to prevail in various other instances of super-acid and sub-acid salts, I thought it not unlikely that this law might obtain generally in such compounds, and it was my design to have pursued the subject with the hope of discovering the course to which so regular a relation might be ascribed.

But since the publication of Mr. Dalton's theory of chemical combination, as explained and illustrated by Dr. Thomson [in his 1807 System of Chemistry], the inquiry which I had designed appears to be superfluous, as all the facts that I had observed are but particular instances of the more general observation of Mr. Dalton, that in all cases the simple elements of bodies are disposed to unite atom to atom singly, or, if either is in excess, it exceeds by a ratio to be expressed by some simple multiple of the number of its atoms.

The facts that Wollaston had observed were two carbonates of potash, two carbonates of soda, two sulfates of potash, all with acid-to-base integral proportions of $2:1$, and three oxalates of potash, with integral proportions of $4:2:1$. A summary of his results (with the modern formula of each of the compounds added) is given in table 10.1.

Wollaston's experiments, concisely but clearly described in the paper, were brilliantly conceived. Whereas Thomson had sought to

Table 10.1
Wollaston's salt analyses (1808)

Name	Base : acid	Compound
Subcarbonate of potash	1 : 1	K_2CO_3
Carbonate of potash	1 : 2	$KHCO_3$
Subcarbonate of soda	1 : 1	Na_2CO_3
Carbonate of soda	1 : 2	$NaHCO_3$
Oxalate of potash	1 : 1	$K_2C_2O_4$
Superoxalate of potash	1 : 2	KHC_2O_4
Quadroxalate of potash	1 : 4	$KHC_2O_4 \cdot H_2C_2O_4 \cdot 2H_2O$
Sulfate of potash	1 : 1	K_2SO_4
Supersulfate of potash	1 : 2	$KHSO_4$

determine the proportions of acid and base present in salts by measuring the amounts that would *combine* with each other in a synthetic reaction, Wollaston chose to determine the proportions present in the salts by *uncombining* them (i.e., by analytical processes). An example is his procedure for the well-known carbonates of potash:[47]

> Let two grains of fully saturated and well crystallized carbonate of potash be wrapped in a piece of thin paper, and passed up into an inverted tube filled with mercury, and let the gas be extricated from it by a sufficient quantity of muriatic acid, so that the space it occupies may be marked upon the tube.
>
> Next, let four grains of the same carbonate be exposed for a short time to a red heat; and it will be found to have parted with exactly half its gas; for the gas extricated from it in the same apparatus will be found to occupy exactly the same space, as the quantity before obtained from two grains of fully saturated carbonate.

This process is illustrated schematically in figure 10.2. In the first experiment, the carbonic acid contained in two grains of carbonate of potash was freed by treatment with muriatic acid, was collected and measured as a volume, and was denoted as V_{gas}. In the second experiment, four grains of carbonate of potash (which, if treated directly with acid, would release twice the volume, $2V_{gas}$, of carbonic acid) had its excess carbonic acid driven off by heating to a quantity of sub–carbonate of potash that, on treatment with muriatic acid, released a volume of carbonic acid also equal to V_{gas}. Thus, a quantity of sub–

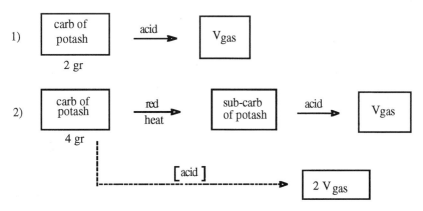

Figure 10.2
Wollaston's analysis of the carbonates of potash.

carbonate of potash, which contains a weight of potash equal to that in four grains of carbonate of potash, releases exactly one-half the volume of carbonic acid gas as four grains of carbonate of potash. In other words, for a fixed weight of potash, carbonate of potash has twice the quantity of carbonic acid as does sub–carbonate of potash.[48]

The most meaningful of Wollaston's results was the compositional relationship of the three oxalates of potash. In the course of his analyses of the two known oxalates of potash (the neutral and the acidic oxalate, salt of sorrel), Wollaston discovered a third salt that contained an even greater excess of acid than that in the acidic salt of sorrel. As shown in table 10.1, the acid content of the new salt was found to be twice that of the superoxalate and four times that of the oxalate of potash. Hence, Wollaston termed it a *quadroxate*. The $1 : 2 : 4$ ratio in acid proportions of the three salts caused him to wonder whether a fourth salt, containing a $1 : 3$ base-acid proportion existed. If a base, B, could combine with an acid, A, to form a sequence of salts of increasing instability, Dalton's rules of atomic combination suggested the sequence BA, BA_2, BA_3, BA_4, and the like, with BA the oxalate of potash, BA_2 the superoxalate, and BA_4 the quadroxalate. After a careful experimental procedure convinced Wollaston that a "tri-oxalate" did not exist (but that a solution of acid and base optimized for its production yielded instead a mixture of the superoxalate and quadroxalate), he proposed that Dalton would explain the results as an example of the B_2A, BA, BA_2 sequence:[49]

> To account for this want of disposition to unite in the proportion of three to one by Mr. DALTON'S theory, I apprehend he might consider the neutral salt as consisting of
>
> | | | 2 particles potash | with 1 acid | $[B_2A]$ |
> |-------------------------------------|--------|--------------------|-------------|------------|
> | the binoxalate as 1 and 1, | or 2 | | with 2 | $[B_2A_2]$ |
> | the quadroxalate as 1 and 2, | or 2 | | with 4 | $[B_2A_4]$ |
>
> in which cases the ratios which I have observed of the acids to each other in those salts would respectively obtain.

After this description of four new, compelling examples of integral multiple proportions and a discussion of their compositional characteristics in terms of atomic theory, Wollaston mused over the impact of atomic theory on crystallography and polyatomic geometry:[50]

> ... I am further inclined to think, that when our views are sufficiently extended, to enable us to reason with precision concerning

the proportions of elementary atoms, we shall find the arithmetical relation alone will not be sufficient to explain their mutual action, and that we shall be obliged to acquire a geometrical conception of their relative arrangement in all the three dimensions of solid extension . . .

But as this geometrical arrangement of the primary elements of matter is altogether conjectural, and must rely for its confirmation or rejection upon future inquiry, I am desirous that it should not be confounded with the results of the facts and observations related above, which are sufficiently distinct and satisfactory with respect to the existence of the law of simple multiples.

Wollaston's short, impressive paper is fundamental to the development of atomic theory for a number of reasons. It presented a series of examples of integral combining proportions together with elegant procedures for their measurement. It gave Dalton's theory the stamp of approval of one of Europe's most respected scientists and noted the potential extension of atomic hypotheses to polyatomic compounds. By combining elements of gas eudiometry with gravimetrical analysis, the paper also established analysis as a key experimental feature of atomic theorizing.[51]

By explicitly noting the independence of his experimental results from atomic conjectures, however, Wollaston's paper fueled the positivistic concerns that encouraged the preference for such terms as *volumes*, *proportions*, and *equivalents* instead of *atoms*.[52] Finally, the paper, when read on the continent, carried atomic theory to an interested but underinformed and occasionally skeptical audience.

JACQUES ETIENNE BÉRARD AND THE FRENCH RESPONSE

Jacques Etienne Bérard was a young man of 18 when he arrived in Paris in 1807 to learn chemistry under the tutelage of Berthollet.[53] Around that time, Berthollet's interest in combining proportions, a phenomenon that had to be accommodated by his affinity theory, had been renewed by the appearance of Thomson's 1807 *System of Chemistry* and of Thomson's and Wollaston's 1808 papers on multiple proportions. Berthollet wrote a lengthy introduction to Riffault's French translation of Thomson's third edition, which appeared in 1809. In it, Berthollet acknowledged the crucial importance of accurate analyses:[54]

[Dalton's hypothesis] provides an explanation of a phenomenon, the cause of which has been very obscure until now; but the more seductive this hypothesis is, the more necessary it becomes to submit it to close examination ...

In order to pass judgement on Dalton's ingenious hypothesis a much more profound discussion would be necessary.... It would be preferable to bring greater accuracy into our experimental results ... than to base our reasoning on hypothetical speculations on the number, the arrangement and the figure of the particles which escape all experience.

Berthollet carried out and published some analyses of salts himself in 1809[55] but assigned a broader investigation to Bérard. In his first publication on the topic, Bérard announced the motivation for his work:[56]

> The accurate determination of the component parts of saline sub-stances is of more importance, because it is employed as the basis of other chemical analyses. Berthollet, who has sought to determine some of these in his late papers, was desirous, that they should be carried to the highest degree of accuracy; and invited me to resume the subject, reiterating the experiments, varying the methods, and taking the greatest care to avoid every source of error. This I have endeavoured to do, and at the same time I extended my observations to a greater number of compounds.

The paper then gives an experimentally rigorous procedure for the analysis of the nitrate of potash, the sulfate of barytes, the muriates and sulfates of potash and soda, and the carbonates and subcarbonates of potash and soda. Interestingly, although Wollaston's analyses of the carbonates is cited, Bérard reports his own proportions as percentages, thus hiding the nearly integral proportions (ratios of 2.01 and 2.09 for the carbonates of potash and soda, respectively) present in some of the salts. Bérard's methods bear testimony to the quality of the work done at Arcueil, for he took great care to ensure that he worked with pure substances, analyzed most of his salts by two or three different methods, did each analysis four times and averaged the results, and checked his results against published values.[57]

Bérard's first publication soon was followed by a second one, which focused on all the known salts of oxalic acid.[41] This report, composed after Bérard had read the complete versions of Thomson's and Wollaston's papers, confirmed Wollaston's work on the oxalates of potash, refuted Thomson's on the oxalate of strontian, and provided three new examples of multiple proportions, each of which exhibited

Table 10.2
Bérard's salt analyses (1810)

Name	Base : acid	Compound
Oxalate of potash	1 : 1	$K_2C_2O_4$
Superoxalate of potash	1 : 1.97	KHC_2O_4
Quadroxalate of potash	1 : 3.92	$KHC_2O_4 \cdot H_2C_2O_4 \cdot 2H_2O$
Oxalate of soda	1 : 1	$Na_2C_2O_4$
Superoxalate of soda	1 : 1.98	$NaHC_2O_4$
Oxalate of ammonia	1 : 1	$(NH_4)_2C_2O_4$
Superoxalate of ammonia	1 : 2.00	$NH_4HC_2O_4$
Oxalate of barytes	1 : 1	BaC_2O_4
Superoxalate of barytes	1 : 2.02	$Ba(HC_2O_4)_2$

integral values. A tabular summary of Bérard's results, again with the modern formulas added, is given in table 10.2.

The values Bérard obtained for the multiple proportions exhibited by the oxalates of soda (1.98 : 1), oxalates of ammonia (2.00 : 1), and oxalates of barytes (2.02 : 1) must have been impressive to anyone who had not already been convinced of the truth of the law of multiple proportions. Nonetheless, Bérard pointed out that his observations could be accommodated within Berthollet's theories of cohesion:[58]

> All the oxalates have not the property of combining with an excess of acid, as my experiments show. It is the force of cohesion of the acid, combined with that of the alkali, which determines the existence of the superoxalates.
>
> In fact, the great number of insoluble salts, which the oxalic acid forms with the bases, tends to prove, that this acid possesses great force of cohesion. To this quality is owing its property of forming with the soluble alkalis salts with excess of acid less soluble than the neutral salts . . .
>
> The conclusions, that may be drawn from the observations I have here submitted to the judgement of the class, are:
>
> 1st. That the soluble oxalates alone are capable of taking up an excess of acid, and forming salts less soluble than the neutral salts.
>
> 2nd. That the property of forming superoxalates depends on the force of cohesion of the acid, combined with that of the alkali.
>
> 3rd. That potash is the only alkali capable of forming a quadroxalate.
>
> 4th. That, in all the superoxalates, the alkali is constantly combined with twice as much acid as in the corresponding neutral oxalate.

Bérard's work, almost totally unknown to historians,[59] brought to a close debate about the existence of multiple combining proportions. Because they were seen by many as little more than a verification of Wollaston's work, his new examples with their remarkable accuracy and precision were largely overlooked by Bérard's contemporaries. However, his experiments, conducted under Bethollet's supervision, were probably an important factor in the evolution of the great chemist's law of mass action.[60]

CONCLUSION

Experimental design and the quantification of scientific phenomena are often, usually uncritically, cited as the foundation of an inductive process that can lead to a law of nature. This chapter has looked at the details of the multiple proportions research of Dalton, Thomson, Wollaston, and Bérard. In this study, the best experimental work (Bérard's)—in the sense of corresponding best with the predictions of atomic theory—was performed after the law of multiple proportions had been conceived (by Dalton), loosely supported (by Thomson), and experimentally established (by Wollaston). Likely, Dalton was able to achieve his nitrous gas–oxygen results only under the impress of his preconceived theory, a suggestion that may apply also to Thomson. Wollaston, on the other hand, had begun to collect his analytical results in the absence of any known theoretical bias but recognized their significance only after learning of earlier work. Finally, Bérard, whose data could genuinely serve as the foundation for a quantized (if not necessarily a particulate) theory of chemical combination, saw his work undervalued by a community that had already accepted the presumed theoretical basis.

Although the proposition that multiple combining proportions was a key experimental component of the construction of Dalton's atomic theory cannot be sustained (nor has it ever been so proposed by a historian other than Meldrum[61]), the accumulation of analytical data by ever-improving experimental techniques became an essential corroborating example of the theory. Furthermore, competing theories of chemical combination, such as that of elective affinities, that could not adequately account for a series of integral proportions became increasingly untenable in the face of the analytical results. Although the creation of "laws of nature" is a complex social process, the experimental results here discussed regarding multiple combining proportions demonstrate that experiments are of critical importance.

NOTES

1. Antoine-Laurent Lavoisier, *Elements of Chemistry*, translated by Robert Kerr (Edinburgh: 1790; Dover reprint, New York, 1965), 130–131.

2. Ibid., 175–176.

3. Ibid., *xxiv*.

4. L. K. Nash, "The Origin of Dalton's Chemical Atomic Theory," *Isis* 47(1956): 101–116; A. W. Thackray, "The Emergence of Dalton's Chemical Atomic Theory: 1801–08," *British Journal for the History of Science* 3(1966): 1–23; and A. W. Thackray, "The Origin of Dalton's Chemical Atomic Theory: Daltonian Doubts Resolved," *Isis* 57(1966): 35–55.

5. See L. A. Whitt, "Atoms or Affinities? The Ambivalent Reception of Daltonian Theory," *Studies in History and Philosophy of Science* 21(1990): 57–89.

6. John Dalton, *A New System of Chemical Philosophy*: II. (Manchester, 1810; facsimile edition undated, London: Dawson & Sons), 222–223.

7. Dalton's lecture and laboratory notes, housed in the library of the Manchester Literary and Philosophical Society, were destroyed in 1940 during the war. The information quoted is taken from H. E. Roscoe and A. Harden, *A New View of the Origin of Dalton's Atomic Theory* (London: 1896; reprinted in 1970, New York: Johnson Reprint).

8. Ibid., 34.

9. Ibid., 38.

10. Ibid., 35.

11. J. Dalton, "Experimental Enquiry into the Proportions of the Several Gases or Elastic Fluids Constituting the Atmosphere," *Manchester Memoirs* (2nd series) 1(1805), 244–258. Reprinted in *Alembic Club Reprints* 2(5–15), (reissue edition, 1969), 8–9. This paper, obviously in a preliminary form, was read to the Manchester Literary and Philosophical Society on 12 November 1802. When published, it contained results from experiments carried out between 1802 and 1805.

12. Ibid., 9.

13. For a discussion of Dalton's contribution to the debate over intermediate versus continuous proportions see K. Fujii, "The Berthollet-Proust Controversy and Dalton's Chemical Atomic Theory 1800–1820," *British Journal of the History of Science* 19(1986): 177–200.

14. See, for example, Dalton, "Experimental Enquiry," *Alembic Club Reprint*, 5, and Freund, *The Study of Chemical Composition* (Cambridge: Cambridge University Press, 1904), 155.

15. J. Dalton, in Roscoe and Harden, *A New View*, 328 refers to the wide vessel as a "common tumbler."

16. W. H. Wollaston, "On Super-acid and Sub-acid Salts," *Philosophical Transactions of the Royal Society of London* 98(1808): 97.

17. J. L. Gay-Lussac, "Mémoire sur la combinaison des substances gazeuses," *Mémoires de Physique et de Chimie de la Société d'Arcueil* 2(1809): 207–234; *Alembic Club Reprints*, vol. 4 (reprint edition, 1949) 8–24.

18. Letter from Thomas Thomson to John Dalton, 13 November 1809, quoted in Roscoe and Harden, *A New View*, 148. Because of its importance in the history of chemistry, Dalton's observation of multiple combining proportions is seen by many as a classic experiment. Both L. K. Nash [*Isis* 47(1956): 105] and J. R. Partington [*A History of Chemistry*, vol. 3 (London: Macmillan & Co., 1962), 791] report difficulties in getting Dalton's numbers in replication experiments. In a paper to be published elsewhere, Katherine D. Watson and M. C. Usselman will report a successful reconstruction of Dalton's experiments.

19. J. Dalton, "Experimental Enquiry into the Proportions," 6.

20. J. Dalton, "On the Constitution of Mixed Gases," *Manchester Memoirs of the Literary and Philosophical Society* 5(1802): 535–602.

21. J. Dalton, "Experimental Enquiry into the Proportions," 8. The other four processes were exposing air to liquid sulfuret of potash or lime; exploding hydrogen gas and air by electricity; exposing air to a solution of green sulphat or muriat of iron in water strongly impregnated with nitrous gas; and burning phosphorus in the air.

22. J. Golinski, *Science as Public Culture* (Cambridge, UK: Cambridge University Press, 1992); especially chap. 8.

23. Roscoe and Harden, *A New View*, 35, 39, 41.

24. Ibid., 44–45.

25. These atomic combinations for the three compounds are given in a notebook table dated 12 October 1803. See Ibid., 45.

26. J. Dalton, "On the Constitution of Mixed Gases," 536.

27. J. Dalton, *A New System of Chemical Philosophy*, 213–214.

28. A. Thackray, *John Dalton* (Cambridge, MA: Harvard University Press, 1972), 100.

29. Thackray, "The Emergence of Dalton's Chemical Atomic Theory," 13.

30. Thomas Thomson, *The History of Chemistry*, vol. 2 (London: Henry Colburn and Richard Bentley 1830–31); reprinted in 1975, New York: Arno Press, 289–291.

31. S. H. Mauskopf, "Thomson Before Dalton: Thomas Thomson's Considerations of the Issue of Combining Weight Proportions Prior to his Acceptance of Dalton's Chemical Atomic Theory," *Annals of Science* 25(1969): 229–242.

32. T. Thomson, "On the Oxides of Lead," *Journal of Natural Philosophy, Chemistry and the Arts* 8(1804): 280–293. In this article, Thomson introduced the terms *protoxide* for the least oxidized compound of a metal, *peroxide* for the most, and *deutoxide, tritoxide*, and the like for the second, third, and additional levels of oxidation.

33. Thomson, *History of Chemistry*, 292.

34. T. Thomson, "On Oxalic Acid," *Philosophical Transactions of the Royal Society of London* 98(1808): 63–95.

35. A. and C. R. Aikin, *A Dictionary of Chemistry and Mineralogy*, vol. 2 (London: Printed for John and Arthur Arch, Cornhill, and William Phillips, George Yard, Lombard Street 1807) 192.

36. Although outside the focus of this chapter, Thomson's atomic digression was perhaps the first to cast Lavoisier's operational definition of compound simplicity in atomic terms and, by so doing, diversified the applicability of atomic theory. He wrote ["On Oxalic Acid," 90], "[w]hen a compound body is decomposed, and resolved into a number of new substances, the products are almost always simpler, or consist of integrant particles, composed of fewer atoms than the integrant particles of the original body."

37. Thomson, "On Oxalic Acid," 70.

38. Ibid., 74.

39. Dalton, "Experimental Enquiry," *Alembic Club Reprint* 2 (1948, original printing); reissue edition, Edinburgh, 1969.

40. Partington, *History of Chemistry*, vol. 3, 719.

41. J. E. Bérard, "Observations on the Alkaline Oxalates and Superoxalates, and Particularly on the Proportions of Their Elements," *Journal of the Nature and Philosophy of Chemistry and the Arts* 31(1812): 28–29; republication of the original article in *Annales de Chimie* 73(1810): 263–289.

42. T. Thomson, *An Attempt to Establish the First Principles of Chemistry by Experiment* (London: Printed for Baldwin, Cradock, and Joy 1825), 12. The reference to Bérard was omitted in the corresponding section of Thomson's 1830 *History of Chemistry*.

43. A reconstruction of Thomson's procedure by Todd Brown and M. C. Usselman has been unable to replicate his results.

44. For an overview of Wollaston's business, see L. F. Gilbert, "W. H. Wollaston Mss. at Cambridge," *Notes and Record of the Royal Society of London* 9(1952): 311–332; and M. C. Usselman, "William Wollaston, John Johnson and Colombian Alluvial Platina: A Study in Restricted Industrial Enterprise," *Annals of Science* 37(1980): 253–268.

45. Thomson, *History of Chemistry*, 293.

46. Wollaston, "On Super-acid and Sub-acid Salts," 96.

47. Ibid., 97.

48. In work to be reported elsewhere, Todd Brown and M. C. Usselman have successfully reconstructed Wollaston's experiments and obtained near-integral proportions.

49. Wollaston, "On Super-acid and Sub-acid Salts," 101.

50. Ibid., 102.

51. Crosland, for example, suggests that the eudiometrical aspects of Wollaston's analyses may have influenced Gay-Lussac's researches. See M. P. Crosland, "The Origins of Gay-Lussac's Law of Combining Volumes of Gases," *Annals of Science* 17(1961): 1–26.

52. Rocke interprets the nomenclature variation as more of a semantic problem than a metaphysical one. See A. J. Rocke, "Atoms and Equivalents: The Early Development of the Chemical Atomic Theory," *Historical Studies in the Physical Sciences* 9(1978): 225–263.

53. M. Crosland, *The Society of Arcueil* (London: Heinemann, 1967), 134, 316.

54. C. L. Berthollet, Introduction to T. Thomson, *Système de Chimie*, 3rd ed., translated by J. Riffault (Paris: mad.Ve Bernard 1809), 21, 27; English translation is taken from M. Crosland, ed., *The Science of Matter* (Harmondsworth, UK: Penguin, 1971), 204–205.

55. C. L. Berthollet, "Sur les Proportions des Élémens de Quelques Combinaisons," *Mémoires de Physique et de Chimie de la Societé d'Arcueil* 2(1809): 42–67.

56. J. E. Bérard, "On the Proportions of the Elements of Some Combinations, particularly of the Alkaline Carbonates and Subcarbonates," *Journal of the Nature and Philosophy of Chemistry and the Arts* 26(1810): 206–207; extracted from *Annales Chim.* 71(1809): 41–69.

57. As editor of the *Journal of Natural Philosophy, Chemistry and the Arts,* Nicholson obviously did not appreciate the significance of Bérard's careful techniques for, after the passage cited in this chapter taken from Bérard's "On the Proportions," he added, "After this introduction, Mr. Berard, in a pretty long paper, gives the detail of his experiments; but as these would occupy much room to little purpose, I shall pass them over, merely giving the tabulated results, with which he concludes."

58. Bérard, "Observations on the Alkaline Oxalates," 32–33.

59. For example, Bérard is not even mentioned in Freund's *The Study of Chemical Composition* (Cambridge: Cambridge University Press, 1904).

60. Although Bérard's results are not mentioned, this issue is discussed in F. L. Holmes, "From Elective Affinities to Chemical Equilibria: Berthollet's Law of Mass Action," *Chymia* 8(1962): 105–145.

61. See Nash, "The Origin of Dalton's Chemical Atomic Theory."

BIBLIOGRAPHY

Crosland, Maurice. *The Society of Arcueil* (London: Heinemann, 1967).

Freund, Ida. *The Study of Chemical Composition* (Cambridge: Cambridge University Press, 1904). Also available as a Dover Reprint, New York, 1968.

Rocke, Alan J. *Chemical Atomism in the Nineteenth Century* (Columbus, OH: Ohio State University Press, 1984).

Thackray, Arnold. *Atoms and Powers* (Cambridge, MA: Harvard University Press, 1970).

Thackray, Arnold. *John Dalton* (Cambridge, MA: Harvard University Press, 1972).

11

ORGANIC ANALYSIS IN COMPARATIVE PERSPECTIVE:
LIEBIG, DUMAS, AND BERZELIUS, 1811–1837
Alan J. Rocke

In his famous biography of his famous mentor, August Wilhelm Hofmann wrote, "The present generation of chemists have not the remotest idea of the difficulties which attended an organic analysis before Liebig invented his bulb-apparatus."[1] As chemical analysis is the necessary starting point for building chemical science, not surprisingly many have dated the rise of organic chemistry to this event. However, Hofmann, the most prominent member of the "present generation of chemists" at the time he made this comment, can have had just as little idea of the difficulties of pre-Liebigian analysis, because he learned the art nearly a decade after Justus Liebig's invention.

An aura of the fabulous has always surrounded this event, as also surrounded the impressive transfer of chemical leadership from France to Germany in the years after 1830, an aura that has been well promoted by such fabulists as Hofmann (not to mention Liebig himself). Liebig's Kaliapparat—the "bulb-apparatus" for measuring the carbon content of the sample—even serves to this day as the iconographic heart of the American Chemical Society's logo. A close examination of the development of organic analysis circa 1830 is revealing, not only regarding the evolution of a particular type of apparatus but also about the ideas surrounding it, influenced by a cosmopolitan international culture at the end of the Romantic era.

Cosmopolitan is one of the least likely words to come to mind when one thinks of Romanticism. Romantic impulses did indeed lend a chauvinist overlay to the views of some German chemists at the beginning of the century. J. W. Döbereiner, resident in Jena, the wellspring of German Romanticism, was not alone in vilifying the French and in attempting to return from Lavoisier's oxygen to the German theory of phlogiston.[2] Ferdinand Wurzer, professor at Marburg, later told Liebig that he could generate mercury in an empty desk drawer and could transmute oxygen into nitrogen in a clay pipe.[3] Liebig's Ph.D. curriculum in Germany led to a full acquaintance with the works of the major phlogistonists but to only a superficial understanding of the chemistry of the oxygenists, despite the ostensible success of the chemical revolution in Germany 30 years earlier. The eminent Swedish chemist Jacob Berzelius had not yet begun to be influential in Germany.[4]

Notwithstanding all the anti-French sentiment and nativist theorizing, much positive French influence also flourished in German culture at this time. During the Napoleonic period, German elites imbibed much from the French occupiers. Some German courts adopted a patina of French culture; when Francophile Prussian naturalist Alexander von Humboldt wrote from Paris to the Hessian capital, Darmstadt, he wrote in French. As we see in what follows, the growth of German science in the 1820s and 1830s, as exemplified and symbolized by Liebig's development of the Kaliapparat, reflected both poles of this German ambivalence toward France.

THE GAY-LUSSAC/BERZELIUS "COLLABORATION" OF 1811 TO 1815

The first substantial steps toward a generally applicable method for determining the percentages of carbon, hydrogen, and oxygen in an organic substance—the prerequisite for calculating its chemical formula—were achieved in France.[5] On the basis of some preliminary experiments of Lavoisier, J. L. Gay-Lussac and L. J. Thenard in their mutual work of 1810-11 provided elemental analyses of 19 different organic substances. They used potassium chlorate as an oxidizing agent in a vertical combustion tube, collected the resulting gases, measured their volumes, and then deduced the respective proportions of carbon, hydrogen, and oxygen that must have been present in the sample.[6] The severe intrinsic difficulties of the method, the details of which need not be specified, and the high quality of the results testify to the superb skills of these chemists.

Berzelius began developing similar methods of organic analysis independently of the French chemists. To slow the combustion and to ensure full oxidation of the sample, he diluted the sample and oxidizer with an admixture of common salt, placed it in a *horizontal* tube, and heated the tube strongly one section at a time, working gradually toward the back of the tube. Berzelius also decided to determine the two true products, water and carbonic acid, by condensed-phase capture (the former using a calcium chloride tube that powerfully absorbs water vapor, the latter by potassium hydroxide solution that attracts the acidic carbon dioxide), and to measure directly by weight rather than indirectly using volume. Berzelius's modification of Gay-Lussac's procedure thus provided higher accuracy, greater simplicity, and a much more direct measurement strategy. Using this method, Berzelius published in 1814 precise analyses of 13 organic compounds.[7]

Gay-Lussac appreciated the merits of Berzelius's approach, and he promptly adopted the horizontal combustion and calcium chloride tube for capturing water vapor, though he retained his volumetric measurements of carbonic acid and nitrogen. However, Gay-Lussac also had further improvements to suggest. One was the use of cupric oxide rather than potassium chlorate, for this substance was far more physically stable and just as good an oxidizing agent; Berzelius followed suit. A second suggestion pertained to the analysis of nitrogenous organic compounds. Gay-Lussac found that for these substances, he was getting inconsistent measurements of nitrogen gas as a combustion product, because the nitrogen tended to be partially oxidized. To solve this problem, he added fresh filings of metallic copper at the front end of the combustion tube. When this copper was heated along with the rest of the tube, it provided a final-stage reducing agent that ensured that elemental nitrogen gas would be collected. The hot copper had no effect on the other products, carbonic acid and water.[8]

To summarize, we see that the first general method for organic analysis was developed between 1811 and 1815, in a series of rapid interacting steps, in what amounted to an unplanned international collaboration between the two finest chemists of their day, Gay-Lussac and Berzelius.[9] This event followed on the heels of the rise of stoichiometry and chemical atomism and was closely connected with that development.[10] By 1815, the method had stabilized to the point that the only substantive difference between Berzelius and Gay-Lussac was in their treatment of carbonic acid. Both collected this product as a gas; Gay-Lussac then absorbed it in liquid potash solution, noting the volume change and calculating the weight from the volume. Berzelius, on the other hand, used a small lye-filled flask floating on the mercury in the receiver, and its gain in weight over 24 hours was the direct measure of carbonic acid absorbed.

These two men were almost exact contemporaries and were then in the prime of their productive years; their personal relationship may be best described as respectful rivalry. Both were working within an essentially French cultural and scientific context, for Berzelius was, as

Söderbaum put it, "a child of the Gustavian period" of the Swedish Enlightenment "and was educated under the dominance of French taste"
and Lavoisian chemistry.[11] For the next 15 years, there was much scru-
tiny, but little lasting improvement, in organic analysis, and the essential elements of the Gay-Lussacian/Berzelian procedure did not change.

LIEBIG AND DUMAS

Liebig's retrospective claim that his scientific education really began
only when he visited Paris was foreshadowed even before his departure. In his grant application, he stated that a period of study in Paris
under "the greatest chemists of our day" was necessary for completing
his education, begun at the Universities of Bonn and Erlangen; his
teacher, Karl Kastner, who initiated the grant request, agreed with this
assessment.[12] Liebig had been studying French since his first semester at
university, because the language was, as he then put it, "necessary for a
scientifically educated man."[13]

Liebig ended up staying for 17 months (from late October 1822
until late March 1824). His initial impressions of Paris were negative,
but within 2 months of his arrival, this 19-year-old German student,
not yet promoted to the doctorate, had made the acquaintance of
Thenard and Gay-Lussac, and he soon applied to these famous chemists
for laboratory space. It seems that Carl Sigismund Kunth (1788–1850),
long-time collaborator of Humboldt, professor of botany at Berlin, and
resident of Paris, provided the introductions;[14] Humboldt himself, the
most prominent German scientist residing in Paris at this time, was then
traveling in Italy.

By this time, Liebig was certain that, however dissipated the
townspeople in the French capital were, its scientists were indeed the
best in the world: they combined an "exceptional mathematical sense"
with a habitual avoidance of all unnecessary hypotheses, and they also
gave superb lectures with dazzling illustrative experiments. As Liebig
wrote at the time to a friend in Hesse, Gay-Lussac and Thenard had
introduced him to a different, more challenging, but altogether more
attractive sort of science. Liebig was thunderstruck by the alteration,
now regarded himself as a mere tyro at this better sort of chemistry, and
spoke fervently of his "metamorphosis" into an aspiring *real* chemist.[15]

In the summer of 1823, with permission from Thenard, Liebig
completed a project in Vauquelin's former laboratory at the École
Polytechnique and, with Thenard's help, wrote up the results. On 28

July 1823, Gay-Lussac read the paper at a meeting of the Académie des Sciences, with Liebig present as a guest.[16] The subject was the fulminates, including an analysis that Liebig conceded as imperfect. Liebig was soon to find that his distrust was justified. This paper, even with its admittedly flawed analytical results, attracted plenty of attention. It was printed in the *Annales de chimie* and was refereed publicly by Gay-Lussac and Dulong, who proffered high praise.[17]

It also impressed another important scientist. As Liebig told the story to Hofmann, after the meeting of the Académie des Sciences at which the paper had been read,[18]

> I was engaged in packing up my specimens, when a gentleman left the ranks of the academicians and entered into conversation with me. With the most winning affability he asked me about my studies, occupations, and plans. We separated before my embarrassment and shyness had allowed me to ask who had taken so kind an interest in me. This conversation became the cornerstone of my future.

The gentleman was Alexander von Humboldt, recently returned to Paris from Italy. Humboldt invited Liebig to dinner with the purpose of introducing Liebig to more intimate acquaintance with his close friend and former collaborator Gay-Lussac. Poor Liebig never showed up, because he had failed to determine the identity of the unknown gentleman.[19] However, the confusion was soon cleared up, and, with Humboldt's help, Liebig had entrée to mentorship and even friendship with Gay-Lussac.[20]

Curiously, 10 months earlier, on his outward journey toward Italy, Humboldt had performed a similar office for a young French chemist, Jean-Baptiste Dumas (1800–1884), then residing in Geneva. Encouraged by Humboldt, Dumas arrived in Paris early in 1823, just 3 months after Liebig. Apparently, and not surprisingly, Liebig and Dumas became acquainted during their 14 months' overlap in Paris, but there seems to be no contemporary reference to their acquaintance. As Gay-Lussac was adopting Liebig as his protégé, Thenard was taking Dumas under his wing.

In Geneva, Dumas's research had been physiological and pharmaceutical; in Paris, influenced by the Arcueil circle, he began to move into chemistry. His first research after his arrival, in collaboration with the pharmacist P. J. Pelletier, was combustion analyses of nine important alkaloids, using the Gay-Lussac procedure. This was Dumas's introduction to organic–chemical research. Their results show reasonable accuracy, though the large molecular weights resulted in some uncertainty in deducing formulas.[21]

A few months after this paper appeared, shortly before Liebig needed to return to Germany, Gay-Lussac invited Liebig to join him in a definitive analysis of the fulminates. This was an unusual step for the French chemist. Liebig later commented that, as far he knew, he was Gay-Lussac's first research student—probably an accurate statement. Although Liebig's collaboration with Gay-Lussac lasted only 6 weeks (in February and March 1824), the two men formed a close bond during this period. He always regarded this period as decisive for his development as a scientist and always looked to Gay-Lussac as his principal mentor. However, this research also marked a stage in the older man's development as well. For one thing, it provided the opportunity for his first extended description of the analytical method he had been using for the previous 9 years. Although the method was by this time very well-known (*très-connu*[22]), the collaborators described it now in detail, really for the first time in this form.

The fulminate analyses were atypical in one respect: Liebig and Gay-Lussac determined nitrogen by a second method, quantitative measurement of hydrogen cyanide released in acid hydrolysis of the compounds. For his general method of determining nitrogen, Gay-Lussac had since 1815 usually applied an accessory combustion, in which the relative amounts of carbonic acid and nitrogen given off were measured. Because this procedure resulted only in measurement of a ratio of volumes from an unweighed sample, it was thereafter routinely denominated, somewhat inappropriately, a *qualitative* determination of nitrogen. The percentage of carbon in the compound was already known from the principal combustion, so the nitrogen–carbonic acid volume ratio could then be used to calculate the percentage of nitrogen in the substance. Dumas and Pelletier, among others, had used this procedure, but, as they later realized, their results were not good, presumably owing to the fact that complex mixtures of nitrogen oxides were inadvertently forming.

During the first few years after his Paris sojourn, Liebig must have regarded himself almost as a nonresident member of the Parisian chemical community. He spoke fluent French; corresponded regularly with Gay-Lussac and other Parisians in their own language; took Pelouze, Regnault, and Jules Gay-Lussac as advanced students and collaborators in Giessen; and regarded the *Annales de chimie*, rather than any of the German journals, as his first outlet for publications.[23] He even maintained good relations with Dumas.

Liebig was certainly regarded as Frenchified by Berzelius and his prize student, Friedrich Wöhler. In Liebig's first paper after his return, Wöhler claimed to smell "a Parisian aroma" of "fast and sloppy" work; Liebig's picric acid paper was, as Wöhler put it, "entirely à la française, that is, he immediately deduces important theoretical conclusions from partial observations and incomplete analyses."[24] Berzelius agreed that Liebig's work showed that he was too often "thoughtless" and "careless" and that he was "fast but incomplete" or "fast but sloppy" in his research.[25]

However, this is nothing compared to Wöhler's and Berzelius' negative early assessment of Dumas. "Dumas hunts after discoveries, as in general all Frenchmen do," wrote Berzelius. "Science [for him] is a public hunting preserve, and therefore the game that is bagged is [his] legal property." Dumas was a schemer ("sehr intriguant") and a "Charlatan"; "instead of discussing difficult questions, this chemical dancing master tries to shine with new explanations."[26] To Berzelius, Wöhler reported documentation of several outright alleged instances of Dumas' plagiarism.[27] When Heinrich Rose visited Paris in 1830, he was highly impressed with all that he saw and all whom he met, including Dumas, but he subsequently learned that at the same time Dumas was flattering him, he had a paper in press that was critical of Rose's work. After this, Rose called Dumas a "Jesuit" (referring not to his Catholic religion but to his duplicitous dealings) and, among the German elite, the epithet stuck.[28]

NEW CHALLENGES FOR ORGANIC ANALYSIS

Celebratory Liebigiana has promoted the notion that no elemental organic analysis was worthy of the name before Liebig's invention. This is clearly not true, as the preceding discussion and other recent scholarship[29] demonstrate. Authors of papers in the 1820s usually omitted all details, merely specifying combustions performed according to the "customary procedure" or the "usual method" (obviously assumed to be well-known to their readers); such phrases referred to the Gay-Lussac-Berzelius standard procedure or to some insubstantial modification thereof. Nor was the procedure confined to only the great masters of the day. Berzelius even asserted, "No particularly great skill of the operator is necessary in order to reach a reasonably reliable result" using his method.[30]

However, Berzelius weakened his point by proceeding to lambaste the sloppy work and hasty conclusions that were threatening to submerge "reliable results" in chemical trash. Furthermore, in the same year during which the preceding words were published (1827), Prout, one of the finest analysts of his or any other day, declared that attaining precision in elemental organic analysis "has always proved a most difficult problem."[31] William Henry agreed, stressing the "considerable skill" required to perform these analyses, even with the latest improvements.[32] Liebig certainly was not satisfied with the state of the art; his papers of the late 1820s repeatedly lamented the uncertainties and difficulties of analyses, his own as well as others'.

The acid test for organic analysis at this time was an exciting and novel series of organic bases called the *alkaloids*, which had resisted easy and precise analysis.[33] One problem was the difficulty of determining nitrogen content accurately, even under the best conditions. Combustions could produce complex mixtures of nitrogen compounds, contrasting starkly with the simplicity of the other two products, carbonic acid and water. Alkaloids presented two further complications. Their nitrogen content was low, commonly around 5%, which meant that analyses required measuring and characterizing tiny amounts of gas; and the results were easily compromised by inadvertent admixture of atmospheric nitrogen derived from the sample tube. Furthermore, the molecular weights of these compounds were high. For instance, Dumas and Pelletier calculated that a morphine molecule contained 107 atoms; Liebig's recalculation a few years later lowered the presumed number to 78, but this was still a very large molecule indeed, and it was typical of the category.[34] The problem is that small uncertainties in measured percentages of carbon, hydrogen, nitrogen, and oxygen, which present no difficulties in determining smaller formulas, lead in larger formulas to significant uncertainties, because with high numbers of atoms, the various percentage intervals corresponding to the alternative formulas are so closely spaced. The net effect of the discovery of alkaloids circa 1820, then, was to pose a new challenge for the art of organic analysis.

Liebig's concern over the deficiencies in analytical methods, especially for nitrogenous compounds, began to become acute, just at the time that he began to form much closer bonds with chemists outside the Parisian orbit. His rapprochement with Wöhler, following their first personal meeting in April 1826, led quickly to an extraordinarily close friendship (already by 1830 they were "Du" and "Du") and his first meeting with Berzelius in September 1830 had a similar effect in his relationship with the older Swedish chemist. One mark of this drift was Liebig's adoption that year of Berzelian atomic weights, used by Wöhler and other (mostly German) Berzelians, abandoning the conventional equivalents preferred predominantly by French and English chemists.

Early in 1830, Liebig published a critique of Prout's new apparatus for elemental organic analysis. He did not like it much, arguing that it was considerably more complicated and no more accurate than the standard approach, but he also used the occasion to point out the deficiencies of the Berzelius/Gay-Lussac method. One difficulty had always been that only very small samples could be burned, on the order of a tenth of a gram, because otherwise the resulting volumes of gases were too large to handle easily—and of course, the smaller the sample, the smaller is the maximum precision attainable. Nitrogen was really the most problematic component, Liebig emphasized; Prout's new apparatus did nothing at all to ameliorate this, because it was designed to analyze only nonnitrogenous compounds. This was the context in which Liebig commented that what was needed was not a new apparatus (meaning the one introduced by Prout) but rather a better method for nitrogen.[35]

This comment should not be taken to indicate that Liebig thought all was well with organic analysis. On the contrary, the foregoing discussion has shown that Liebig was much troubled by the inaccurate and cumbersome character of many aspects of the art, especially (but not solely) nitrogen determinations. In September 1830, Liebig published an analysis of camphor, in which he once more lamented that only approximate numbers were so far achievable, owing to the relatively large molecular weight and the volatility of the substance.[36] The relevant published literature of the 1820s contains many such incorrect, discordant, or insecure analyses, even from the hands of the elite. In sum, in the 1820s, highly accurate and secure results were achievable, but not always, and not with all kinds of compounds. This situation changed suddenly in the fall of 1830, when Liebig invented what he had months earlier declared to be unnecessary: a new apparatus.

Introducing the Kaliapparat (1830–1831)

On 19 December 1830, Wöhler wrote Berzelius somewhat cryptically: "I am curious to learn [Liebig's] new method of organic analysis. He can apply it to very large quantities of sample."[37] Berzelius did not have to remain curious very long, for 3 weeks later, in his first letter to the Swedish chemist, Liebig wrote,

> I have felt compelled to invent a new apparatus for my analyses, which permits one to subject to combustion not just a few tenths of a gram of sample, as is now customary, but any arbitrarily large quantity. The carbonic acid is captured in a specially constructed vessel in which the absorption is complete, and in which the acid can be weighed directly and without the slightest loss.

Liebig proceeded to say that he had thoroughly tested the new method by using it on well-known compounds, such as racemic acid and urea, and asserted that it made organic analysis just as simple and precise as inorganic analysis:[38]

> You will tell me that a father does not easily scold his child, and so I have praised the apparatus more than it deserves. I am therefore very curious to hear your opinion, after you read the description of it, which would be much too tedious for a letter, in Poggendorff's journal.

The article appeared in the January 1831 issue.[39] Liebig began by pointing out the general limitations of organic analysis: small samples mean lower accuracy of results, but large samples yield unwieldy amounts of gas; for small formulas, this does not matter, but in large-weight compounds, such as alkaloids, the limitation becomes so severe as to make an accurate analysis impossible. Furthermore, the determination of nitrogen content is beset with special problems, including the impossibility of excluding all traces of atmospheric nitrogen and the difficulty of collecting pure nitrogen gas from the combustion.

Liebig's principal solution to these interconnected difficulties was to devise a means by which to capture in condensed phase and to measure gravimetrically both carbonic acid and water in a single operation and then, if required, to carry out a second "qualitative" volumetric determination of nitrogen. By collecting condensed carbonic acid and water, Liebig could increase sample size dramatically, thus increasing precision by the same factor; moreover, the gravimetric measurement strategy was simpler and more direct than were the previous versions. Of course, since 1811, product water vapor had always been captured in condensed phase and by the same method Liebig used: a calcium chloride tube. However, all analysts had hitherto collected effluent carbon dioxide as a gas, and no one had attempted to capture it directly in a condensed phase.

To do this, Liebig developed his potash apparatus, a triangular piece of glassware connected to the end of the combustion train, in which five bulbs were blown. Three of the bulbs, arrayed in a horizontal line at the base of the triangle, held a potassium hydroxide solution that strongly attracted the carbon dioxide and condensed it in the form of potassium carbonate; the other two bulbs, flanking and situated above the base line, prevented overflow of the bubbling solution during the course of the combustion.

The apparatus was carefully weighed before the analysis began. After the combustion was complete, the operator broke the upturned tip off the back of the combustion tube and, from the front end of the Kaliapparat, he sucked "for a short time a certain portion of air" through the train; this ensured that any residual product carbonic acid and water vapor would end up in the Kaliapparat and the calcium chloride tube, respectively.[40] Then the potash apparatus was reweighed, the increase in weight being equal to the carbonic acid released by the sample. With this device, Liebig could capture almost any quantity of carbonic acid, so he could burn almost any quantity of sample. The increase in precision and ease of measurement was dramatic.

However, Liebig was aware that his debts to previous analysts were deep; he modestly commented in introducing the discussion, "[T]he only thing that is new about this apparatus is its simplicity, and the complete reliability it affords."[41] He knew that Berzelius had long been weighing carbonic acid rather than measuring its volume, as Gay-Lussac did, but *collecting* it volumetrically, as all workers had hitherto done, imposed serious restrictions. He wrote Berzelius,

> After I had begun to concern myself preferentially with organic analysis, I quickly came to the conclusion that only your method of determining carbon by the *weight* of carbonic acid promised entirely secure results in all circumstances, and since then all my efforts have been devoted to making this process more easily accessible; this was the way my apparatus came to be.

Liebig went on to point out that a reliable gravimetric determination of carbon could serve as a control on the "qualitative" determination of nitrogen (the separate measurement of the nitrogen–carbon dioxide volume ratio to calculate the percentage of nitrogen in the compound), for if calculations based on this ratio differ from those based on the gravimetric measurement, the purity of the evolved nitrogen gas cannot be trusted. "This control has never before been used, and for this reason it is possible that very many earlier analyses are faulty; here I do not exclude my own analysis of picric acid, in any case I plan to repeat it."[42]

A contingent instrumental reason also required the sample-size limitations, hence precision limitation, of the Gay-Lussac/Berzelian procedure. Any process that required volumetric collection of evolved gases required a closed system ending in a pneumatic trough, and mercury was usually necessary as the fluid, because so many gases are at least partially soluble in water. The weight of the mercury generally places the entire combustion train under pressure; this pressure can be minimized by clever arrangements,[43] but it cannot be eliminated. It was also necessary to heat the combustion tube strongly to produce complete combustion. Hot glass and fragile couplings under pressure are a recipe for inconvenience, or even disaster. One could, and did, wrap the combustion tube in sheet-steel to prevent blowholes, but this procedure had its own disadvantages and, in any case, such palliatives had limits.

The introduction of Liebig's Kaliapparat eliminated the need for a pneumatic trough and allowed for the entire process to take place at atmospheric pressure. The tube could also be heated more strongly, ensuring complete combustion, for it no longer mattered if the glass softened. Finally, Liebig's technique of sucking air through the tube at the end of the combustion to collect residual combustion products could be performed only in such an "open" system; until Liebig's elegant solution, the problem of residual gases had always posed a difficult dilemma for analytical accuracy. In sum, Liebig's move to an "open" atmospheric-pressure system resulted in several important advantages.

Liebig argued for the superiority of his new method partly by demonstrating the large number of accurate analyses that could be performed by semiskilled hands. Immediately after describing the apparatus and procedure, he stated that "Herr Hess, one of my students, has at my suggestion undertaken an analysis of racemic acid, as his first assignment of this kind." The percentage composition determined by this unpracticed student could be compared to that just published by none other than Berzelius himself, and Liebig judged it "exact."[44] He then proceeded to report analyses of two additional well-studied substances (urea and cyanuric acid) and of 13 alkaloids, the ultimate test of organic analysis. The quality of this surprisingly large number of analyses was very high, measured by both contemporary historical and modern standards.[45]

Frederic L. Holmes has studied the role of Liebig's students in the immediate aftermath of the invention of the Kaliapparat. He concludes that the new method "enabled [Liebig] almost at once to accelerate the pace of his personal research ... [and he] noticed that students could learn very quickly to produce reliable results with the new apparatus;" in general, the method "quickly changed research practices in the field at large." Nonetheless, Holmes portrayed the innovation as consisting of only "small modifications" and "a further refinement" of an existing method that Liebig himself found to be satisfactory; he doubted that Liebig anticipated in advance any profound effect of the apparatus on the field or that the method would soon become so routine that any student could produce publishable results.[46] The preceding discussion suggests, however, that Liebig was well aware of the importance of the invention and that it may well have placed good analyses within the grasp of unpracticed hands.

There is some real ambiguity in the story. Part of it is due to Liebig's introduction of his method with the words, "There is nothing new about this apparatus except" simplicity and reliability. In one sense, this is literally accurate, but I believe that in a larger sense, Liebig knew otherwise. German scientists in publications were often inclined to the rhetorical device known as *litotes*: understatement calculated for effect, expressing exaggerated (but ultimately insincere) modesty. We find similar rhetoric in other landmark German chemical papers, such as Wöhler's synthesis of urea and Kekulé's structural and benzene theories.[47] The much more flamboyant French tended to follow rhetorical strategies very different from those of the Germans, and this literary and stylistic difference was the source of innumerable misunderstandings in the course of the nineteenth century. We return to this point at the end of the chapter.

THE EARLY RESPONSE (1831–1833)

Liebig sent a French version of his paper to Gay-Lussac for inclusion in the *Annales de chimie*. As Liebig had found multiple occasions in it to criticize Dumas's and Pelletier's 1823 paper, Gay-Lussac gave the paper to his younger colleague for review, and it was published in the fall of 1831 with Dumas's response immediately following.[48] The chemical rumor mills had been churning; Dumas was reported to have been preparing to "one-up" Liebig, but many were betting on Liebig.[49] In the event, Dumas's response was surprisingly mild—and inconsistent. Analysis of nonnitrogenous organic compounds, Dumas averred, was the "simplest of operations," and no improvements were necessary. As for Liebig's arrangement for determining nitrogen, Dumas claimed that his own (very similar) method was distinctly superior, and he provided many details. Both men had taken the basic idea—combustion of the sample followed by in situ reduction of the nitrogen oxide mixture— from Gay-Lussac's 1815 paper. Virtually all of Dumas's comments related to the nitrogen method.

Despite his cavalier dismissal of the need for any innovations, Dumas declared in his very first paragraph that Liebig's method was "destined without any doubt to change the state of organic chemistry in the very near future." Reconciling the contradictory judgments contained in this response is difficult; they provide an indication of how frustrating Berzelius and many Germans found reading Dumas's papers.

Early in October 1831, Dumas sent Liebig a copy, hot off the press, of the issue of the *Annales de Chimie* that contained Liebig's article and his own response. Dumas told Liebig that he had discovered that he had erred in critiquing Liebig's method for determining nitrogen; he had looked over Liebig's paper too fast and had not understood (or even properly read) what was there. He could now see, Dumas wrote apologetically, that their nitrogen methodology was essentially the same, and he promised that he would correct his "blunder" by writing a note to be inserted in the next issue. A reason for Dumas's propitiation of Liebig is revealed in this letter: he felt that he badly needed Liebig's influence to be elected to a new vacant position in the Académie des Sciences. Judging from Liebig's published comments, Dumas feared that his countrymen had gained the false impression that Liebig valued Robiquet's scientific work over Dumas's, which might prove fatal to his chances to gain entry to the Académie:[50]

> In the current state of chemistry in France, it is Germany that determines the opinion. . . . Of all the chemists I have met, it is you whose character and ideas inspire the most attachment in me . . . From Germany, one cannot gain an idea of my position here. I am the only one in Paris who reads your papers. I cannot talk chemistry with anyone, for no one stays current with what is happening in the science.

Liebig responded with grace and cordiality.[51] The following month, Dumas published, as promised, a long footnote that corrected the errors and misleading impressions in his earlier response. He conceded some flaws in his alkaloid article of 1823, but, despite his conciliation of Liebig, continued to maintain the superiority of his own procedures.[52]

Dumas in fact almost immediately adopted the Kaliapparat.[53] In the late autumn of 1831, he was still measuring carbonic acid by volume,[54] but, in a paper published early in 1832, he wrote that "for some time" he had been using the Liebig device "with full success."[55] Dumas may have been influenced by Jules Gay-Lussac, the son of the famous chemist, who studied with Liebig in the fall of 1831 and learned the new method there. Jules Gay-Lussac and Pelouze became the first Frenchmen to publish a paper (actually just a note) in which the Kaliapparat was employed for analysis.[56]

In a long article published in March 1834, Dumas discussed all available methods for organic analyses, giving decided preference to those which weigh carbonic acid rather than measure its volume. The method "that merits preference in every respect is that which uses the ingenious absorption apparatus of M. Liebig." The potash-bulb device "simplifies organic analysis to such a degree, and gives results so precise, that it can be regarded as one of the most precious acquisitions which analytical chemistry has made for many years."[57] As Dumas went, so went Parisian chemistry; the Gay-Lussac/Berzelius method of volumetric capture of carbonic acid died in the early 1830s.

Berzelius read Liebig's paper in April 1831 and studied it with great care. He just had time to insert a review of the article into his annual report for calendar year 1830, just then being printed, and he praised the new apparatus and analyses highly.[58] His corresponding letter to Liebig was entirely focused on the data itself, not on the apparatus (which he presumably had not yet attempted to construct). He was particularly interested in Liebig's critique of Dumas and Pelletier:[59]

> The unreliability of the French analyses, which is brought out so strikingly in your work, is a damned curious thing. For when one reads the paper of Dumas and Pelletier, considers the [nitrogen/carbonic acid] controls and the lovely agreement which they achieved in everything, it is clear that wherever Dumas is present, the results have been helped along with the pen, and when he couldn't figure out the correct answer by calculation he took a false one as his model.

In Liebig's wide-ranging response to Berzelius's letter, he touched on some other measurements recently published by Dumas, namely on vapor densities:[60]

> As little as I trust Dumas' work, the calculations of this tightrope dancer seldom fail to meet the mark; I have become convinced ... that Dumas is right; it always irks me that this fellow, in spite of his unclean, impossible and sloppy manner of working, nonetheless shakes masterpieces from his sleeve, for which, to be sure, his pen deserves the most credit.

In 1832, Liebig reported to Berzelius that he had reduced the size of the Kaliapparat to the point that it could be weighed on any balance; he offered to send one to Berzelius. Berzelius gratefully accepted the gift, commenting, "You are at the moment certainly the greatest master in the art of carrying out precise organic analyses which we now have." Due to various postal difficulties, it was more than 9 months before the gift arrived. Liebig had sent two Kaliapparate, of which only one survived the journey; however, Berzelius noted that this made little difference, as he could now blow as many as he liked, using the one surviving specimen as a model.[61] Some 18 months later, Berzelius wrote Wöhler,[62]

> We are using Liebig's apparatus daily for these [investigations]. It is a magnificent instrument. By means of small insubstantial modifications we are now to the point that the result that one arrives at cannot be incorrect, and that one neither obtains water in excess nor loses any carbonic acid.

By this time, Wöhler had long since adopted the device. In November 1831, Wöhler visited Liebig's lab in Giessen for two weeks of intense collaborative work. Liebig taught him the use of the new apparatus, which Wöhler reported to Berzelius as "superb" (*ganz vortrefflich*), and they carried out many analyses together. This was "a side of chemistry to which I was hitherto a complete stranger."[63] After his first experience of working directly with Liebig, Wöhler had even more respect for his friend:[64]

> He is, by the way, the best and most honest fellow in the world, and in chemistry has virtually unrivaled zeal. The days with him passed like hours, and I count them among the happiest of my life. His organic apparatus seems quite splendid to me; he is moreover a

master of organic analysis and performs it with a pedantic exactitude. But as regards inorganic analysis, such as filtrations, use of the lamp, etc., one spots the imperfect French methods. He uses neither a filter stand, nor good filters, nor usually the lamp. He knew no better, but was immediately ready and happy to come over to the Swedish flag.

The collaboration was so welcome to both men that Wöhler returned to Giessen 2 months later for another 2-week stint and then again for 4 weeks in July and August 1832.[65]

In April 1831, Liebig became co-editor of Geiger's *Magazin der Pharmacie* (renamed *Annalen der Pharmacie* in January 1832) and began to use it as his personal publication organ. The first issue of the renamed journal contained no fewer than three contributions that mentioned analyses performed in the new manner.[66] Even before this, Liebig students had published in other journals articles with analyses using the Kaliapparat,[67] and the flow of such articles, by Liebigians and others, continued in the new journal.[68]

Robert Bunsen must have learned the method when he spent some days in Giessen just at the time (summer 1832) when Liebig was working with Wöhler on the oil of bitter almonds. The community of Berzelians in Berlin—Heinrich and Gustav Rose, Gustav Magnus, and most especially Eilhard Mitscherlich—quickly adopted it. The relevant French chemists also converted, as we have seen. The English were slow to move, because they were not yet oriented toward organic chemistry in general or organic analysis in particular: Prout, who published his last analysis in 1827, was still dismissive about the Kaliapparat 10 years later, which illustrates how out of touch many English chemists had become in this area.[69] Perhaps the last paper in which the older method is known to have been used was written by Pelletier in 1833.[70]

SUBSEQUENT DEVELOPMENT OF THE METHOD (1833–1837)

Other than his reduction of the size of the Kaliapparat to facilitate weighing, Liebig retained the original apparatus and method virtually unchanged throughout the 1830s; this is clear not only from the essential identity of the two-page description of it in January 1831 with the detailed monograph that appeared 6 years later, but also from his explicit statements and those of his friends.[71] By March 1833, Liebig was prepared to expand his claims about the accuracy achievable with his apparatus.[72] In fact, he made the "very unfortunate discovery" that most of the analyses performed before the Kaliapparat was introduced were inaccurate, his own as well as others', for he found that the volumetric procedure, for mysterious reasons, tended to give a slightly larger value for carbonic acid than the Kaliapparat did. Among other consequences, he was now forced to admit that his pre-1830 analyses of turpentine, uric acid, indigo, picric acid, and camphor were wrong, and Dumas's were right.[73]

In exculpation of their teacher on this occasion, Liebig's students, Blanchet and Sell, commented that only then did science possess an analytical method—Liebig's—that always gave the same results, no matter who undertook the analysis.[74] Liebig himself often emphasized this point as well, only taking care to remark that one must always operate on fully purified and homogeneous samples, that analysis was merely a tool of the scientific investigator, and that no number of mere analyses strung together constituted a proper research program.[75]

We have seen that Dumas converted enthusiastically to the use of the Kaliapparat (circa February 1832). However, Dumas, like Liebig, always emphasized the urgent need for an accurate method for nitrogen, which had so confounded the best analysts of his day. By early 1833, he had devised a modification of the existing procedure that constituted a signal advance.[76] Dumas placed lead carbonate in the back of the combustion tube, with the sample and oxidizer arranged as usual. Application of a vacuum followed by strong heating of the carbonate resulted in a surge of carbonic acid gas through the tube, which flushed out all vestiges of residual air. The tube was again placed under vacuum, and the combustion was carried out as usual. The nitrogenous gases produced in the combustion were reduced by passing over hot activated metallic copper. Finally, another surge of carbonic acid from heated carbonate ensured that any residual sample nitrogen was flushed out of the tube into the collector. The collector was a graduated bell jar in a pneumatic trough, the mercury of which was topped by a layer of potassium hydroxide solution. The potash absorbed all the carbonic acid from the carbonate and absorbed carbonic acid and water vapor from the sample, leaving (in principle, at least) the nitrogen alone to be measured volumetrically.

By this procedure, Dumas was confident both of eliminating even the smallest traces of contaminant nitrogen from residual air in the sample and of quantitative collection of pure nitrogen from the sample. He emphasized that this was the first nitrogen method that yields results "with a rapidity and certainty at least equal to that now obtainable" for carbon and hydrogen (by which he meant with the Kaliapparat). Previous methods have been "radically inexact," Dumas asserted, rather contradicting his confidence of two years earlier.[77] Dumas's nitrogen method could be, and almost immediately generally was, coupled with the Kaliapparat procedure to provide an excellent two-step analytical method for elemental analysis of nitrogenous organic compounds. In a paper of April 1834, Dumas himself wrote,[78]

> The procedure to follow in all research on the composition of the organic bases has been very nicely traced in the remarkable paper in which M. Liebig described his precious condenser; here it will suffice to repeat some examples from this paper.

However, carrying out the Dumas nitrogen method in its original form was not particularly simple; despite the schematic clarity, eliminating the contamination of sample with air and product with oxides was difficult. For these reasons, the method did not displace its competitors until it was gradually simplified. The systematic errors of the method were still troublesome as late as the beginning of the twentieth century, even though by then it was long since acknowledged as the best general method of its type.[79]

As early as the mid-1830s, Liebig's Kaliapparat was widely employed by European analysts, especially in Germany and France. Berzelius provided a fine summary of the "state of the art" in his classic textbook of chemistry. In a section on organic analysis written circa 1835 and published 2 years later, he characterized Liebig's procedure as leaving "nothing to be desired" and stated that it had brought elemental analysis to such a state of simplicity that the procedure was "one of the easiest operations," the previous methods being no longer of any but historical interest. In fact, the simplicity and low skill demands of the Kaliapparat led Berzelius to predict a swell of very bad analyses: bad not because of the analytical procedure itself, which was unproblematical even in unpracticed hands, but because the preliminary steps of identification and purification were not properly carried out. In other words, Berzelius was suggesting that the challenging and problematical character of the earlier methods had the beneficial concomitant that only sophisticated and conscientious chemists performed analyses at all; this internal protection no longer applied. In contrast to Berzelius' extravagant praise of Liebig, he simply listed Dumas's nitrogen method among several alternatives, including Liebig's own version.[80]

Despite Berzelius's praise of Liebig, publication of this section of his textbook caused consternation in Giessen. As Liebig had published nothing in detail on the method—his 1831 paper had been sketchy — Berzelius had been forced to turn to a detailed discussion that had just appeared in Eilhard Mitscherlich's textbook.[81] However, Mitscherlich and Liebig were enemies, and Berzelius's innocent adoption of a number of Mitscherlich's modifications gave Liebig a bad case of heartburn. (Berzelius had been Mitscherlich's mentor; he was aware that Liebig had taught Mitscherlich the Kaliapparat method directly and assumed that Mitscherlich's modifications had been devised by Liebig; he also underestimated the depth of enmity between the two men.) Liebig's secretiveness may well have been intentional, to retain a degree of monopoly on the method, but he could now see that his tactic (if such it was) had become self-defeating.

These events led Liebig to decide that it was time for a full account of the method from his own pen, and he was given the appropriate occasion when he had the opportunity to write an article on "Analysis, organic" for a chemical handbook he was helping to edit. The article appeared in the third fascicle of the first volume of the handbook, published in late summer 1837, but an offprint was available as early as April of that year. Liebig assured the publisher that the fascicle would sell very well, because he was, as he put it, giving away all his secrets in it.[82] Liebig's article was really a small monograph, a classic of analytical chemistry. It is filled with descriptions of ingenious auxiliary devices and procedures, all designed to increase simplicity, reliability, and accuracy. For example, hygroscopic moisture in the sample was removed by an artful arrangement of drying tube, sample in a sand bath, and water-filled three-necked bottle that served as a device to provide continuous suction. Using a siphon from the bottle, a continuous stream of perfectly dry air could be made to pass effortlessly through the tube containing the heated sample, the only operation necessary to continue the process being to replenish the water in the bottle by means of a funnel installed in the third neck.

Another example was the sucking of air through the combustion tube at the end of the combustion, to remove and collect residual water vapor and carbonic acid. After the back tip of the combustion tube was snipped off, an 18-inch-long glass tube was placed over the broken tip, to ensure that carbonic acid from the charcoal fire was not drawn through the tube. A quantity of air was sucked through, estimated to be equal to the volume of the combustion tube plus drying tube.

Liebig urged that this be done by mouth, rather than by mechanical means (as Berzelius had recommended), both for an analytical purpose and for convenience:

> When the combustion has been thorough, no taste is perceived in the air drawn through. When it has been imperfect, a more or less distinctly empyreumatic taste is perceived. The latter circumstance is not always proof that the analysis has failed; for it very often happens that two analyses agree perfectly, in one of which an empyreumatic taste is perceived, in the other not. This proves how minute a quantity of matter suffices to communicate a taste to the air.

Liebig readily conceded that this procedure added a little extra moisture from atmospheric water vapor to the calcium chloride tube, hence to the final measurement of product water and hence to the determination of hydrogen. However, the average excess could readily be measured by "blank" tests; it amounted to approximately 0.6 mg of hydrogen. This amount could then be subtracted from the final result to achieve a higher accuracy. However,[83]

> In publishing the weights obtained in any analysis, we must not make the above deduction, but give the numbers as they occur, since the amount of excess, due to hygrometric water, furnishes to the reader a valuable means of judging the accuracy of the determination of the hydrogen. It is only in calculating the composition with a view to discover the formula that we are to make the above correction.

Liebig discussed nitrogen determination in detail, including the methods introduced by Gay-Lussac, by Liebig and Gay-Lussac, and by Dumas. Although French chemists had generally decided that the Dumas method for nitrogen was superior, Liebig and his countrymen, as well as Berzelius, were not yet convinced. Liebig not only gave no indication that Dumas's method was the best, he avoided mentioning any name in connection with the procedure.

Throughout this monograph, Liebig exhibited a sophisticated but also highly pragmatic (and essentially instinctive) understanding of error analysis. Systematic errors, once identified, were compensated for or eliminated by one of the following means: adjustment of the procedure to reduce or remove the problem; no action at all, if the perturbing effect could be shown to be sufficiently small; no action at all, if approximately equal compensating systematic errors were found to exist; or compensation by post hoc calculation, controlled by blank combustions or other tests. His methodology was also pragmatic, for he always had simplicity, ease of operation, and rational use of the analyst's time in mind; wherever he could argue that a more complex or difficult procedure made no *practical* difference in results, he discarded it for the simpler one. At the start of the monograph, Liebig commented, "In what follows we hold to the rule of Berzelius, the most experienced chemist of our own, and probably also of all times; and of two equally good methods, we prefer the simple to the complicated one."[84]

Despite these flattering words, Liebig had some unpleasant surprises for Berzelius in this work. Berzelius, as we have seen, adopted many of Mitscherlich's small modifications, alterations that had led Mitscherlich to write at times of "his" method. In his monograph, Liebig, who had been feuding with Mitscherlich constantly over the preceding few years, rejected all these changes, sometimes with not-very-gentle language. He also took a shot at Berzelius regarding the apparently trivial point about how the combustion tube was to be connected to the calcium chloride drying tube. After describing the Berzelian modification without comment early in the monograph, he came back to the point later; Berzelius's preference, he stated, "can only be accounted for on the supposition that he has never examined" Liebig's alternative. "To make a good analysis with Berzelius' arrangement is an operation which requires the hand of a master; and we must always consider it a fortunate event when the analysis is safely over." The end effect of adopting Berzelius' modification, therefore, would be to confine such analyses "to a proportionally small number of experimenters."[85]

Liebig attempted to head off problems with his older friend by writing him a disarming letter:[86]

> ... I have had to depart from your details in many respects, which you will not be able to hold against me when you consider the fact that the multitudinous analyses that are made throughout the year in my laboratory are the best touchstone for the less secure and less well proven experiences of other workers.

Liebig's ploy did not work, and Berzelius was wounded. He wrote back,[87]

> I come now to a matter that is very unpleasant for me. It concerns your excellent Instructions for the Analysis of Organic Compounds. Of the little that I have done for this sort of analysis, you have totally rejected everything, excepting only the use of a combustion tube, as

either defective or unnecessary, and you have thereby publicly entered the lists against me, all while we have engaged in the friendliest correspondence. You could have indicated your doubts to me, I could then have provided you explanations, but you have preferred to put me in the embarrassing position of having to publicly defend myself against so valued a friend as you. My dear Liebig, I say this without the slightest trace of resentment: you must stop being a chemical executioner . . .

Now it was Liebig's turn to be upset. He told Berzelius that he had never meant to make Berzelius the target of his scorn, but rather "my bitterest and most vicious enemy . . . [Mitscherlich] who is doubly dangerous since his cowardice does not allow him to attack me directly, but only secretly." He said the same thing to Wöhler, pointing out that it was Mitscherlich who wrote in his textbook "of *his* method, *his* apparatus . . . how could I have quietly watched such a performance and thus lost the question rather than won?" Liebig offered the strongest possible protestations of his regret for his unintended insult and of his regard for Berzelius.

Berzelius responded with a friendly letter armed with a "sermon" about how to conduct scientific controversies as gentlemen, to which Liebig answered,[88]

Your letter of 3 April gave me great joy, my sins are thus forgiven and I hope also forgotten. Your sermon was not preached in vain, but sometimes the devil knows what my pen has in mind. I press your fatherly hand with sincere love and admiration, and say with you, "friendship for ever."

Berzelius printed his "public defense" of his views, with a discussion of his differences with Liebig, in the seventh volume of his textbook.[89] Another critic was Russian chemist G. H. Hess, who contested one of Liebig's error analyses. In his reply to Hess, Liebig also publicly defended himself against Berzelius. The "many thousands" of analyses performed in his laboratory since 1831 had given Liebig the empirical warrant to speak with authority about which procedures work well and which do not. Berzelius's older method was excellent, he averred, for a master such as Berzelius or Chevreul, who did not mind spending a week or two on each analysis, but neither chemist used that method any longer, for obvious reasons. The question at issue was no longer what method gave the highest attainable accuracy, but what method gave high accuracy with fast, simple, and reliable procedures. A single investigation sometimes required as many as 60 to 100 individual analyses, and something like 400 were being performed annually in the Giessen laboratory. If absolute perfection were sought, such investigations would not be slowed, they would be rendered absolutely impossible.[90]

Until 1837, Liebig's method had been communicated largely by direct means, one chemist showing another the procedure in chains of transmission that led back to Liebig himself. After the publication of this monograph, anyone with minimal chemical expertise could fashion the devices and perform the operations from the detailed instructions readily available in bookstores. The essential superiority of the method and its now openly accessible details ensured that it would be employed anywhere chemistry was done. The method can be found in German as well as non-German textbooks throughout most of the rest of the century, little altered from Liebig's description of 1837.[91]

The one major European country where the method was initially slow to spread was England, largely because English chemists generally failed to participate in the ferment in chemistry (especially organic) of the 1820s and 1830s. The respected London chemist J. F. Daniell demonstrated unintentionally how out of touch the English had become by describing the Gay-Lussac/Berzelius method as standard, with no mention of Liebig or Dumas, in his 1839 textbook.[92] In that same year, William Gregory declared that "for a good many years previous to 1836, no organic analyses were published by any British experimenter;" the following year, two English chemists working in Berlin averred more generally, ". . . [b]ut little of what is done abroad, especially in Germany, seems to find its way into England, or, at least, until after the lapse of some years."[93]

These sentiments were shared by many foreign observers. On the approaching death of Humphry Davy, Berzelius commented, "England could not bear to lose Davy and [William] Wollaston simultaneously; her science is on a much lower level than many other countries. Faraday and Turner, despite all their virtues, are far from being able to replace the deceased." Regarding Edward Turner's new textbook, Berzelius wrote two months later,[94]

> What abominable rubbish it is; no knowledge of anything that was not done in England by Englishmen, and with the imperious tone that accompanies true ignorance . . . It is strange to see how far the French and English chemical literature falls short of the German. I would never have believed it.

The deficiencies began to be addressed in the late 1830s, largely owing to Liebig's impetus. Several British chemists studied in Giessen starting circa 1835, including Thomas Thomson, Jr., W. C. Henry, R. Kane, William Gregory, Lyon Playfair, and John Stenhouse. On returning to their homeland, their professional efforts (writing of books and papers, editing of journals, teaching of students, and the founding of the Chemical Society and the Royal College of Chemistry) resulted in a quickening of English chemistry, especially in the organic realm. Liebig's influence in Britain was also much fostered by his visit to England, Ireland, and Scotland for 2 months in the summer of 1837.[95]

Thomas Thomson described Liebig's method in his *Chemistry of Organic Bodies* of 1838. He wrote, "Professor Liebig has published this article as a separate pamphlet. Were any person to favour us with an English translation of it, he would contribute essentially to promote the prosecution of vegetable chemistry in Great Britain, and would be conferring an important favour on the British chemical public."[96] Liebig's student Gregory supplied this desideratum (prefaced with an apology for its appearance 2 years after the original). This English version was widely read. It was echoed shortly thereafter by Thomas Graham's treatment of the Liebigian method in his widely influential textbook.[97]

In the same year in which this last account appeared, Gregory published a polemic about the sad state of British chemical education, pointing out that British students were generally going to study with Liebig or Wöhler to find there what they could not obtain at home.[98] However, Gregory and several of his German-educated chemical friends had started a process that soon raised the level of British chemistry markedly.

A French translation of Liebig's monograph also shortly appeared; this was published together with a critical essay by the peripheral and iconoclastic French organic chemist F. V. Raspail, but Raspail's opprobrium was directed not to the analytical method, which in fact he praised highly, but rather to Liebig's theoretical inclinations.[99] Thereafter, as in England, the Kaliapparat method entered the mainstream French chemical literature. We have seen that Dumas himself recommended it whole-heartedly in his textbook of 1835; in the 1840s, the other principal French chemical textbooks strongly urged its use. The fact that two of Liebig's research students, Regnault and Pelouze, were authors of perhaps the two most prominent French textbooks of the day did not hurt.[100] In Germany, no doubt ever was voiced about the immediate dominance of the Liebigian method. Carl Löwig wrote in 1844 of the "new epoch" thereby introduced into chemistry, and such rhetoric was typical of German textbook writers of this period.[101]

The Kaliapparat soon acquired virtually mythical proportions among chemists. Liebig's students were in the habit of sporting five-bulbed lapel pins as a badge of honor.[102] Many chemists in the nineteenth century cherished their "potash bulbs," sometimes employing a single apparatus for 20 years or more.[103] This mythical success had a sound foundation. Twenty years ago, a repetition of organic analyses using a modern reproduction of the Kaliapparat and following Liebig's 1837 instructions produced "astonishingly accurate results".[104] Although organic analyses today are performed by specialized commercial laboratories rather than by individual research chemists, the method is essentially the same as Liebig's.

Conclusions

In summary, all previous writers have rightly emphasized the simplicity, reliability, and accuracy of the method, but not enough attention has been paid to considering these attributes in their particular historical setting. For example, we have seen how Liebig's "open" atmospheric-pressure procedure had certain advantages over the closed mercury pneumatic trough. An interesting analogy can be drawn here to the invention of the first commercially successful steam engine, Thomas Newcomen's atmospheric engine of 1710, which displaced Thomas Savery's failed high-pressure device. In each case, the earlier device actually operated and could even be made to work well in special cases. However, in each case, the minimum technical requirements exceeded the skill levels, technological capabilities, and financial resources of all but a few sets of unusual circumstances. By contrast, the atmospheric devices of Newcomen and Liebig were well attuned to the demands of the day. They shared the virtues of simplicity, inexpensiveness, reliability, safety, and accessibility to "ordinary" levels of skill.

Liebig's innovation also represented a complete transition to gravimetry and thus to an essentially *chemical* methodology. Gay-Lussac was a chemist in the fullest sense, but he made his initial reputation in the physics of gases and of heat; he taught physics in the Paris Faculty of Sciences from its origin in 1809 until 1832, and served in the physics section of the Académie des Sciences. Eudiometry was more natural to him than was gravimetry; Crosland speculates that we may see here "a late strand of French Cartesianism," in which "extension" was instinctively regarded as the proper measure of matter.[105]

Nonetheless, we have seen that a fully chemicalized and gravimetrized procedure—carbonic acid captured in condensed phase by a chemical reaction, the formation of carbonate, rather than physically as a gas—had essential advantages. Another very curious circumstance is that Liebig's chemical procedure had such advantages over the physical method as regards accuracy and precision; in this century, it was a turn back to physics that provided a quantum leap in analytical accuracy. This is yet another sense in which the analogy to Newcomen holds, for high-pressure steam engines in the spirit of Savery eventually displaced atmospheric devices, and the high-pressure principle is used today in the most efficient engines of the type (modern steam-electric power plants).

Liebig's method fit its context well in other respects, for the apparatus was theoretically more at home in Germany than in France. The Kaliapparat method developed hand in hand with chemical atomism. For one thing, the inability to arrive at secure atomistic formulas for the alkaloids was a major reason for Liebig's unhappiness with earlier analytical methods and provided the proximate incentive for developing a new one. For another, the emergence of the phenomenon of isomerism just at this time posed another sort of challenge for elemental analysis; atomic theory provided a route out of the conundrum. Chemical atomism of the Berzelian type influenced German chemists quickly and decisively. Already by Liebig's Paris years, it was becoming influential and, by the time of Liebig's own conversion to Berzelian atomic weights—in the very year he invented the apparatus —it was dominant in Germany. The French response to Berzelian atomism was more muted and ambivalent. Although a prevailing school of chemical atomists lived in Paris by the 1820s, these ideas were vitiated by continuing rhetoric of "volumes" as a putative synonym for "atoms," and by a strong antimetaphysical tendency that made atomistic pronouncements seem suspect in tone and epistemological character. It is reasonable, therefore, to believe that there may have been a greater sense of urgency in Liebig's mind than in Dumas's to improve analytical methods.

This last point underlines Bernadette Bensaude-Vincent's reference to Bachelard's view of instruments as "materialized theories."[106] The Kaliapparat may also be regarded as fitting comfortably into a characteristic institutional context as well (a point made at length by

Holmes[107]). From his student days, Liebig planned a career in which he would instruct *groups* of students in pharmaceutical and chemical practica. The Kaliapparat method was superbly suited to this pedagogical strategy, which interacted strongly with his research program and developed into a characteristic German style of group-oriented research. French chemists did not develop research groups of the German type until the late nineteenth century and so were not able to exploit the method in the way that Liebig and other German chemists did. Furthermore, Dumas's method for nitrogen, which has some parallels to Liebig's for carbon, had intrinsic limitations and never achieved the kind of mass-production simplicity that use of the Kaliapparat afforded.

We have intimated that Liebig's analytical innovations were connected with his drift away from the Parisian orbit in the late 1820s. The change was particularly evident by 1832, when he published a harsh denunciation of the French chemical community, lambasting them for arrogance, chauvinism, provincialism, rhetorical bombast, and scientific thievery. The viciousness of these judgments was slightly ameliorated by Liebig's argument that much of the behavior he described was an inevitable product of the *structure* of the French scientific establishment. Because university lectures were free, young French scientists had no opportunity to earn a livelihood early in their careers, as German Privatdozenten could; furthermore, the monopolistic power of the Académie des Sciences, "the source of all remunerative positions," led to an unseemly scramble for success. This was why French scientific papers all seemed so arrogant and self-promoting, Liebig thought.[108]

By contrast, German scientists might be said to have worn their sobriety and modesty on their sleeves, vaunting it and at times almost overplaying the card, which may be regarded as another form of arrogance, less forgivable because it is concealed. To a German scientist used to German rhetoric, the literature from Paris made the French sound considerably more brash and presumptuous than they were. When German scientists visited Paris and spoke directly with their French counterparts, their impressions were invariably different and much more favorable.[109]

The battles over elemental organic analysis circa 1830 provide an interesting window on wider aspects of chemistry, science, and European culture. Liebig borrowed essential elements of French culture and French chemistry, some of which remained with him for the rest of his life, but he added other elements as well, including German and Berzelian. His continuing ambivalence toward the French reflected the eclectic character of his education and his culture.

NOTES

1. A. W. Hofmann, "The Life-Work of Liebig," in *Erinnerung an vorangegangene Freunde*, vol. 3 (Braunschweig: Vieweg, 1888), 195–305, on 229. The other standard biographies are Jacob Volhard, *Justus von Liebig* (Leipzig: Barth, 1909); F. L. Holmes, in *Dictionary of Scientific Biography*, 8(1973): 329–350; and William Brock, *Justus von Liebig: The Chemical Gatekeeper* (Cambridge, UK: Cambridge University Press, 1997).

2. J. W. Döbereiner to Goethe, 7 December 1812, in *Die Chemie in Jena zur Goethezeit*, ed. H. Döbling (Jena: Fischer, 1928), 162–163; and Döbereiner, "Chemische Bemerkungen und Versuche," *Journal für Chemie und Physik* 59(1818): 318–327, on 321–322.

3. J. Liebig, "Eigenhändige biographische Aufzeichnungen," in *Justus von Liebig: In eigenen Zeugnissen und solchen seiner Zeitgenossen*, 2nd ed., ed. Hertha von Dechend, (Weinheim: Verlag Chemie, 1963), 13–27, on 19. Wurzer yielded his chair to Bunsen in Marburg in 1839; by this date, Liebig had spent 15 years as professor in nearby Giessen. Regarding the role of nationalism in the debate over the oxygen theory, see M. Crosland, "Lavoisier, the Two French Revolutions, and 'The Imperial Despotism of Oxygen,' " *Ambix* 42(1995): 101–118.

4. Liebig, "Biographische Aufzeichnungen," 20–22; Liebig to his parents, 17 December 1821, in *Briefe von Justus Liebig nach neuen Funden*, ed. Ernst Berl (Giessen: Liebig-Museum, 1928), 31, where he reports purchasing an early, partial, and fairly deficient German translation of Berzelius's textbook.

5. Regarding the history of organic analysis, see for example M. Dennstedt, *Die Entwickelung der organischen Elementaranalyse* (Stuttgart: Enke 1899); J. R. Partington, *A History of Chemistry*, 4 (London: Macmillan, 1964), 234–239; and F. L. Holmes, "The Complementarity of Teaching and Research in Liebig's Laboratory," *Osiris* [2] 5(1989): 121–164. Despite these fine examples, the history of organic analysis has not been thoroughly explored.

6. J. L. Gay-Lussac and L. J. Thenard, *Recherches Physico-Chimiques*, vol. 2 (Paris: Deterville 1810–1811), 268–350. Although elemental analysis is important in its own right, for it is the principal criterion for chemical identity, analysis became much more avidly pursued after the introduction of the atomic theory into chemistry.

7. J. Berzelius, "Experiments to Determine the Definite Proportions in Which the Elements of Organic Nature Are Combined," *Annals of Philosophy* 4(1814): 323–331, 401–409; 5(1815): 93–101, 174–184, and 260–275; method described on 401–408.

8. Gay-Lussac, "Recherches sur l'acide prussique," *Annales de chimie* 95(1815): 136–231, on 181, 184–186; idem, "Observation sur l'acide urique," ibid., 96(1815): 53–54.

9. A good source for the method in its ultimate form is Berzelius, *Lehrbuch der Chemie*, 2nd ed., 3 : 1 (Dresden: Arnold, 1827), 157–174. This volume was Berzelius's first full treatment of organic chemistry in his famous textbook, and this German version, edited and translated from the Swedish manuscript by Friedrich Wöhler, was the editio princeps, for the Swedish "first edition" followed behind the German. A good French description is C. Despretz, *Elémens de Chimie Théorique et Pratique*, vol. 2 (Paris: Méquignon-Marvis, 1830), 742–757. F. L. Holmes was the first to point out that the Gay-Lussac-Berzelius interaction produced a "standard apparatus ... developed in part by Gay-Lussac and in part by Berzelius" (Holmes, "Liebig's Laboratory," 135).

10. See A. Rocke, *Chemical Atomism in the Nineteenth Century: From Dalton to Cannizzaro* (Columbus: Ohio State University Press, 1984), 21–123.

11. H. G. Söderbaum, "Berzelius und Hwasser, ein Blatt aus der Geschichte der schwedischen Naturforschung," in *Studien zur Geschichte der Chemie*, ed. J. Ruska (Berlin: Springer, 1927), 176–186, on 177; J. Berzelius, *Autobiographical Notes* (Baltimore: Williams & Wilkins, 1934), 16–38, 123–128, 179–180; Anders Lundgren, "The New Chemistry in Sweden: The Debate that Wasn't," *Osiris* [2] 4(1988): 146–168; and E. Melhado and T. Frängsmyr, eds., *Enlightenment Science in the Romantic Era: The Chemistry of Berzelius* (Cambridge: Cambridge University Press, 1992), passim. Lundgren makes the important point that factors internal to Sweden strongly conditioned the early and generally favorable response to French oxygenist chemistry.

12. Liebig to Grand Duke Ludwig, [10 April 1822]; Kastner to Grand Duke Ludwig, 12 April 1822, in Berl, *Briefe*, 34–35.

13. Liebig to his parents, 19 November 1820 and 27 October 1821, ibid., 12 and 29.

14. "Ich habe auch Herrn Prof. Knuth [*sic*—possibly a transcription error] aufgesucht ... durch ihn habe ich den berühmten ... Blume der Chemiker Gay-Lussac ... [words missing due to tears in the letter]" (Liebig to his parents, 1 January 1823, in Berl, *Briefe*, 43); "Durch die Güte des Herrn Professor Knuth [*sic*], Mitarbeiter des Herrn v. Humboldt, hatte ich Gelegenheit, einer Sitzung der königlichen Akademie beizuwohnen, ich hatte hier die Freude, die Bekanntschaft von Vauquelin, Gay-Lussac und anderen vortrefflichen Männern zu machen ..." (Liebig to Schleiermacher, 17 January 1823, ibid., 46). How Liebig contrived to gain an introduction to Kunth in the first place is not known.

15. Liebig to Schleiermacher, 17 January 1823, ibid., 45; Liebig to August Walloth, 23 February 1823, ibid., 49. On Gay-Lussac, see M. Crosland, *Gay-Lussac: Scientist and Bourgeois* (Cambridge, UK: Cambridge University Press, 1978). John Heilbron has recently and perceptively explored the quantitative culture in late eighteenth-

century French science: *Weighing Imponderables and Other Quantitative Science Around 1800* (Berkeley: University of California Press, 1993), esp. 23–33, 141–149. Maurice Crosland has enriched the literature about eighteenth- and nineteenth-century French scientific culture with a number of important essays, conveniently collected in *Studies in the Culture of Science in France and Britain Since the Enlightenment* (Aldershot: Variorum, 1995).

16. Liebig to his parents, 12 June, 30 June, n.d., 17 July, and 6 August 1823, ibid., 58–64; Liebig, "Sur l'Argent et le Mercure Fulminans," *Annales de Chimie et de Physique* (hereafter *Annales de Chimie*) [2] 24(1823): 294–317. On the last page of this article, Liebig expressed his gratitude to Thenard and to H. F. Gaultier de Claubry, Thenard's préparateur at the École Polytechnique, who had shared his laboratory space with Liebig.

17. In *Annales de chimie*, [2] 24 (1823), 421.

18. Hofmann, "Life-Work of Liebig," 233.

19. In his reminiscences, Liebig said that Humboldt had returned just days earlier, that for this reason most of his friends did not yet realize he was back, and that this was why he failed to determine the man's identity. In fact, Humboldt returned to Paris some 5 months before this date: *Briefe Alexander von Humboldt an seinen Bruder Wilhelm* (Stuttgart: Cotta, 1880), 90.

20. Volhard, *Justus von Liebig*, 46–50. Liebig's reminiscence is in a long book dedication, written 17 years after the event, later (apparently) supplemented orally to his student and biographer, Volhard. Only a single sentence about this is contained in Liebig's surviving correspondence: "Herr von Humboldt, the famous traveler, even came up to me [at the Académie des Sciences], and conversed nearly an hour with me" (letter of 6 August 1823, in Berl, *Briefe*, 64). Regarding the relevance of Humboldt's known homosexuality and Liebig's possible inclinations in that direction, see Pat Munday, "Social Climbing Through Chemistry: Justus Liebig's Rise from the Niederer Mittelstand to the Bildungsbürgertum," *Ambix* 37(1990): 1–19, on 7. Liebig's affair with Platen is described in Xavier Mayne, *The Intersexes: A History of Similisexualism* (Privately printed, ca. 1908; Arno reprint, 1975), 609–611, and in Brock, *Justus von Liebig*, 21–26.

21. Dumas and Pelletier, "Recherches sur la Composition Élémentaire et sur Quelques Propriétés Caracteristiques des Bases Salifiables Organiques," *Annales de Chimie* [2] 24(1823): 163–191.

22. Liebig and Gay-Lussac, "Analyse du fulminate d'argent," *Annales de Chimie*, [2] 25(1824): 285–311, on 294.

23. For instance, although Liebig published his papers about picric acid (described in the next section) in German in Schweigger's *Journal für Chemie* and in Poggendorff's *Annalen der Physik*, his corresponding French-language articles in the *Annales de Chimie* appeared earlier, presumably owing to Liebig's habit of sending them directly to Gay-Lussac as enclosures in letters.

24. Wöhler to Berzelius, 11 December 1825; 17 May 1828; 18 May 1829; 14 February and 26 March 1830; in O. Wallach, ed., *Briefwechsel zwischen J. Berzelius und F. Wöhler* (Leipzig: Engelmann, 1901), 1, 101, 218, 258, 287, and 291.

25. Berzelius to Wöhler, 13 January 1826; 9 April and 9 July, 1830; ibid., 106, 292, 304.

26. Berzelius to Wöhler, 18 July and 22 November 1826; 9 April 1827; 22 January 1831; and 24 January 1832; ibid., 132–133, 155–156, 170–172, 335, and 396.

27. Wöhler to Berzelius, 11 March 1827; ibid., 168.

28. Related in Wöhler to Berzelius, 7 November 1831; ibid., 319.

29. F. L. Holmes, "Liebig's Laboratory," 121–164.

30. Berzelius, *Lehrbuch*, 173–174.

31. Prout, "On the Ultimate Composition of Simple Alimentary Substances," *Philosophical Transactions of the Royal Society* 117(1827): 355–388, on 357.

32. William Henry, *Elements of Experimental Chemistry*, suppl. to 2nd American ed., from 9th London ed. of 1823 (Philadelphia: Desilver, 1823), 122–128. Henry added: ". . . [s]ome practice in [these operations] is necessary to enable a person, who is even conversant in the general processes of chemistry, to obtain accurate results. A single experiment should never be depended upon; but the analysis of each substance should be several times repeated, and a mean taken of those which do not present any very striking disagreement . . ."

33. John E. Lesch, "Conceptual Change in an Empirical Science: The Discovery of the First Alkaloids," *Historical Studies in the Physical Sciences* 11(1981): 307–328.

34. Dumas and Pelletier, op. cit.; Liebig, "Über einen neuen Apparat zur Analyse organischer Körper, und über die Zusammensetzung einiger organischen Substanzen," *Annalen der Physik* [2] 21(1831): 1–47, on 18. The present formulation of the molecule, using a two-volume rather than Liebig's four-volume formula, is approximately one-half as large and contains 40 atoms.

35. Liebig, "Über die Analyse organischer Substanzen," *Annalen der Physik* [2] 18(1830): 357–367.

36. Liebig, "Über die Zusammensetzung der Camphersäure und des Camphers," *Annalen der Physik* 20(1830): 41–47, on 43, 45.

37. Wallach, *Briefwechsel*, 327.

38. Liebig to Berzelius, 8 January, 1831, in J. Carrière, ed., *Berzelius und Liebig, Ihre Briefe von 1831–1845*, 2nd ed. (Munich: Lehmann, 1898), 4–5.

39. Liebig, "Über einen neuen Apparat," 1–47.

40. Ibid., 6. Liebig left these directions vague, probably intentionally. Too little air sucked through the train would fail to deposit all remaining combustion

products, but too much would lead to excess product, owing to the inadvertent collection of naturally occurring atmospheric water vapor and carbon dioxide. The right amount of sucking could be learned only by experience or expert advice.

41. Ibid., 4–5.

42. Liebig to Berzelius, 14 September, 1833, in Carrière, *Berzelius and Liebig*, 71.

43. The Liebig/Gay-Lussac collection method outlined in their collaborative article of 1824 was designed to do exactly this.

44. Liebig, "Über einen neuen Apparat," 7.

45. A comparison of Liebig's, Dumas's, and modern results is revealing. For morphine, for example, Dumas and Pelletier arrived at the formula $C_{15}H_{20}NO_{2.5}$; Liebig's was $C_{17}H_{18}NO_3$; the modern formula is $C_{17}H_{19}NO_3$. For the sake of comparison, Dumas's and Liebig's formulas have been stated using modern atomic weights.

46. Holmes, "Liebig's Laboratory," 139–142.

47. A. Rocke, *The Quiet Revolution: Hermann Kolbe and the Science of Organic Chemistry* (Berkeley: University of California Press, 1993), 171, and 239–241; idem, "Hypothesis and Experiment in the Early Development of Kekulé's Benzene Theory," *Annals of Science* 42(1985): 355–381, on 364–368.

48. Liebig, "Sur un Nouvel Appareil pour l'Analyse des Substances Organiques; et sur la Composition de Quelques-Unes de ces Substances," *Annales de Chimie*, [2] 47(1831): 147–197; J. B. Dumas, "Lettre de M. Dumas à M. Gay-Lussac, sur les Procédés de l'Analyse Organique," ibid., 198–213.

49. Berzelius to Wöhler, 4 October and 10 November 1831, in Wallach, *Briefwechsel*, 372 and 379.

50. Dumas to Liebig, n.d., postmarked 5 October 1831, Liebigiana IIB, Bayerische Staatsbibliothek, Munich. "Dans l'état où est la chimie en France, c'est l'allemagne qui forme l'opinion.... De tous les chimistes que j'ai rencontrés, vous êtes celui dont le caractère, les idées m'inspirent le plus d'attachement ... On ne se fait pas en allemagne une idée de ma position. Il n'y a que moi à Paris qui lise vos mémoires. Je ne puis trouver personne à qui parler chimie, car personne ne se tient au courant de ce qui se passe dans la science."

51. Liebig to Dumas, 23 October [1837], Archives de l'Académie des Sciences, Paris.

52. Dumas, untitled footnote, *Annales de Chimie*, [2] 47(1831): 324–325n.

53. Holmes has a brief but trenchant discussion in "Liebig's Laboratory," 142. More details are revealed in a letter (n.d., postmarked 23 January 1832, Liebigiana IIB) in which Dumas told Liebig that he intended in future to perform analyses in Liebig's fashion because of the ease and rapidity of the method.

54. Dumas, "Recherches sur la Liqueur des Hollandais," *Annales de Chimie* [2] 48(1831): 185–198.

55. "Depuis quelque temps ... avec un plein succès ...": A. Dumas, "Sur l'Esprit Pyro-Acétique," *Annales de Chimie* 49(1832): 208–210.

56. Jules Gay-Lussac and T. J. Pelouze, "Sur la Composition de la Salicine," *Annales de Chimie* [2] 48(1831): 111. The issue date was September 1831, but the actual date of appearance may not have been before the end of the year.

57. A. Dumas, "De l'Analyse Élémentaire des Substances Organiques," *Journal de Pharmacie* [2] 20(March 1834): 129–156; republished verbatim in *Traité de Chimie Appliquée aux Arts*, vol. 5 (Paris: Béchet, 1835), 3–30, esp. 26–28.

58. Berzelius, untitled review in *Jahresbericht* [for 1830] 11(1832): 214–215.

59. Berzelius to Liebig, 22 April 1831, in Carrière, *Berzelius and Liebig*, 7.

60. Liebig to Berzelius, 8 May 1831; ibid., 11. J. R. Partington mistranslated an archaic form of the word *nevertheless* in this phrase (*demongeachtet = demungeachtet*) as "with the devil's help": *A History of Chemistry*, vol. 4 (London: Macmillan, 1964), 339.

61. Liebig to Berzelius, 6 November and 22 December 1832, and 30 May 1833; Berzelius to Liebig, 27 November 1832; 15 January, 21 May, and 30 August 1833; in Carrière, *Berzelius and Liebig*, 43, 46, 49–51, 60, 66, and 68 (quotations on 50–51 and 68).

62. Berzelius to Wöhler, 20 March 1835, in Wallach, *Briefwechsel*, 609.

63. Wöhler to Berzelius, 24 November 1831, in Wallach, *Briefwechsel*, 381.

64. Wöhler to Berzelius, 1 December 1831; ibid., 387. Berzelius wrote to Liebig (13 December 1831), "I am happy to learn that you have made Wöhler into a proselyte for organic analyses. He was always before somewhat disinclined towards this kind of work;" Carrière, *Berzelius and Liebig*, 19.

65. Wöhler to Berzelius, 17 January and 19 August 1832, in Wallach, *Briefwechsel*, 399 and 448.

66. C. H. Pfaff and J. Liebig, "Über die Zusammensetzung des Caffeins," *Annalen der Pharmacie* 1(1832): 17–20; Wöhler and Liebig, "Über die Zusammensetzung der Schwefelweinsäure," ibid., 37–43; Pelouze and J. Gay-Lussac, "Über die Zusammensetzung des Salicins," ibid., 43. All three papers emphasized the use of the Kaliapparat. Regarding Liebig's takeover of the journal, see Ulrike Thomas, "Philipp Lorenz Geiger and Justus Liebig," *Ambix* 35(1988): 77–90.

67. "Herr Hess" was mentioned in Liebig's landmark article (see preceding); also Carl Oppermann, "Einige vergleichende Versuche mit dem sogenannten Baum-wachs und Bienenwachs," *Magazin für Pharmacie* 35(1831): 57–64.

68. Carl Ettling, "Beiträge zur Kenntniss des Bienenwachses," *Annalen der Phar-macie* 2(1832): 253–267; Wöhler and Liebig, "Untersuchung über das Radikal der

Benzoesäure," *Annalen* 3(1832): 249–282; R. Blanchet and E. Sell, "Über die Zusammensetzung einiger organischer Substanzen," *Annalen* 6(1833): 259–313; and A. F. Boutron-Charlard and T. J. Pelouze, "Mémoire sur l'Asparamide et sur l'Acide Asparamique," *Annales de Chimie* [2] 52(1833): 90–105.

69. William Brock, *From Protyle to Proton: William Prout and the Nature of Matter* (Bristol: Hilger, 1985), 18–20.

70. J. Pelletier, "Untersuchung über die elementare Zusammensetzung mehrerer näheren Bestandtheile der Vegetabilien," *Annalen der Pharmacie* 6(1833): 21–34.

71. In a letter to Wöhler of 9 February 1837, Liebig stated that the only change made from the original apparatus of 1831 was the use during the final aspiration of a diagonal glass tube inserted over the broken end of the combustion tube, to prevent CO_2 contamination from the charcoal fire [A. W. Hofmann, ed., *Aus Justus Liebig's und Friedrich Wöhler's Briefwechsel in den Jahren 1829–1873* (Braunschweig: Vieweg, 1888), 1, 99]. Wöhler wrote to Berzelius (12 February 1837) that it was his understanding that Liebig still used the apparatus "in its original configuration and simplicity" (Wallach, *Briefwechsel*, 674).

72. Holmes ("Liebig's Laboratory," 143) states that Liebig made no claims regarding higher accuracy until this time. I had no greater success than did Holmes in finding an explicit statement of this kind from Liebig earlier than 1833. Nonetheless, Liebig must have been conscious that his procedure was intrinsically more precise than the volumetric approach. Liebig, along with everyone else, emphasized the advantage of the method in its capability of handling much larger sample size, and the attainable precision was well understood to increase (*ceteris paribus*) proportionally with sample size. As described above, the crucial test was the alkaloids, which *required* a more precise method to arrive at secure results. Liebig's confidence in January 1831 that he had successfully cracked this problem demonstrated his conviction regarding the greater precision of the method, even if he declined to render that claim explicit. In 1839, he argued once more that the issue was not precision but rather ease and reliability (fn. in *Annalen* 26[1838]: 193n.), pointing out that the old method could, with skill and care, be used to derive values that stood the test of time. However, Liebig knew that this depends on what substances are being analyzed, and presumably he never would have denied that the higher-molecular-weight compounds required the Kaliapparat for trustworthy results.

73. Liebig to Berzelius, 15 March, 30 May, and 14 September 1833, in Carrière, *Berzelius and Liebig*, 52, 62, and 71; Berzelius to Liebig, 30 August 1833, Carrière, *Berzelius and Liebig*, 67. Berzelius commented acidly, "It's good to have a confirmation of Dumas' analyses, for it is impossible to rely on D.'s results, since he so often lets himself be misled by theoretical predictions, and never publicly confesses his errors."

74. Blanchet and Sell, "Zusammensetzung," 304–305.

75. Liebig, editorial note, *Annalen der Pharmacie* 22(1837): 50–52; Liebig, "Über die vorstehende Notiz des Hrn. Akademikers Hess in Petersburg," *Annalen der*

Pharmacie 30(1839): 313–319. Holmes has rightly emphasized this point in his discussion in "Liebig's Laboratory."

76. Dumas, "Recherches de Chimie Organique," *Annales de Chimie* [2] 53(1833): 164–181. The paper was read on 5 August 1833, but Dumas commented that he had earlier taught it to Pelouze and Boutron-Charlard, who used it in their paper on aspartic acid (read 11 March 1833). This is the first publication of Dumas's innovative method for nitrogen; most citations in the secondary literature incorrectly cite Dumas's letter to Gay-Lussac published in the *Annales de Chimie* in 1831.

77. Ibid., 171–172.

78. Dumas, "Détermination du Nombre d'Atomes qu'une Matière Organique Renferme," *Journal de Pharmacie* [2] 20(1834): 185–223, on 211.

79. Victor Meyer and Paul Jacobson, *Lehrbuch der organischen Chemie*, 2nd ed., vol. 1 (Leipzig: Veit, 1907), 21–26.

80. Berzelius, *Lehrbuch der Chemie*, 3rd ed., vol. 6 (Dresden: Arnold, 1837), 49, 55, 60–63. That this section was written in 1835 can be verified by reading the relevant passages in the Berzelius-Wöhler correspondence, for Wöhler was still serving Berzelius as translator and editor. As was the case for the second edition, the German third was the *editio princeps* of this important work.

81. Mitscherlich, *Lehrbuch der Chemie*, 2nd ed., vol. 1 (Berlin: Mittler, 1834), 205–210. According to the preface, this volume had been printed in November 1832.

82. Liebig, "Analyse, org.," in *Handwörterbuch der reinen und angewandten Chemie*, vol. 1 (Braunschweig: Vieweg, 1836–1842), 357–400 (third fascicle, 1837); idem, *Anleitung zur Analyse organischer Körper* (Braunschweig: Vieweg, 1837); idem, *Instructions for the Chemical Analyses of Organic Bodies*, translated by W. Gregory (Glasgow: R. Griffin, 1839). For precise publication information, see M. and W. Schneider, "Das 'Handwörterbuch' in Liebigs Biographie," in *Orbis pictus: Kultur- und Pharmaziehistorische Studien* eds. W. Dressendörfer and W.-D. Müller-Jahnke (Frankfurt: Govi-Verlag, 1985), 247–254; the Liebig citation is from his letter to Eduard Vieweg of 11 January 1837 (251–252). The *Handwörterbuch*, a landmark scientific publication, was sold piecemeal in fascicles over the course of many years, which has led some scholars to misdate the various volumes and the articles within them.

83. Liebig, "Analyse, org.," 361, 369, 378; *Instructions*, 5, 17, 28. The translation, here and later, is Gregory's English translation, slightly modified by comparison with the German original.

84. Liebig, "Analyse, org.," 360; *Instructions*, 4.

85. Liebig, "Analyse, org.," 364, 379–380; *Instructions*, 9, 29–31.

86. Liebig to Berzelius, 25 February 1837, in Carrière, *Berzelius und Liebig*, 126.

87. Berzelius to Liebig, 20 February 1838, in Carrière, *Berzelius und Liebig*, 146.

88. Liebig to Wöhler, 2 March 1838; Liebig to Berzelius, 7 March and 15 April 1838; Berzelius to Liebig, 3 April 1838; in Carrière, *Berzelius und Liebig*, 147–154.

89. Berzelius, "Nachtrag zum VI. Band: Ueber die Analyse organischer Körper durch Verbrennung," *Lehrbuch der Chemie*, 3rd ed., vol. 7 (Dresden: Arnold, 1838), 610–630.

90. Liebig, editorial note, *Annalen der Pharmacie* 26(1838): 192–1994n.; Liebig, "Über die vorstehende Notiz," 313–319. Liebig made the same claims, with the same data, in his letter to Wöhler of 2 March and his letter to Berzelius of 7 March 1838 (Carrière, *Berzelius und Liebig*, 147–149).

91. For example, C. R. Fresenius, *Anleitung zur quantitativen chemischen Analyse*, 3rd ed. (Braunschweig: Vieweg, 1853), 342–391; and George Fownes, *A Manual of Elementary Chemistry, Theoretical and Practical*, 10th ed. (Philadelphia: Lea, 1874), 448–457.

92. J. F. Daniell, *An Introduction to the Study of Chemical Philosophy* (London: Parker, 1839), 316–317. A fuller discussion is found in his second edition (1843, 606–610), in which Liebig's method was introduced as "the more usual process." However, Prout's method was also described, with no clear preference given.

93. W. Gregory, "Translator's Preface" to *Instructions*, iii; William Francis and Henry Croft, letter of 13 December 1840 from Berlin, *Philosophical Magazine* [3] 18(1841): 202.

94. Berzelius to Wöhler, 1 May and 10 July 1829, in Wallach, *Briefwechsel*, 253 and 267.

95. See Volhard, *Justus von Liebig*, 131–49; and W. H. Brock and Susanne Stark, "Liebig, Gregory, and the British Association," *Ambix* 37(1990): 134–147.

96. T. Thomson, *Chemistry of Organic Bodies* (London: Bailliere, 1838), *vi*.

97. T. Graham, *Elements of Chemistry* (London: Bailliere, 1842), 698–708.

98. William Gregory, *Letter to the Right Honorable George, Earl of Aberdeen . . . on the State of the Schools of Chemistry in the United Kingdom* (London: Taylor & Walton, 1842), esp. 18–22 and 28–29. See also R. Bud and G. K. Roberts, *Science versus Practice: Chemistry in Victorian Britain* (Manchester: Manchester University Press, 1984).

99. Liebig, *Manuel pour l'Analyse des Substances Organiques*, translated by A. J. L. Jourdan (Paris: Baillière, 1838).

100. J. Pelouze and E. Fremy, *Cours de Chimie Générale*, (Paris: Masson, 1848–50), "The simplest of all analytical methods, and one which has replaced those we have just discussed, is due to M. Liebig ... [It] is as simple as it is exact; it has much contributed to the very great progress made by organic chemistry in recent years" (vol. 3, 49–60); V. Regnault, *Cours Élémentaire de Chimie*, 2nd ed., vol. 4 (Paris: Masson, 1849–50), 7–17. Regnault's work was translated into German and was

edited by Adolf Strecker; it was perhaps the most popular organic chemical textbook in Germany in the 1850s.

101. Löwig, *Chemie der organischen Verbindungen*, 2nd ed., vol. 1 (Braunschweig: Vieweg, 1846) 141 (preface dated September 1844).

102. O. P. Krätz and C. Priesner, *Liebigs Experimentalvorlesung* (Weinheim: Verlag Chemie, 1983), 11.

103. A student of Adolphe Wurtz (himself a student of Liebig) wrote, "I myself used one given me by Wurtz for twenty years [ca. 1851–71], and it almost broke my heart when it finally cracked in my hands:" Scheurer-Kestner to Hofmann, in Hofmann, *Erinnerung*, vol. 1, 224–225.

104. Willi Conrad, *Justus von Liebig und sein Einfluss auf die Entwicklung des Chemiestudiums und des Chemieunterrichts an Hochschulen und Schulen* (doctoral dissertation, Technische Hochschule Darmstadt, 1985), 17. A new series of experiments using a reconstructed Kaliapparat and following Liebig's 1837 instructions has been carried out under the direction of Professor Melvyn Usselman; a report on this subject is in preparation. The combustions fully confirm the accuracy and simplicity of the method.

105. Maurice Crosland, *Gay-Lussac: Scientist and Bourgeois* (Cambridge, UK: Cambridge University Press, 1978), 92–114, 117–118 (quotation on 93).

106. Is the Kaliapparat an *instrument* or an *apparatus*? It actually is both. It is an apparatus in that it is a laboratory contrivance designed to accomplish a procedure or task; but it is an instrument in that the outcome of the procedure is a precise numerical measurement. It would be interesting to consider how large this dual category "instrument-apparatus" is.

107. Holmes, "Liebig's Laboratory."

108. Liebig, "Bemerkungen zu vorhergehenden Abhandlung [von Thenard]," *Annalen der Pharmacie* 2(1832): 19–30, esp. 19–22.

109. See, for example, Wöhler's fascinating comments on his trip to "Babylon," in his letters to Berzelius of 13 and 27 October 1833, in Wallach, *Briefwechsel*, 526–538.

12

Chemical Techniques in a Preelectronic Age: The Remarkable Apparatus of Edward Frankland
Colin A. Russell

One of the masters of chemical technique in the nineteenth century was Edward Frankland (1825–1899).[1] The illegitimate son of a noted lawyer, he was raised by his mother in Garstang and then in Lancaster, England. After attending a number of schools that mostly failed to provide any kind of scientific education, he was apprenticed to a pharmacist in Lancaster as a substitute for the medical education that his mother could not afford. Here the foundations were laid for neatness and scrupulous accuracy in dispensing, and the continuous use of his manual skills brought them to a high level of proficiency. Meanwhile, he learned the rudiments of scientific theory from popular lectures, at the Mechanics Institute and in a laboratory equipped for young would-be students by a philanthropical doctor in the town.

Through the good offices of this doctor and the local member of Parliament, Frankland then moved to London to work in the chemistry laboratory of Lyon Playfair. Here he began his lifelong friendship with another assistant, Hermann Kolbe, and worked (still with Playfair) at the Putney College of Civil Engineering. After a preliminary visit to Bunsen's laboratory at Marburg during the summer of 1848, he returned to England in some haste, having received a call to teach at Queenwood College in Hampshire, a strange "progressive" institution that left him astonishingly free to develop practical teaching of science in almost any way he liked. Within a year, he was back at Marburg, this time staying for the year needed to start and complete a doctoral degree. This was granted in 1849, largely on the basis of his discoveries of a new kind of compound that he called "organometallic" and of a sub-class known as *dialkylzincs*.

Currently, this is one of the most important branches of chemistry, founded by Frankland and since employed in a multitude of organic syntheses. The great reactivity of these compounds opens up endless possibilities for alkylation and other reactions, whereas the study of their structure is important for both organic and inorganic chemists. In fact, Frankland's organozinc compounds were too reactive, for which reason the full exploitation of organometallic compounds had to await the discovery of their more amenable analogs, the Grignard reagents, in 1900.

After a brief spell back at Putney College, he obtained the first chair in chemistry at the new Owens College in Manchester. He was there from 1851 to 1857. He spent the rest of his career in London, serving at a variety of institutions (sometimes several at once), including St Bartholomew's Hospital, the Addiscombe Military Academy, the Royal Institution, and the Royal College of Chemistry (later Imperial College).

In this chapter, several of Frankland's technical innovations are described. Notably, they sprang from a wide range of chemical interests: civil and military; synthesis and analysis; academically obscure and prominently visible; above all, pure and applied. This breadth of interest was rarely encountered elsewhere at that time and may go some way to explaining why he was so successful. A final section attempts to draw some further general conclusions both as to the reasons for his success and as to the origins and reception of his techniques. An interesting enquiry would be how far they apply to scientific procedures in general. Meanwhile, the following are some of the techniques by which he transformed the face of chemical laboratory practice in the years before black boxes, automation, and other by-products of the electronic age.

USE OF THE REFLUX CONDENSER

In 1846, Frankland and Kolbe wanted to test the theory that alkyl cyanides were nitriles of carboxylic acids. This was part of a wider hunt for "radicals" that they (and relatively few others) believed to be constituent parts of organic molecules. What was required was a demonstration of the presence of "alkyl" radicals in a range of compounds, including the famous aliphatic acids. Were their nitriles in fact just alkyl cyanides? If so, the acids too must contain alkyl. To establish this hypothesis, they subjected ethyl cyanide (bt 97°C) to boiling, concentrated aqueous potash, a reaction that we now know requires many hours to complete. At first, they seem to have used a distillation apparatus, repeatedly returning the distillate to the retort until no odor of the cyanide was perceptible.[2] After evaporation, followed by acidification of the solid residue, the contents of the flask were redistilled. The

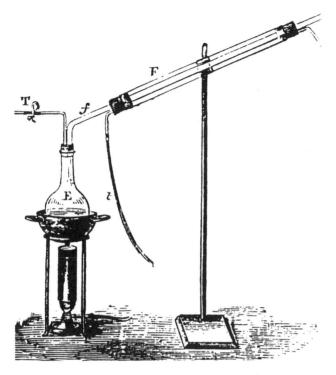

Figure 12.1
An early illustration of a Liebig condenser in the semireflux position, enabling volatile reagents to be recondensed and returned to the flask. [From E. Frankland, *How to Teach Chemistry: Hints to Science Teachers and Students*, ed. G. Chaloner (London: J. & A. Churchill, 1875).]

distillate showed all the characteristics of aqueous propionic acid, thus confirming their theory as to the nature of nitriles.[3] However, this method was not applicable to the more volatile methyl cyanide (bt 82°C), and here Frankland employed a "simple contrivance" consisting of a Liebig condenser whose lower end was bent downward to an obtuse angle, thus enabling the condensed nitrile to return repeatedly to the flask (figure 12.1).[4] This "condensed refrigerator" he claimed to be the first use of a condenser in the reflux position.[5] He had in fact been anticipated by Mohr 10 years previously,[6] but was apparently unaware of that. One wonders why such an extremely simple device had not been in common use for a long time. It probably reflects as much the primitive state of organic manipulation in the early nineteenth century as Frankland's skill at laboratory techniques.

Employment of a reflux condenser is now so commonplace that finding an area of organic chemistry to which it is never applicable is difficult. Subjecting volatile liquids to this technique permits heating them without intermission for an indefinite period of time, the low-boiling components being continuously returned to the flask for further reaction.

It has been employed for simple reactions, such as ester formation and hydrolysis; acylation of amines; aromatic nitrations; preparation of acetoacetic ester; and use in syntheses, the Wittig reaction, the Friedel Crafts reaction, and a host of others. By this means, Grignard was able consciously to follow Frankland's example and discover the alkylmagnesium halides named after him, by heating magnesium, alkyl halide, and ether under reflux conditions. In conjunction with the later Soxhlet extractor, the technique opened up a powerful but simple method of using volatile solvents to extract organic materials from natural products as leaves, bark, and the like.

HEATING UNDER PRESSURE

At an early stage of his chemical career, Frankland was infected with a passion to isolate the radical "ethyl." It was all very well to know that it might be present in acids, nitriles, and so on. The next step must surely be actually to isolate it, to collect it in some way, and to analyze it eudiometrically (by explosion with excess oxygen). Such thinking was inescapably in the spirit of Berzelius, whose radical theory of organic chemistry was still surviving despite several devastating blows from its opponents. In Kolbe he found a kindred soul, and the latter was in fact able to produce what appeared to be a hydrocarbon "radical" from his electrolysis of fatty acid salts.

While still at Queenwood College, in 1847/48, Frankland made the momentous decision to try to release "ethyl" from ethyl iodide by treating it with the very electropositive metal potassium. A strong evolution of gas ensued when the temperature approached the boiling point of ethyl iodide. It seemed hopeful, but the liquid naturally tended to boil away, so he tried again, this time in a closed vessel to prevent ebullition. The vessel was a glass tube sealed at each end. However, the gas produced was not "ethyl" but a strange mixture that seemed to contain hydrogen and "methyl" (perhaps better called *ethane*). So, he determined to replace the vigorous reagent potassium with the less electropositive element zinc, thus hoping to avoid any decomposition of his expected ethyl.

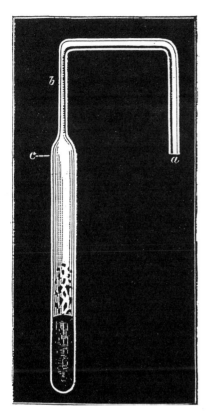

Figure 12.2
Reaction vessel for allowing zinc and alkyl iodides to react. After introduction of the reagents, the tube was heated locally at *c*, and the upper part drawn out into a narrow tube (*b*), which then was sealed at *a*. [From E. Frankland, *Experimental Researches in Pure, Applied and Physical Chemistry* (London: van Voorst, 1877), 69.]

This time he heated the reagents together from the beginning in a sealed tube and made the wholly unexpected discovery of the remarkable zinc alkyls. To prepare these, he took a tube, some 30 mm long and made from hard Bohemian glass. Metal was introduced, and one end was sealed. It was then alternately warmed and cooled, with the open end dipping into ethyl iodide, thus introducing the liquid into the tube. The latter was first evacuated of air by a pump, and the open end then was sealed in a flame (figure 12.2). The sealed tube was then placed in an oil bath and heated for several hours, almost the first time such a technique had been used. Because his only eudiometer at Queenwood had exploded, the tube had to be left unopened and was in fact transported to Marburg. Later, in the following year, he opened the tube under water with spectacular and unexpected results: "A greenish blue flame, several feet long, shot out of the tube."[7]

Several years later, in Manchester, he determined to exploit the reactivity of the zinc alkyls to the full. However, severe limits restricted the amounts that could be made by heating sealed tubes in an oil bath, with the constant danger of explosion. At approximately this time, he developed a lifelong friendship with engineer James Nasmyth. Out of their discussions emerged designs for digesters capable of conducting chemical reactions under considerable pressure. Using iron digesters supplied by Nasmyth, he found that the sealed tubes of reagents were kept from bursting by the external pressure of steam in the vessel.[8] This latest Mancunian technology of high-pressure vessels led to the production of diethylzinc on a large scale. Figure 12.3 illustrates one such apparatus made of copper.

This was another example of Frankland's technical versatility. With the product, he examined many new reactions of diethylzinc (e.g., that with nitric oxide in the apparatus in figure 12.4, where a piston within a condensing syringe [D] forces the gas into the digester). This semiautomatic apparatus was necessary, as the reaction could take several weeks for completion.

Such digesters proved to be of great value to other chemists, including Hofmann[9] and Williamson. Apologizing for a delayed delivery date to the latter, Nasmyth expressed the hope to Frankland that he would "keep a keen look out to commercialise ... and turn them to profitable results" for then, "with such powerful tools as money can furnish," he would be able "to set out as pioneer of that glorious band who are soon in the van of discovery."[10] He was unlikely to have turned away from such a prospect.

The use of sealed tubes is extremely simple, needing only a hard-glass tube and a steam bath or (for higher temperatures) an oil bath. On the other hand, it does require a certain confidence and skill in glass blowing to ensure that the seal is able to withstand the great internal pressures generated.

The procedure enjoyed a certain vogue in the late nineteenth and early twentieth centuries, although the inherent dangers of heating under great pressures deterred many and encouraged a search for alternative techniques. It has the great advantage of enabling liquids to be heated considerably above what would be their boiling temperatures at normal pressures, with no interference from atmospheric contamination or need to employ a reflux condenser. The technique

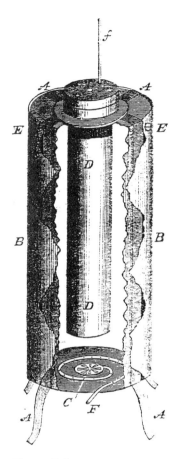

Figure 12.3
Reactions under pressure. A sealed copper digester is placed in the copper oil-bath
(*DD*), heated by a gas burner (*f*), and enclosed in a sheet-iron cylinder (*BB*).
[From E. Frankland, *Experimental Researches in Pure, Applied and Physical Chemistry*
(London: van Voorst, 1877), 194.]

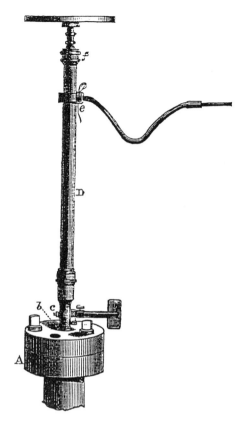

Figure 12.4
Apparatus for introducing a gas (such as nitric oxide) into a copper digester containing dialkylzinc. *A* is the head of the copper digester, *D* is a syringe with a tightly fitting piston, and *c* is a stopcock. [From E. Frankland, *Experimental Researches in Pure, Applied and Physical Chemistry* (London: van Voorst, 1877), 213.]

was formerly employed in the Carius method for estimation of halogen. It has been used synthetically for various condensation reactions (as that between aldehydes and active methyl compounds as 2- and 4-methyl quinolines).

GAS COLLECTION

As part of his quest for "radicals," Frankland wanted to prepare "ethyl" from ethyl cyanide and potassium. Assuming it was gaseous, he devised the apparatus in figure 12.5 to collect the gas in bulk and also in separate sample tubes.

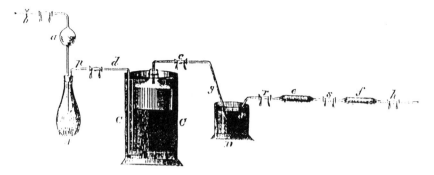

Figure 12.5
Production of "ethyl" (butane) by the action of potassium on ethyl cyanide. See text for details. [From E. Frankland, *Experimental Researches in Pure, Applied and Physical Chemistry* (London: van Voorst, 1877), 50.]

Potassium was in flask *A*, ethyl cyanide in bulb *a*, and mercury in *D*, and the whole apparatus beyond *B* was filled with water. The nitrile was introduced by opening tap *b* and, after displacement of the air, the hydrocarbon gases were collected in *B*. Sample tubes *e* and *f* were filled by opening *h*, thus allowing gas from *B* to displace water in the tubes, which then were isolated, removed, and stored.

The technique was chiefly noteworthy for the added precautions to expel air before collection, to remove the nitrile (by standing the gas in the water in *B* [see figure 12.5] for some hours) and (later) to dry the gas samples and to equilibrate with atmospheric temperature before weighing. The preparation of large volumes of gases is no longer a major feature of organic chemical practice, chiefly because the hydrocarbon products are far better obtained from petroleum and partly because they are no longer identified with "radicals." Further, it cannot be claimed that the specifics of this apparatus were of importance to posterity, but their employment emphasized in general terms several paramount needs in organic chemical practice, especially those of product purification and accurate measurement of volumes and weights.

DISTILLATION IN INERT ATMOSPHERES

Distillation in inert atmospheres, commonplace today, makes one of its earliest appearances with Frankland's apparatus for distillation of the highly reactive dialkylzincs under hydrogen (figure 12.6). *D* is a

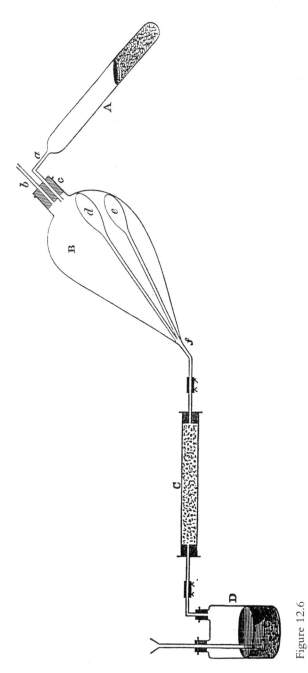

Figure 12.6
Distillation of the highly reactive dialkylzincs under hydrogen. See text for details. [From E. Frankland, *Experimental Researches in Pure, Applied and Physical Chemistry* (London: van Voorst, 1877), 174.]

hydrogen generator, C is a calcium chloride tube, A is the container for dimethylzinc, and B is a receiver containing several bulbs that can be filled with the condensed liquid. After flushing the apparatus with hydrogen, the tubes at b and f were sealed, and A was warmed while B was cooled. The entire operation must have been a most hazardous procedure, though the danger was perhaps less than allowing the organometallic compound to have access to either air or water. Probably in organometallic chemistry this procedure is still most employed, though the inflammable hydrogen is now more commonly replaced by pure dry nitrogen or (occasionally) argon.

WATER ANALYSIS

Frankland's development of new methods of analyzing water supplies brought him greatest fame in his lifetime and also performed his most direct service to the community. The problem was complex as, in the 1860s, the microbiological causes of waterborne diseases were ill understood. The approach of Frankland was to measure nitrogen concentration as an indication of previous sewage contamination and also to determine the ratio of organic and inorganic nitrogen in the water.

The old methods all left much to be desired, and his collaborator, H. E. Armstrong, observed of Frankland:[11]

> Thorough in everything he did, he soon became dissatisfied with the methods, especially the determination of organic impurities indicative of sewage contamination. He decided to revise them all. At the close of the summer session of 1866, he did me the great honour to propose that I should carry out the work for him. The method of combustion analysis *in vacuo* we devised was made known, the following year, in a lecture by Frankland at the Royal Institution; the work generally was described at the Chemical Society in February 1868.

The biggest problem, apparently intractable, was that "No process has yet been devised by which the amount of organic matter in water can be even approximately estimated."[12] After nearly 2 years' work, they found an answer.

The new method was a development of Liebig's classic technique for estimating carbon and nitrogen in organic compounds. The essential feature of this variation (figure 12.7) was oxidation of the evapo-

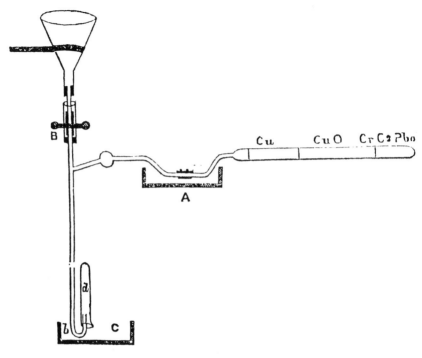

Figure 12.7
Estimation of carbon and nitrogen in water residues, with oxidation by lead
chromate and measurement of the volumes of the resultant gases, CO_2 and N_2.
[From E. Frankland, *Experimental Researches in Pure, Applied and Physical Chemistry*
(London: van Voorst, 1877), 574.]

rated residue with lead chromate in an evacuated tube, the resultant
gases (CO_2 and N_2) being estimated volumetrically. A refined form of
apparatus for the latter procedure was on general sale (figure 12.8). To
avoid interference from carbonates, the preliminary evaporation was
first conducted with "sulphurous acid," thereby expelling CO_2, and
also liberating nitrogen from nitrates and nitrites (a "remarkable reaction" that "could scarcely have been predicted"). The evacuation was
effected, before and after combustion, by the mercury fall-pump
described by Sprengel in 1865.[13]
    While this work was in progress, an alternative attack on the
problem had been developed by Frankland's former student, J. A.
Wanklyn (now professor of chemistry at the London Institution) and
two of his collaborators.[14] They treated the water first with alkali to
expel ammonia from ammonium salts and urea and second with alka-

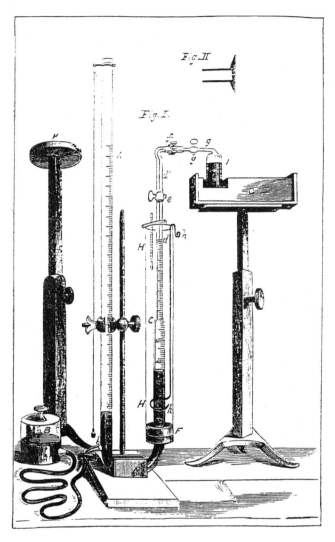

Figure 12.8
Modified apparatus for "analysis of gases incident to water analysis." [From E.
Frankland, *Experimental Researches in Pure, Applied and Physical Chemistry* (London:
van Voorst, 1877), 594.]

line potassium permanganate, claiming the organic material to be quantitatively oxidized to ammonia, which may then be estimated colorimetrically using Nessler's solution. This method was clearly simpler than that of Frankland and Armstrong but was less accurate. They pointed out that in Wanklyn's technique, even for albumen, oxidation is never complete,[15] and Wanklyn later admitted that only two-thirds of the organic material is so oxidized.[16]

In retrospect, clearly the Frankland technique had several advantages. First, it was extremely accurate. To read Frankland's papers on the subject is to be transported into a world of painstaking attention to detail, of careful standardization of reagents and calibration of apparatus, of blank control experiments, of rigorous procedures for quantitative transfer, of flawless vacuum techniques, even of meticulous instructions for "reading" Nessler tubes. It is a world largely forgotten today, except perhaps by those with long memories stretching back into a preelectronic era. Second, the method was capable of amendment and adjustment as experience demanded. Those (including Frankland's critics) who saw this as a sign of inadequate technique completely missed the point that *all* scientific procedures need constant updating, though often only in minor matters of detail. It was, as Odling said, "a new and very refined process of water analysis."[17]

However, it also had a number of disadvantages. First and foremost among these was the sheer difficulty of the techniques involved. For a Frankland or an Armstrong, this was not a problem, nor was it likely to be so for the numerous assistants whom they had trained. However, for lesser mortals, it presented serious difficulty, though this can be exaggerated. American chemist J. W. Mallet, writing many years later and after some simplifying improvements, described it as "a process of great delicacy, and quite satisfactory in its details with proper precaution and in trained hands."[18]

Two further and related problems were the time taken for a complete analysis (up to 2 days) and the financial cost (in terms of time, apparatus, and chemicals). Finally came the critical question as to what exactly was being measured. One analyst put his finger on the problem thus:

> Most chemists are, we believe, convinced that, assuming the organic matter to be once inside the combustion tube free from an admixture of nitrates, its carbon and nitrogen can be estimated with an extraordinary degree of accuracy, by means of Frankland and Armstrong's

process. The real questions, however, are, firstly: does the process enable us to get the organic matter dissolved in a litre of water into our combustion tube undiminished in quantity, and freed from the large excess of nitrates, with which it is often associated, and, secondly, can we make accurate allowance for any ammonia which may be present in the water?[19]

By comparison with the Frankland-Armstrong process, the alternative introduced by Wanklyn and his colleagues was both faster (taking perhaps 3 hours) and cheaper. On the other hand, it suffered from the defect that albumen is not quantitatively decomposed into ammonia under the test conditions and moreover, that urea, supposedly eliminated as ammonia by preliminary treatment with sodium carbonate, does not give 100% conversion. For these reasons, Wanklyn introduced his correction factor of two-thirds. This stratagem, together with his failure to conduct his own tests of the rival process, led to much acrimony in the pages of *Chemical News*.[20] The results obtained by this method were almost always lower than those obtained by Frankland and raised the critical question as to whether albuminoid substances were in fact the right subjects for analysis.

By 1872, some claimed that most leading analysts had rejected Frankland's process in favor of Wanklyn's, their number including Thomas Way, Angus Smith, W. A. Miller, A. Voelcker, and Henry Letheby. Crookes, the editor of *Chemical News*, remarked, "[w]e scarcely know a single chemist of reputation who approves of Dr. Frankland's water analysis."[21] Four years later, Frankland himself conceded the popularity of the ammonia process.[22] Nevertheless, one should not exaggerate the monopolistic position claimed for the latter.

Thus, the Nottingham Public Analyst, E. B. Truman, assured Frankland in 1874 that he was using his method for "hundreds of analyses."[23] J. W. Thomas of Cardiff wrote that, though he had temporarily gone over to Wanklyn's method to keep fees down, he hoped to return to the Frankland method as soon as possible,[24] and Odling told Frankland "you may always refer to me as a chemist habitually using your process of water analysis," even though they might occasionally disagree over the conclusions.[25] An undated note in Frankland's handwriting discloses that his process was in use by Bischof, Campbell Brown, H. Brown, Donkin, Hill, Moulting, Odling, O'Sullivan, Russell, Tate, Thomas, Truman, and Williamson.[26] A copy-letter to Sir Hugh Owen, also undated but probably dating in the early 1890s, names Hill, Mills, Odling, and Tidy as users of his own process that, however, is "tedious, laborious and costly." However, the ammonia process, though "utterly untrustworthy," is used by "the profession at large."[27]

In the end, Frankland's water analysis techniques were replaced by those of microbiology. Their influence on subsequent chemistry was trivial. Yet they had an importance for Victorian society that is difficult to exaggerate. Despite the protestations of the chemists employed by the water companies, it seems very likely that urbanized Britain might have suffered something approaching a pandemic through waterborne diseases. Only Frankland's method was capable of identifying water that had been in contact with previous sewage contamination, and this was the critical question.

True, his conclusions were sometimes over-severe (as when nitrate concentrations were abnormally high through seepage of innocuous fertilizers), but it was preferable to err on the side of caution. Water condemned by Frankland was at best a risky commodity and at worst a lethal cocktail of virulent microorganisms. Understandably, his water analysis caused Frankland to became best known by the public at large.

The development of these techniques affected the chemical community in a variety of ways. First and most obviously, they divided it. Chemist was ranged against chemist as testimonies were given, often in a court of law, for and against an individual water company. Of course, the reason was that no real agreement existed as to the true etiology of the outbreaks of typhoid, typhus, cholera, and the like. Second, they were intimately connected with the rise of the chemical profession in Britain. As chemists performed chemical analyses of waters that had once been the domain of doctors, engineers, or even the Astronomer Royal, so their professional awareness slowly increased. It is no coincidence that the first president of the new incarnation of this awareness, the Institute of Chemistry, should be none other than Edward Frankland himself.

COMBUSTION APPARATUS

By the early 1860s, a renewed interest was aroused in combustion generally. Once it had been at the very heart of Lavoisier's revolution, but its prominence had receded with the advent of electrochemistry and the growth of organic chemistry. Now the demands of applied chemistry were bringing it back into fashion. After a successful ascent of Mont Blanc in 1859, Frankland conducted a series of experiments on combustion of candles at low and high altitudes. Possibly through his connection with a military college at Addiscombe, he turned his attention to the variation of burning times for shell fuses at different altitudes (and therefore pressures). It had been shown that, unlike Frankland's candles, time fuses took much longer to burn at high altitudes than at low levels.[28]

To examine this phenomenon, he constructed an apparatus that embodied many familiar features; a large iron cylinder with an air pump at one end and at the other a 6-foot-long gas pipe connected to a mercury pressure gauge and capable of receiving the fuse. An iron ball was suspended by a thread at the far end of the fuse, so when combustion eventually reached that end, the thread would burn, and the ball would drop onto an iron plate below. Seemingly this was a rare case in Frankland's work where the critical signal was acoustic. Again, it was an astonishingly simple solution to a complex problem. He concluded that each diminution of 1 inch of barometrical pressure causes a retardation of 1 second in a 6-inch or 30-second fuse. The relevance of this work to gunnery soon became evident to the British army, especially as they operated on mountainous terrain. Yet his discovery of a relation between pressure and combustion rate was to foreshadow the programs of combustion research conducted by the oil companies before and after World War II. Ironically true is that crucial to the complex mechanisms were the formation and reaction of exactly those transient entities sought so diligently but never obtained by Frankland: the alkyl radicals.

CONCLUSION

Even a cursory examination of Frankland's new techniques reveals several obvious reasons for their success.

Availability of Materials
When Frankland returned to England in the summer of 1848, he brought with him, for the benefit of the students at Queenwood, a very large quantity of apparatus and chemicals made in Germany. What could he have imported?

First was Bohemian glass, a commodity prized for its use in chemical manipulations. It is low in lead and low in soda, having a higher melting temperature than the usual alternative, soda glass. To manufacture it into apparatus, a blowpipe was of course required. From this material he constructed his sealed tubes, for example. Another commodity was *caoutchouc* (rubber) that formed the essential joining between adjacent glass vessels. It was made by fastening a sheet around a glass tube and keeping it in boiling water for some time.[29]

For consumable raw materials, Frankland usually made his own chemicals, as ethyl iodide, acetoacetic ester, and the like. He used his considerable network of contacts to acquire all kinds of raw materials, as in the case of 6-inch fuses provided by his friend, F. A. Abel, at Woolwich Arsenal.

Technology Transfer

Frankland was a master at transferring a technique from one field to another. He seems to have learned these techniques from people rather than from books.

His use of the water-cooled condenser was probably learned from his brief sojourn at Liebig's laboratory at Giessen in 1848. To turn its end through an angle of perhaps 120 degrees was a simple adaptation of apparatus from one role (distillation) to another (refluxing). To turn Nasmyth's massive digesters to his own purposes of heating under pressure was not merely the effective invention of the pressure cooker but a successful example of scale-up technology. Learning from Kolbe, and still more from Bunsen, of the latter's techniques of gas analysis, enabled Frankland to develop his own technique of eudiometry, and he was making such equipment by the early 1850s. In 1853, he and W. J. Ward devised a compact apparatus incorporating several features of an apparatus of Regnault (such as water cooling of the eudiometer tube).

However, in the apparatus for water analysis, Frankland most obviously brought together a number of hitherto unrelated techniques, especially two. First, the estimation of carbon and nitrogen was basically an adaptation of the classic technique of Liebig as applied to organic compounds in relatively large quantities. Second, on the scale of water analysis developed by Frankland, the quantities were far smaller, and the apparatus had to be evacuated of air before use. His genius lay in a development recorded by Armstrong: "I recollect, when he suggested that we should make the attempt, that he referred me to the account given by Graham of the use he had made of the Sprengel mercury pump in studying the diffusion of oxygen through india-rubber."

Experimental Skill

One cannot read about Frankland without a growing awareness of his astonishing personal skills at the bench. Armstrong did not doubt their origin: "His real education was probably obtained during his apprenticeship in a local druggist's shop. The work he did there taught him to use his hands." Indeed, the tidiness and self-discipline in a pharmacist's profession gave invaluable lessons in the general manipulation of chemicals, crucially important for anyone working in synthetic organic chemistry.

Writing of his work with Frankland on water analysis, Armstrong observed,

> The experience was invaluable, as I was thrown much on my own resources, though sufficiently aided at critical times ... The task was not entirely simple. A few days later, seeing that I was in difficulties at the blowpipe table, he came to my aid and made the first Sprengel fall-tube for me. We were the first after Graham to use the pump. If the traditions of the early Frankland school had been kept alive and the few students with hands [sic] had been regularly tutored in the methods of gas analysis, the vacuum combustion process would now be a preferred process, I believe.[30]

Frankland himself asserted that he acquired considerable skills in glass blowing at Marburg, so much so that he promised to return to Giessen to use his glass-blowing skills for the benefit of that famous laboratory.

Favorable Social Environment

A survey of the conditions in which Frankland worked leaves no room for doubt as to the favorable environment in which he did his best work. In his early years, as at Queenwood, the conditions were not naturally propitious, though it was one of the few schools in the country to take science at all seriously, and Frankland was given a largely free hand as to how he worked. Also, through his drive and ambition, he made the best of what he had. The apparatus he invented was of the simplest kind, and he had ample space in which to use it. In the German universities of Marburg and Giessen he was, of course, in contact with a tradition far more favorable to scientific enquiry than in Britain. In Manchester, on the other hand, conditions were much less satisfactory (heavy teaching load, poor salary, etc.), and research took second place to more remunerative consultancy work.

He came into his own in London, with the unparalleled traditions and good facilities of the Royal Institution, and there he did his best work in organic synthesis. When later he took up his post at the Royal College of Chemistry and shortly afterward became heavily involved in water analysis for the government, he was in a position to make his conditions favorable. Thus, he was rarely in the teaching laboratory; he relieved himself of as many routine chores as possible and eventually set up his own laboratory. Success bred success; by the end if his life, he was in receipt of a very large annual income and well able to spend money on expensive equipment. As was supposed to be the case of Humphry Davy, the very elitism of his social position led to an elitism of his science. Consequently, one could say that his techniques were not applicable to ordinary people. One of them put it thus:

> The Frankland process is quite within the reach of the manipulative skill of any fairly-trained chemist, but it requires practice, and probably pretty constant practice. It cannot be taken up off-hand, and even tolerable results obtained at once. From the hands of a person without proper laboratory training, its results are utterly valueless. It is hence better adapted to regular use in the examination of many samples of water in a large public laboratory than to an occasional use by a private individual.[31]

In other words, it was not even financial but professional elitism that led to a rejection of his methods. For the president of the world's first institute for professional scientists, that is irony indeed.

Two final considerations may place Frankland's technical advances in the widest context. The first concerns the surprising fact that, today, most of this is completely unknown to modern chemists and, in fact, to not a few historians. The apparent eclipse of such a reputation has much to do with Frankland's life style, preference, and extreme reluctance to promote himself in a public way. These have been discussed elsewhere.[32]

Additionally, Frankland never had the good fortune to have a technique, a reaction, or a piece of apparatus named after him. Though generations of chemists have heard of Kipp, Mohr, Victor Meyer, Liebig, Hofmann, and other inventors of laboratory ware, Frankland has not been so commemorated. One reason is that some of his techniques hardly outlived their inventor, but another is more curious. His effective introduction of the reflux condenser and sealed glass tube was so simple, so obvious in retrospect, and so widely applicable that most chemists used them almost as second nature, scarcely bothering to wonder where they came from. They could not even be ordered from the laboratory stores by name, for each was a totally ordinary piece of equipment used slightly differently or adapted by simple glass blowing on the bench. If one *ordered* a "sealed tube," it would be of no use!

Much the same fate befell what was once known widely as "Frankland's notation," though it was again so obvious, so simple, and so universal that it became simply "graphic formulae." We frequently use such devices to this day. Thus, in one sense, Frankland was a victim both of his own success and of the elegant simplicity of his procedures. However, that should not distract historians from a proper evaluation of work that transformed the face of chemistry, particularly the realm of organic syntheses. Despite his past obscurity, it is entirely justifiable to hail Frankland as one of the great founders of synthetic organic chemistry.[33]

The second point is more subtle. The very fact of Frankland's extraordinary experimental success, coupled with the very limited nature of his own chemical training, seems to have stirred him to the invention of one further technique, though this was not chemical but pedagogic: the revolutionary idea that students should learn their chemistry not just from books but from lecture-demonstrations and personal laboratory experience themselves. Of course, he was not the only one to have such ideas, but he was one of the few to translate them into effective practice. His opportunity came when, in 1865, he succeeded Hofmann not only to the chair at the Royal College of Chemistry but to the post of examiner to the Department of Science and Art. Here he had a unique opportunity to fashion the curricula, examination papers, and chemical education of thousands of young people who studied in the evenings for their own improvement.[34]

To bring his dream to reality, he instituted in 1869 a scheme of teacher training in which members of the profession were welcomed for one week to the laboratories of the Royal College of Chemistry. He also wrote a book, *How to Teach Chemistry*, which featured 109 experiments that should be demonstrated to the students.[35] Some of these involved the techniques he had developed: reactions in inert atmospheres, refluxing, gas collection, and so on. Moreover, so popular was his scheme that a number of chemical suppliers competed with each other to provide the apparatus "recommended by Dr. Frankland."

His is a remarkable story. Hugely important advances were made in teacher training and in laboratory instruction. Frankland's name was everywhere in the popular literature of chemistry. His distinctive techniques were no longer the prerogative of an elite of rich chemists but the experience of the common man. In both this democratization of chemical practice and in the introduction of important new procedures, chemistry in Britain gained a very great deal.

NOTES

1. Much of this material will be found in my book, *Edward Frankland: Chemistry, Controversy and Conspiracy in Victorian England* (Cambridge, UK: Cambridge University Press, 1996).

2. E. Frankland, *Autobiographical Sketches from the Life of Sir Edward Frankland*, 2nd ed., ed. MNW and SJC (London: privately published, 1902), 169.

3. E. Frankland, "On the Chemical Composition of Metacetonic Acid, and Some Other Bodies Related to It." *Mem. Chem. Soc.* 3(1847): 386–391; *Experimental Researches in Pure, Applied and Physical Chemistry* (London: van Voorst, 1877), 29.

4. E. Frankland, *Experimental Researches*, 41.

5. Ibid., 29.

6. C. F. Mohr, *Annalen* 18(1836): 232.

7. E. Frankland, *Sketches*, 186.

8. E. Frankland, "Researches on Organometallic Bodies—Zincethyl," *Philosophical Transactions of the Royal Society of London* 145(1855): 259–275.

9. Hofmann to Frankland, 28 April 1855 (Raven Frankland Archive, Open University microfilm 01.03.0274).

10. Nasmyth to Frankland, 22 January 1856 (Raven Frankland Archive, Open University microfilm 01.02.0279); regarding Nasmyth (1808–90) see S. Smiles, ed., *J. Nasmyth, An Autobiography* (London: 1885), and *DNB*.

11. *Chemistry & Industry*, 53(1934): 462; the paper was read on 16 January and further discussion ensued on 6 February 1868.

12. E. Frankland and H. E. Armstrong, "On the Analysis of Potable Waters," *Journal of the Chemical Society*, 21(1868): 87.

13. H. Sprengel, *Journal of the Chemical Society* 18(1865): 9–21.

14. E. T. Chapman, M. H. Smith, and J. A. Wanklyn, *Journal of the Chemistry Society* 20(1867): 445–454.

15. E. Frankland and H. E. Armstrong, "On the Analysis of Potable Waters," *Journal of the Chemistry Society* 21(1868): 83.

16. J. A. Wanklyn, *Journal of the Chemical Society* 20(1867): 591–595.

17. W. Odling, testimonial for Armstrong, 1 December 1870 (Imperial College Archives, Armstrong papers, first series, 335).

18. J. W. Mallet, Supplement no. 19, *National Board of Health Bulletin* (Washington, D.C., 27 May 1882), 4.

19. Anonymous, *Analyst* (note 39).

20. For example, editorial comments in September and October 1868, *Chemical News* 18(1868): 151, 153, 165.

21. Editorial, "Water Analysis," *Chemical News* 25(1872): 157.

22. E. Frankland, *Journal of the Chemical Society* 29(1876): 825–851, on 847.

23. E. B. Truman to Frankland, 25 June 1874 (Raven Frankland Archive, Open University microfilm 01.04.0409).

24. J. W. Thomas to Frankland, 22 December 1876 (Raven Frankland Archive, Open University microfilm 01.04.0415).

25. W. Odling to Frankland, 27 April (Raven Frankland Archive, Open University microfilm 01.04.0418).

26. Frankland, undated memorandum (Raven Frankland Archive, Open University microfilm 01.04.0421).

27. Frankland to H. Owen, n.d. (Raven Frankland Archive, Open University microfilm 01.07.0848). As far away as New Zealand and as late as 1896, Dunedin Public Analyst A. G. Kidston-Hunter was arguing the diagnostic importance of albuminoid nitrogen and ammonia in water, while earlier estimations of oxidizable organic matter probably employed permanganate oxidation. If so, Frankland's techniques do not appear to have made much progress in that part of the Empire [R. J. Wilcock, "Water Chemistry," in *Chemistry in a Young Country*, ed. P. P. Williams (Christchurch: New Zealand Institute of Chemistry, 1981), 195–205].

28. J. Mitchell, *Proceedings of the Royal Society* 7(1855): 316–318.

29. E. A. Parnell, *Applied Chemistry* vol. 2 (London: Taylor & Warren, 1844), 249n.

30. H. E. Armstrong, *Pre-Kensington History of the Royal College of Science and the University Problem* (London: Royal College of Science, 1920), 7.

31. J. W. Mallet, *National Board of Health Bulletin*, suppl. 19, 27 May 1882, 4.

32. C. A. Russell, *Edward Frankland*, chapter 17 and elsewhere.

33. G. Porter, "The Chemical Bond Since Frankland," *Proceedings of the Royal Institution* 40(1965): 384–396, on 387.

34. C. A. Russell, *Edward Franklin*, chapter 10.

35. E. Frankland, *How to Teach Chemistry: Hints to Science Teachers and Students*, ed. G. Chaloner (London: J. & A. Churchill, 1875).

BIBLIOGRAPHY

Brock, W. H. "Edward Frankland," in *Dictionary of Scientific Biography* (New York: Scribner's, 1970–80).

Frankland, Edward. *Autobiographical Sketches from the Life of Sir Edward Frankland*, 2nd ed., MNW and SJC, (London: privately published, 1902).

Frankland, Edward. *Experimental Researches in Pure, Applied and Physical Chemistry* (London: van Voorst, 1877).

Russell, Colin A. *Edward Frankland: Chemistry, Controversy and Conspiracy in Victorian England* (Cambridge, UK: Cambridge University Press, 1996).

Russell, Colin A. "Edward Frankland," in *New Dictionary of National Biography* (in preparation). Several of the letters and other manuscript documents cited in this chapter are in an archive privately owned by Mrs. Raven Frankland and are at present accessible only in microfilm form at the Open University. Other deposits of Frankland papers, though immensely important, are not directly relevant to the present subject.

13

Bridging Chemistry and Physics in the Experimental Study of Gunpowder

Seymour H. Mauskopf

> It were indeed to be wish'd that our art had been less ingenious, in contriving means destructive to mankind; we mean those instruments of war which were unknown to the ancients, and have made such havoc among the moderns. But as men have always been bent on seeking each other's destruction by continual wars; and as force, when brought against us, can only be repelled by force, the chief support of war must, after money, be now sought in chemistry.[1]

> The most important era in the history of powder is undoubtedly to be found in the improvements and alterations made during the last three decades during which period the combined scientific knowledge and engineering skill of those whose names will be always remembered in connection with this work, have removed gunpowder from the ranks of ordinary explosives, and have made it an absolutely reliable propellant.[2]

Chemistry has always been the quintessential "mixed" science, as much devoted to the creation and improvement of material products as to the elucidation of the natural laws that govern material behavior. The very early division of chemistry into "pure" and "applied" testified to this realization on the part of at least one eighteenth-century practicing chemist.[3] Much of my recent research has been focused on eighteenth- and early nineteenth-century chemists' involvement with the improvement of one such product: gunpowder.

In this chapter, I concentrate on somewhat later research on gunpowder carried out in the 1850s by the chemist, Robert Bunsen and his Russian student, Leon Schischkoff, and in the 1860s and 1870s by the English gunnery expert Andrew Noble and the chemist Frederic Abel. Their work was concerned less with "improving" gunpowder, which had generally meant increasing its ballistic force, than with precisely determining this force.

Before discussing the research of these protagonists, I establish a framework for relating their research to earlier experimentation on bal-

listic force. The determination of the ballistic characteristics of gunpowder and, thereby, the prediction (and improvement) of its ballistic
performance, have always been of fundamental importance to both the
manufacturers and consumers of gunpowder, particularly, of course, to
the military consumers.

However, in fact, not one experimental means for measuring
ballistic force but several existed. The most straightforward was incorporated in a variety of instruments developed in the eighteenth and
nineteenth centuries: mortar-eprouvettes, ballistic pendulums, and elec-
troballistic chronographs. I shall term it *mechanical* because all these
instruments depended on the mechanical behavior of the projectile: the
distance it traveled, the impact it had on a target, the recoil it produced
in the test instrument, or the velocity it achieved, as measured by successive rupture of electrical circuits.[4]

At the same time during which these devices were being used
(and improved), other means for determining ballistic force were also
being developed. I argue that these can be analyzed in terms of what I
label *research approaches* to the study of gunpowder (*chemical* and *physical*)
and *research styles* (*laboratory* and *field*). The chemical approach was concerned with the chemical analysis of gunpowder, especially the nature of
the explosion reaction and its products.[5] The physical approach had
more to do with what artillerists call *interior ballistics*, in which the challenge was to measure such parameters as temperature, pressure, and rate
of burn.[6]

By laboratory and field research styles, I distinguish between
research in which gunpowder was treated as an ordinary laboratory
substance and was studied under normal laboratory conditions and
research in which the behavior of gunpowder was studied under conditions approximating its military use, particularly in heavy cannon.
These approaches and styles were exemplified respectively by the
research of Joseph-Louis Proust (chemical approach and laboratory
style) and of Benjamin Thompson, Count Rumford (physical approach
and field style), concurrently conducted in Spain and Germany in the
1790s.

The research of Bunsen and Schischkoff was published as "*Chemische Theorie des Schiesspulvers*" ("A Chemical Theory of Gunpowder") in
1857; it was heralded throughout the rest of the nineteenth century as a
landmark in the study of gunpowder, primarily for its unprecedented
detailed analysis of the chemical products of gunpowder explosion.
However, in terms of my analysis, a deeper significance is seen in

Bunsen and Schischkoff's research: it bridged, for the first time, the two research *approaches*, chemical and physical. This was made possible through the use of thermodynamics, novel to chemists at this time. By means of this, Bunsen and Schischkoff were able to use their chemical data to determine theoretical values for such physical parameters as temperature and pressure of gunpowder explosion.

If Bunsen and Schischkoff bridged the two research approaches, their research *style* remained emphatically laboratory. However, in what was probably the most ambitious research program on the ballistic characteristics of gunpowder ever carried out, Nobel and Abel both adopted the thermodynamic perspective of Bunsen and Schischkoff and situated it in a research style that, in turn, bridged laboratory and field.

By way of background, I give a brief overview of Proust's and Rumford's research. Then I focus on the chapter's theme: the research of Bunsen and Schischkoff and of Noble and Abel.

BACKGROUND: PROUST AND RUMFORD

Both Proust and Rumford carried out their research under government patronage, Proust as professor of chemistry at the Royal Artillery School in Segovia, Spain (1785–1798), and Rumford as military adviser to the Duke of Bavaria in 1793. Proust's research had the practical objective of determining scientifically how to make the "best" powder, which, for the eighteenth century, meant the powder with the greatest ballistic force. However, he carried out his research as a laboratory chemist, basing his theories and methods particularly on those of his predecessor, Lavoisier.[7]

Proust's research program was to determine the ballistic force of gunpowder by an indirect, chemical approach. Following (and modifying) Lavoisier, Proust defined ballistic force as a function of the volume of produced gases and the speed of reaction; he employed the techniques of pneumatic chemistry both to collect and to identify the gaseous products of explosion. In another expression of his chemical orientation, Proust conducted a comprehensive series of parameter-variation experiments in which the principal parameters were the proportions of the constituents and the types of charcoal. For each case, he determined the volume and the nature of the gaseous products, the nature of the solid residues (less systematically), and the speed of the reaction.

In these experiments, Proust made no attempt to mimic the conditions under which gunpowder was actually fired. Although he exploded relatively large quantities of ingredients (e.g., 12 gm of salt-peter to 60 gm of charcoal to test speed of burn of charcoal from dif-ferent wood sources), he used ordinary laboratory apparatus to carry out these burns. Proust's experimental style was exemplary of what I mean by *laboratory*.

By contrast, Count Rumford attempted to "ascertain the force of gunpowder by *actual measurement*, in a direct and decisive experiment." I consider his experiments physical in approach and field in style.[8] The physical approach Rumford adopted was to measure the *pressure* of gunpowder at the moment of firing. His point of departure was the research of Benjamin Robins, published half a century earlier[9], espe-cially two of Robins's conclusions about explosion pressure: (1) that the pressure of fired powder was a linear function of its density and (2) that maximum pressure (i.e., when explosion took place in a container completely full of powder) was equal to 1000 atmospheres. Rumford's field style—and challenge—was to measure this pressure directly by constructing an apparatus capable of withstanding the extreme pressure and temperature of explosion and of providing reliable quantitative results.

Rumford certainly attempted to meet this challenge in the appa-ratus he devised (figure 13.1). Its core was a strong "hammered iron" barrel containing a bore in which the powder was put. This was stop-pered by a steel hemisphere that had a reinforced seal but could be lifted by the pressure of the exploding powder. The entire apparatus was attached to a wooden scaffold on which weights could be balanced atop the gunpowder apparatus. These weights were graduated up to an 8081-lb cannon. Pressure was measured by the weight that could just be lifted off the stopper of the powder bore by the explosion. Rumford also did parameter variation, but his variable was the "density" of the powder exploded (i.e., the ratio of the weighted amount to powder used to the maximum weight that would fill the powder chamber). He exploded progressive amounts of powder (one, two, three grains, etc), all of the same composition.[10] Although Rumford used quantities of powder smaller than those used by Proust, his apparatus was designed to produce—and sustain—much higher pressures.

Rumford did indeed find a nonlinear relationship between den-sity of powder and pressure, and from his pressure-density function, he extrapolated a maximum pressure, very much higher than that of Robins, of more than 29,000 atmospheres.[11] Even this figure he con-

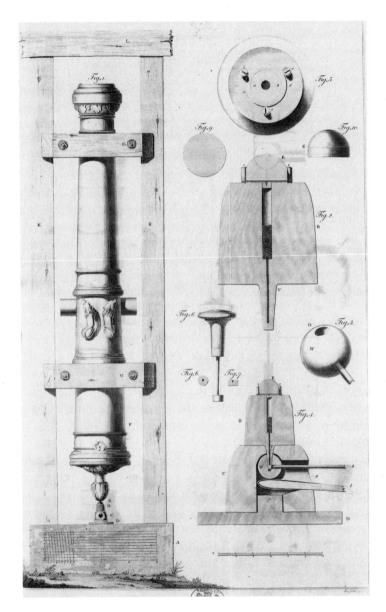

Figure 13.1
Rumford's apparatus for determining the pressure of fired gunpowder. The apparatus on the right can also be seen under the cannon on the left. [From G. Piobert, *Traité d'Artillerie Théorique et Pratique* (Paris: Bachelier, 1847).]

sidered to be much too low; he thought that he had data indicating maximum pressure of more than 100,000 atmospheres![12]

The research of Proust and Rumford had contrasting fates in the nineteenth century. Proust's chemical studies of gunpowder explosion were, in my opinion, the most ambitious for the eighteenth century, but they proved to lack cash value. His chemical determination of "best" powder composition merely underscored the verity of the traditional French military powder formula of 6(saltpeter) : 1(sulfur) : 1(charcoal). His laboratory-based recommendation concerning the best wood source for charcoal was turned down by the French Gunpowder Administration because it was not borne out when subject to field ballistic tests using the standard mortar-eprouvette.[13] Proust's analyses were appropriated in the 1820s by chemists such as M.-E. Chevreul and J.-B. Dumas, who repackaged them in quantitative stoichiometry in which the variety of gaseous and solid products that Proust had in fact identified was simplified to nitrogen and carbon dioxide gases, with potassium sulfide as the solid product.[14] This simple stoichiometry remained standard until the publication of Bunsen and Schischkoff's explosion-products analysis. Proust's own studies virtually disappeared from view.

Despite its stated expectation of utility, Rumford's research appears to have had as little success as Proust's in realizing practical consequences. Yet, unlike Proust's, Rumford's research did not fall into obscurity. During the first half of the nineteenth century, it was cited as definitive[15] and, after that, as the fountainhead of an important research tradition in munitions.[16]

Its continued citation stemmed, I believe, from practical and social contexts.[17] The practical context pertained to the perceived challenges to artillerists of bursting cannon arising from more powerful gunpowder,[18] augmented in midcentury by the development of very powerful cannon.[19] The eighteenth-century research objective of maximizing gunpowder's ballistic force gave way to the new goal of controlling its development in the gun barrel. Here, physical considerations seemed much more relevant than those of a chemical nature.

The social context had to do with the emergence of a new type of artillery researcher in the early nineteenth century: the École Polytechnique–trained artillerist–military engineer–researcher, exemplified by the author of the most important scientifically oriented artillery textbook of the first half of the nineteenth century,[20] Guillaume Piobert (1793–1871).[21]

In his own research, Piobert had led the principal investigations into the cause of the bursting cannon, and he had gone on to perform pioneering experiments on virtually all the physical parameters of gunpowder explosion, carefully relating his research to the actual processes that transpired in guns and cannon.[22] These included the study of the density-pressure relationship of gunpowder explosion; hence Piobert's interest in and acclaim for Rumford's research.[23]

As for chemistry, not only was Proust's research neglected; more generally, the first half of the nineteenth century was barren of chemical research in munitions (excepting the discoveries of guncotton and nitroglycerine in the 1840s, materials whose potential, though immediately recognized, was not realized for some time.)[24] The only chemical research of any significance (as measured by citation in contemporary and later literature) was carried out by J.-L. Gay-Lussac in 1823 as a function of his governmental appointment as a munitions expert.[25]

BUNSEN AND SCHISCHKOFF, NOBLE AND ABEL

This situation changed dramatically in 1857, however, when the well-known German chemist, Robert Bunsen, and his Russian student, Leon Schischkoff, published their "A Chemical Theory of Gunpowder." The authors themselves harked back to Gay-Lussac's analysis "published thirty years ago" as their predecessor.[26]

As the title suggests, the heart of the paper was a detailed chemical analysis of the products of gunpowder explosion, both solid and gaseous. However, their research went beyond chemical analysis to treat physical aspects of explosion that had never been directly addressed by chemists: the temperature, the pressure, and most novel, the work produced by it. This research brought together, for the first time, what I have called the chemical and the physical traditions of research. However, the research style was emphatically that of the laboratory; it was carried out "for only one type of powder (sporting and rifle powder), and at atmospheric pressure," although the authors promised that their method "could, with certain slight modifications, be applied to the study of gunpowder combustion under other circumstances."[27]

The paper appears to have little or no antecedent in Germany; Germany seems to have been as barren of chemical investigations of gunpowder as was the rest of Western Europe. This is borne out by the data (or lack thereof) in the most compendious survey of gunpowder research and industrial development of the late nineteenth century.[28] What seems likely, moreover, is that the impetus for the research came less from Bunsen than from Schischkoff, a Russian artillerist with an interest in the chemistry of munitions.[29]

However, the synthetic physicochemical approach of the research probably came from Bunsen. The year of the publication of this paper also marked the appearance of Bunsen's treatise on gasometry; the two seem to be closely linked.[30] Moreover, Bunsen was unusual among chemists in the 1850s for his interest in physics.[31] The particular aspect of physics highlighted in this paper was thermodynamics and the related interest in "thermochemistry" that was just beginning to develop in the 1850s. For instance, Bunsen and Schischkoff made use of some of the thermochemical data compiled by Favre and Silbermann, who had amassed the most comprehensive data on heats of reactions in the 1850s.[32]

The point of departure for Bunsen's and Schischkoff's chemical investigation of gunpowder explosion was the disparity between the prediction of the standard simplified stoichiometric analysis of gunpowder explosion[33] of the volume of gaseous products and the actual results that they (and their predecessors) had obtained: the standard stoichiometry predicted much too great a volume.[34]

A thin stream of the powder[35] was made to fall into a glass bulb suspended over a burner in which the powder was combusted (figure 13.2). The solid products remained in the bulb; the smoke collected in a long tube connected to the bulb. The gases passed through an enclosed, narrower tube connected to an aspirator and issued into glass ampules that could be closed off and hermetically sealed. The gases were chemically analyzed by standard mid-nineteenth-century techniques (e.g., carbon dioxide and hydrogen sulfide absorbed by potash; oxygen absorbed by pyrogallate of potash).[36]

Collecting and analyzing the solid and gaseous products, Bunsen and Schischkoff produced a chemical scenario for gunpowder explosion far more complex than what had been standard. This was particularly true for the solid residues, which not only were more varied but made up a greater part of the products by weight than had the traditional, simplified stoichiometric prediction.[37] Conversely, gases made up only 31.38% by weight of the products (instead of the predicted 59.22%), and the volumetric ratio of carbonic acid gas (carbon dioxide) to nitrogen was much lower than that predicted.[38] This weight in turn

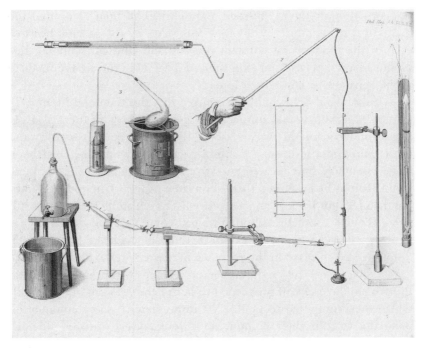

Figure 13.2
The apparatus of Bunsen and Schischkoff. [From R. Bunsen and L. Schischkoff, "On the Chemical Theory of Gunpowder," translated by E. Atkinson. London, Edinburgh, and Dublin Philosophical Magazine and Journal of Science (4th series) 15(1858): 489.]

was correlative with the volumetric production of 193.1 cc of gases, considerably smaller than the predicted 330.9 cc of gases.[39]

In addition to reconciling the disparity between prediction and experiment, the detailed analysis of the products of gunpowder explosion gave the chemists a purchase on determining theoretical values for parameters necessary to determine the force of the explosion reaction: the temperature, the pressure, and the work. The remainder of the paper was devoted to these determinations.

The key to these was the calorimetric measurement of the heat produced by gunpowder explosion; to the best of my knowledge, Bunsen and Schischkoff were the first to obtain this calorimetric data directly.[40] Their modest and simple apparatus, carefully described in their text, bespoke the laboratory style of their gunpowder research.[41] From their data, they determined an adjusted value of 619.5 calories (modern terminology).[42] Bunsen and Schischkoff then computed the theoretical temperature by dividing this value by the specific heats of each of the products (at constant pressure and constant volume); this produced a temperature of explosion of 2993°C in the air and 3340°C in the apparatus at constant volume.[43]

Bunsen and Schischkoff recognized that this theoretical temperature of explosion would immediately decrease through thermal dissipation; indeed, for this very reason, "[I]t cannot be measured by the usual thermometric means."[44] And, in fact, prior estimates had been made essentially by educated guesswork. For example, in 1742, Benjamin Robins had assumed that "the Flame of fired Gunpowder is not less than red-hot Iron,"[45] and a century later Guillaume Piobert was still citing Robins' estimate, although he had come to think that an upper limit of 2400°C was more realistic because of evidence that metallic fusion had taken place in the presence of gunpowder explosion.[46]

Using the computed temperature along with the other data, Bunsen and Schischkoff proceeded to derive the pressure of explosion, which they computed to be 4373.6 atmospheres.[47] They commented pointedly that, if their computation were correct, "many previous assumptions as to the pressure exerted in guns must depend on very faulty premises, for the best artillery works give this pressure at from 50,000 to 100,000 atmospheres."[48] Their own citation here was to Piobert's discussion of Rumford in his *Traité d'Artillerie*.[49] Employing their computed pressure and using a formula derived from Clausius, Bunsen and Schischkoff made their final computation: the theoretical work of explosion (assuming a thermally insulated system).[50]

"These very important researches" gave rise to a number of research programs, some purely chemical and others devoted to determining physical data of temperature, pressure, and work as well.[51] The importance of the latter was particularly enhanced by the changes in large gunnery taking place at the time the paper of Bunsen and Schischkoff was published: first, the introduction of much larger cannon during the Crimean War and American Civil War, including breech-loaded, rifled cannon.[52] These developments, in turn, exacerbated the long-standing problem of the ballistically powerful gunpowder. An English researcher, recounting attempts to determine pressure in large guns in the 1860s, noted what amounted almost to a crisis over the estimation of explosion pressure engendered by its theoretical determination by Bunsen and Schischkoff.[53]

The enormous discrepancies between the 1000 atmospheres estimated by Robins and the 100,000 atmospheres of Rumford will not
have escaped you; and even coming to quite recent dates, the difference of opinion between authorities like Piobert on the one hand,
and Bunsen and Schischkoff on the other, are quite startling enough
to show you the difficulties with which the subject is enveloped.

The particular research was carried out by a Committee on Gunpowder and Explosive Substances, set up in 1869 by the secretary of state
for war and under the presidency of Colonel Charles Wright Younghusband, superintendent of the Royal Gunpowder Factory. It included
the artillerist and gunnery expert and manufacturer Andrew Noble
and the chemist Frederick Abel. The establishment of government-
sponsored munitions-testing bodies, such as this committee, no doubt
testified to the concern of the British government in this period of
rapid change in ordnance (and tumultuous international political
events) with maintaining military parity with its rival nations.[54] The
style of the research was distinctly *field* in my typology, and its findings
had to do mainly with the detailed study of the interior ballistics of
different powders and guns. They confirmed the general range of the
pressure estimate of Bunsen and Schischkoff, as contrasted with the
"extravagant estimates which have frequently been made."[55]

The career of Andrew Noble (1831–1915) represented an inter-
section of military, industrial, and research interests. He was educated
at the Royal Military Academy and received a commission in the
Royal Artillery in 1849. Abroad until 1858, he returned to England at
just the time that military ordnance was undergoing the fundamental
change just described. Having earlier shown an interest in mathematics
and chemistry, Noble became expert at experimental research on ordnance and munitions. In 1860, he was gazetted out of the army to join
in partnership with a major ordnance innovator, William George
Armstrong, of whose company Noble eventually became chairman
(1900).[56]

By the time of the publication of this paper, Noble had been at
work for 3 years on an even more ambitious set of gunpowder investigations. This time, he had a collaborator: munitions chemist Frederick
Abel (1826–1902). Abel was unusual—indeed, unique—among important mid-nineteenth-century chemists in pursuing a life-long career
in military chemistry. A charter student (and one of the most esteemed)
of W. A. Hofmann at the Royal College of Chemistry, Abel held the

positions of ordnance chemist at the Woolwich Arsenal and professor of chemistry at the Royal Military Academy there (succeeding Faraday) from 1854 until his retirement in 1888. He devoted most of his research to guncotton (nitrocellulose), focusing particularly on the challenges of purifying and stabilizing it so that it could be stored, transported, and deployed militarily. However, his research extended to all aspects of munitions. Even after retirement, he continued to be active in munitions chemistry, developing the important smokeless powder, "cordite," with James Dewar.[57]

The research on gunpowder by Noble and Abel, begun in 1868, was published in two installments in the *Philosophical Transactions of the Royal Society* (1876, 1880) under the simple title, "Research on Explosives". Like Noble's earlier work, this research had the practical military objective of determining the most suitable powder for heavy ordnance, "which is still continually increasing in size" and was carried out in tandem with a military committee appointed by the secretary of state for war.[58] However, the publication under the aegis of the Royal Society of London also underscored the underlying scientific objectives of this research. The hybrid objectives were mirrored in the mixed methods employed by Noble and Abel, combining laboratory and field styles. Moreover, the research was linked to the two long-standing investigative programs that this chapter features: the physical determination of the pressure of fired gunpowder and the chemical analysis of the explosion reaction.[59] Thus, though the study of Bunsen and Schischkoff was a first synthesis of research approaches, this research was a synthesis not only of these approaches but of the styles of research delineated in this chapter.

The doubly synthetic nature is brought out clearly in the set of eight "objects of experiments" that Noble and Abel set for themselves. Some dealt with the determination of chemical parameters (the products of gunpowder combustion and the volume of the permanent gases produced);[60] others with physical parameters (the determination of pressure, temperature, and work).[61] However, still others turned to seek interaction between physical and chemical parameters: for example, did variation in the physical characteristics of the powder—density and size of powder grains—or in the pressure of firing influence the chemical reaction of explosion?[62] Moreover, in their research methodology, Noble and Abel explicitly combined laboratory and field styles. Thus, the sixth object was "to compare the explosion of gunpowder fired in a close vessel with that of similar gunpowder when fired in the

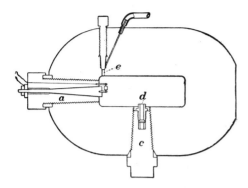

Figure 13.3
The apparatus of Noble and Able. [From Lawrence L. Bruff, *A Text-Book of Ord-nance and Gunnery*, 2nd ed. (New York: John Wiley, 1903).]

bore of a gun,"[63] and they determined the chemical products of gun-powder as "fired under circumstances similar to those which exist when it is exploded in guns and mines".[64]

To do this, the investigators made use of an "explosion apparatus" that Noble had devised for his earlier investigations. This consisted of a pair of vessels made of "mild steel ... of great strength, carefully tempered in oil," in the chambers of which the powder was fired. The chamber of the larger vessel could hold a kilogram of powder.[65] Great care was taken to secure the firing plug as tightly as possible. A crusher gauge for measuring pressure and an escape tube leading to a mercury trough for collecting the gaseous products were also built into the apparatus (figure 13.3). The powder was secured in its chamber and was exploded by a battery detonator: "[T]he only audible indication of the explosion is a slight click."[66]

Somewhat ingenuously, the authors observed, "The difficulties we have met with in using the apparatus are more serious than might at first sight appear." The obvious danger was the force of the explosion itself, possibly destroying the apparatus (and the experimenters). However, even when the apparatus behaved satisfactorily, difficulties arose in getting at the explosion products, because of their fusion to the screws and metal surfaces of the container.[67]

Aside from the need to wait a bit after the explosion (5–15 minutes), collection and analysis of the gaseous products was straight-forward. The solid products were another matter. They were found in fused deposits on the bottom and sides of the cylinder in "an exceed-ingly hard and compact mass, which always had to be cut out with steel chisels."[68] Although some characteristics were common to the solid residues of all the experiments,[69] what more impressed was their physical and chemical variability. Moreover, their analysis was also complicated by their refractory physical state, necessitating protracted pulverization, and their tendency to absorb oxygen and water from the air, necessitating the provision of a nitrogen atmosphere in which to carry out the pulverization.[70]

As in Noble's earlier experiments, a number of different powders were used, all standard English military powders,[71] except one pebble powder of Spanish provenance. Elaborate tables were given of the powder composition and the analyses of the explosion products. These were tabulated not only according to the type of powder but also to its density (i.e., proportion of firing chamber occupied).[72] Moreover, the analyses of Bunsen and Schischkoff and of some of their continental successors were also tabulated. These tabulations are the most elaborate of which I am aware up to this time.

The profiles of the explosion products obtained by Noble and Abel showed some important contrasts with that of Bunsen and Schischkoff, particularly for the solid products.[73] For example, Bunsen, Schischkoff, and the other continentals obtained double the amount of potassium sulfate and less of potassium carbonate and potassium sulfide than did Noble and Abel. To the English investigators, these contrasts derived from the difference in experimental methods used. In the laboratory method of Bunsen and Schischkoff, the explosion took place under much lower pressure and for a duration considerably longer (to explode all the powder) than that in the more cannonlike firing chamber used by Noble and Abel. Therefore:[74]

> We feel warranted by the results of our experiments, in stating, with confidence, that the chemical theory of decomposition of gunpowder, as based upon the results of Bunsen and Schischkoff, and accepted in recent text-books, is certainly as far from correctly representing the general metamorphosis of gunpowder as was the old and long-accepted theory, according to which the primary products were simply potassium sulphide, carbonic anhydride [CO_2], and nitrogen.

As things turned out, however, Noble and Abel had nothing to substitute in place of the chemical paradigm of Bunsen and Schischkoff. Indeed, the conclusion Noble and Abel drew was pessimistic as far as it concerned defining any specific chemical profile of the explosion reaction:[75]

The variations in the composition of the products of explosion furnished in close [sic] chambers by one and the same powder under different conditions as regards pressure, and by two powders of similar composition under the same conditions as regards pressure, are so considerable that no value whatever can be attached to any attempt to give a general chemical expression to the metamorphosis of a gunpowder of normal composition.

This, combined with the their recognition of the long-standing view that variation in the physical properties of the powder[76] were indeed the principal determinant of the ballistic behavior of the powder, led Noble and Abel to conclude, "A minute examination into the nature of the products of explosion of powder does not necessarily contribute directly to a comprehension of the causes which may operate in modifying the action of fired gunpowder."[77]

In the remainder of the paper, Noble and Abel followed the schema of Bunsen and Schischkoff in addressing the determination of physical parameters: pressure, temperature, and work performed. Their synthetic union of traditions and research styles was evident here in the contrasts of their investigations to those of the German researchers. In particular, the experimental apparatus, mimicking the conditions under which powder was exploded in gun barrels, enabled Noble and Abel to build up an elaborate empirical database for some of these parameters. For example, instead of merely calculating a "theoretical pressure," as Bunsen and Schischkoff did, Noble and Abel amassed data correlating powder "density" and explosion pressure up to levels approaching densities of unity, much as Rumford had done, but they determined a maximum pressure of only 6350 atmospheres.[78]

Noble and Abel compared their own experimental pressure-density data with that actually observed in cannon and with various theoretically generated pressure-density correlations for the firing of projectiles.[79] The military data were supplied by the government Committee on Explosives; pebble and rifled large-grain powders were used in a 10-inch 18-ton gun.[80] Noble and Abel made an exceedingly detailed report and analysis of these data, not only supplying the requisite comparative pressure-density data but a very elaborate study of the changes in pressure in the gun as a fired projectile moved down the gun barrel.[81]

Their procedures for determining the other physical constants was similar to those they used for determining pressure: they used the general physical and thermodynamical paradigm of Bunsen and

Schischkoff, but interposed their own empirical data and amended it where they felt it necessary. Thus, although they accepted the method for determining theoretical temperature used by Bunsen and Schischkoff,[82] they were skeptical about the way these latter had calculated the specific heats of the products. This was because Noble and Abel thought that they had reason to believe that the nongaseous products were actually liquified rather than solid at the moment of explosion,[83] as Bunsen and Schischkoff had assumed, and "the specific heat is largely increased in passing from solid to the liquid state."[84]

Thus, instead of accepting the simple calculation of Bunsen and Schischkoff, Noble and Abel devised a much more elaborate algorithm to determine explosion temperature. Moreover, it deployed the empirical data that they had collected, in particular data correlating powder "density" and explosion pressure. Using this and the gas law, they determined a theoretical temperature of 2231°C.[85] That this temperature was reasonable, they believed, was indicated by the behavior of platinum wires and thin sheets in their apparatus, elements that seemed to show that the explosion temperature was slightly higher than the melting point of platinum but well under that necessary to volatilize that metal.[86]

The hypothesis that the nongaseous products of explosion were liquified also informed their calculation of the theoretical work of the reaction. Work was performed only by the released permanent gases, but during the firing of a gun, these gases were able to secure additional heat from the liquified residue, which acted as a heat reservoir. From this model, Noble and Abel derived functions for the temperature, pressure, and work produced in the firing of gunpowder.[87]

Calculating the work for various values of explosive expansion (up to 20-fold), Noble and Abel extolled the results:[88]

> [They are] of very considerable importance. They enable us to say by simple inspection what is the maximum work that can be obtained from powder such as is employed by the British Government in any given length of gun. To make use of the table, we have only to find the volume occupied by the charge (gravimetric density = 1) and the number of times this volume is contained in the bore of the gun.

The climax of this long memoir[89] was, indeed, the tabulation of

> first, the total work realized per lb. of powder burned for every gun, charge, and description of powder in the English service; second, the

maximum theoretic work per lb. of powder it would be possible to realize with each gun and charge; and third, the factor of effect with each gun and charge—that is, the percentage of the maximum effect actually realized.

Inspection of the energy data of these tables allowed deduction of the velocity of any standard projectile in any standard large gun and at any point along the gun bore (figure 13.4).[90]

Noble and Abel's studies of the chemical and physical character-istics of gunpowder must surely be the most ambitious and elaborate ever carried out for this material. They were recognized as such by contemporaries[91] and by subsequent writers on the subject.[92] Yet, ironically, within 10 years of publishing the first of their "Researches on explosives," a "complete revolution in gunpowder manufacture" was under way, brought about by the development of the first smoke-less nitrocellulose powder. The very object of Noble and Abel's studies, black powder, was "washed away over-night, and now belongs to his-tory alone,"[93] at least in regard to its military use.

However, if the material object of these studies was soon to be superceded, the general manner in which they were carried out was to be readily applicable to all munitions and was paralleled in other research programs. Indeed, the principal criticisms directed at Noble and Abel's research came from a rival and even more elaborate program across the Channel. This Parisian-based research group was formed in the wake of the Franco-Prussian War; its leader was chemist Marcellin Berthelot, but he was really only *primus inter pares* among such col-leagues as Émile Sarrau and Paul Vieille.[94] To try to narrate and delin-eate the very varied and important research carried out by this group would take us much too far afield. Suffice it to say that it, too, melded the laboratory and field styles of research and the chemical and physical traditions. The melding of the styles was no doubt a reflection of the long involvement of alumni of the *École Polytechnique* directly in study and even production of munitions. Both Sarrau and Vieille were *polytechniciens*.[95]

The bridging of physics and chemistry was more comprehensive, systematic, and theoretical in the work of this group than in that of the English researchers, and it was intimately connected with the devel-opment of Berthelot's thermochemistry. Yet, as with Noble and Abel, the point of departure for the thermochemistry of munitions was the work of Bunsen and Schischkoff. Berthelot himself paid tribute to the

TABLE 21.—*Giving, with the data necessary for calculation, the work per lb. of powder realised, the total maximum theoretic work, and the factor of effect for every gun and charge in the British Service.*

Nature of gun	Bore — Diameter (Inches)	Bore — Length (Calibres)	Charge — Nature	Charge — Weight (Lbs. oz.)	Projectile — Weight (Lbs.)	Projectile — Velocity (Feet per sec.)	Gas — Total volumes in bore	Gas — Final density	Total (Foot-tons)	Energy of powder — Realised per lb. of powder (Foot-tons)	Energy of powder — Calculated maximum (Foot-tons)	Percentage realised
38 tons	12	16·5	P.	110 0	700	1430	7·342	·1362	9932	90·3	97·0	93·1
35 tons	12	13·5	P.	110 0	700	1300	6·007	·1665	8209	74·9	90·2	82·7
25 tons	12	12·0	P.	85 0	600	1300	6·910	·1447	7036	82·9	94·9	87·3
			R.L.G.	85 0	495	1353	6·910	·1447	6334	74·5	94·9	78·6
			R.L.G.	55 0	495	1142	10·679	·0936	4479	81·4	108·9	74·8
25 tons	11	13·2	P.	67 0	600	1130	8·765	·1141	5797	86·4	102·8	84·1
			R.L.G.	67 0	495	1271	9·765	·1141	5549	82·3	102·8	80·6
			R.L.G.	50 0	495	1140	11·750	·0351	4464	89·3	111·3	80·0
			P.	85 0	535	1315	5·355	·1708	6419	75·5	89·2	84·7
18 tons	10	14·5	P.	85 0	535	1315	5·355	·1700	6419	75·5	89·2	84·7
			R.L.G.	70 0	535	1217	7·109	·1407	5498	78·6	95·3	82·1
			P.	70 0	535	1217	7·109	·1407	5498	73·9	95·3	82·1
			R.L.G.	70 0	400	1364	5·367	·1704	5164	73·9	99·4	82·6
			P.	44 0	400	1340	8·867	·1704	4984	71·2	99·4	79·7
12½ tons	9	13·9	R.L.G.	60 0	400	1125	9·334	·1071	3513	79·3	104·7	76·3
			P.	40 0	400	1293	6·344	·1461	4676	77·9	94·5	82·4
			R.L.G.	50 0	250	1117	10·269	·0974	3463	86·6	107·9	80·3
			P.	50 0	250	1420	5·742	·1742	3498	70·0	88·6	79·1
			R.L.G.	43 0	250	1420	5·742	·1742	3498	72·0	93·6	77·1
9 tons	8	14·3	P.	30 0	250	1336	6·683	·1496	3096	72·0	93·6	77·1
			R.L.G.	35 0	180	1336	6·683	·1496	3096	82·1	105·2	73·2
			P.	35 0	180	1192	9·566	·1045	2465	71·3	90·9	73·4
			R.L.G.	30 0	180	1413	6·136	·1630	2493	71·3	90·9	76·3
			P.	30 0	180	1413	6·136	·1630	2493	73·7	96·0	76·3
			R.L.G.	20 0	130	1330	7·154	·1398	2209	84·5	109·1	77·6
			R.L.G.		130	1330	7·154	·1398	2209		96·0	
			R.L.G.		130	1163	10·724	·0932	1689			

Figure 13.4

Table providing the work realized per pound of powder, the total maximum theoretical work, and the factor of effect for every gun and charge in the British

Service. [From Andrew Noble and Frederick Able, "Researches on Explosives," in *Artillery and Explosives*, ed. A. Noble (London: John Murray, 1906).]

importance of their work in the preface to the massive third edition (1883) of his *Sur la force des matières explosives d'après la thermochemie*. Indeed, what he says could serve as a summation of this chapter:[96]

> The old treatises, of which the work of Piobert constitutes one of the most carefully done models, were concerned mainly with problems of ballistics ... But the authors remained indeterminate about the true causes of [ballistic] force both from a chemical as well as a mechanical point of view.
>
> In 1857, MM. Bunsen and Schischkoff, in a most remarkable work, set the problem of black powder on a rational basis by measuring the volume of gas and the heat released and by endeavoring to infer the mechanical effects from these. But these two fundamental data were measured only in an empirical fashion and without the authors having attempted to deduce these data from the powder's initial composition or from knowledge of the products of the explosion.

The theme of this chapter has been the evolution of research methods to determine the ballistic force of gunpowder. I have tried to show that the very research definition of ballistic force was a highly protean concept, itself determined by scientific backgrounds and research objectives. Proust, the chemist, experimentally defined ballistic force rather differently from Rumford, the physicist and military scientist; from cursory examination of their experimental techniques and instruments, one might hardly guess that they were attempting to determine the same thing. Although more clearly Bunsen and Schischkoff and Noble and Abel were studying the "same thing," they too used very different instrumentation to do so because of their different research objectives (research styles). Thus, a discussion of research techniques and instrumentation without situating them in the kind of contexts suggested here is impossible.

Sometimes, indeed, even conceptually separating experimental techniques and instruments from theories is difficult. A case in point is the strategy by which Bunsen and Schischkoff bridged the physical and chemical approaches: the use of thermochemistry. Was this just a new theoretical perspective, or might it also be viewed as a novel experimental strategy? It provided the rationale for their carrying out an unprecedently detailed chemical analysis of the products of gunpowder explosion, and it mandated their deployment of the calorimeter. The combined chemical and calorimetric data enabled Bunsen and

Schischkoff to calculate physical parameters, such as temperature and explosion pressure, factors that had previously been beyond the reach of experimentation, especially experimentation carried out using ordinary laboratory apparatus.

In the broad context, the evolution of research methods analyzed in this chapter represents a case study of how the "mixed" science of chemistry not only maintained its dual pure and applied nature in the nineteenth century but, indeed, enhanced its effective synergy between theory, experimentation, and application during the course of that century. As in other cases, such as metallurgy and pharmaceuticals, this enhancement was fostered both by the linkage of chemistry with other sciences and by the establishment of close connections between laboratory science and product development. For better or for worse, the epigraph that "the chief support of war must, after money, be now sought in chemistry" was far more true at the end of the nineteenth century than when it was coined a century and a half earlier.

NOTES

1. H. Boerhaave, *Elementa chemiae*, translated by T. L. Davis, "Chemistry in War: An 18th-Century Viewpoint," in *Army Ordnance, 5*, no. 30 (May-June, 1925), 783. This appears to be Davis's own translation.

2. Vivian B. Lewes, *Cantor Lectures on Explosives and Their Modern Development* (Delivered Before the Society for the Encouragement of Arts, Manufactures and Commerce) (London: Printed by William Trounce, 1895), 3–4.

3. Johann Gottschalk Wallerius in *Bref om Chemiens rätta Beskaffenhet, Nytta och Wärde* (1751), quoted in Christoph Meinel, "Theory or Practice? The Eighteenth-Century Debate on the Scientific Status of Chemistry," *Ambix* 30(1983): 126. The idea of "pure" versus "applied" or "mixed" appears to have originated in mathematics. Wallerius subsequently amended his definition so as to include both pure and applied chemistry under the rubric of "science." The terms became instantly very popular with Scandinavian and German writers.

4. S. H. Mauskopf, "Explosives, Instruments to Test Ballistic Force of," in *Instruments of Science: A Historical Encyclopedia*, R. Bud and D. J. Warner, eds. (New York: Science Museum, London, and National Museum of American History, Smithsonian Institution, in association with Garland Pub., 1998), 234–236.

5. It also dealt with determining the purity of gunpowder's constituents, the consistency of the final product, and the development of new explosives.

6. "Interior ballistics treats of the formation, temperature and volume of the gases into which the powder charge in the chamber of a gun, is converted by combustion, and the work performed by the expansion of these gases upon the gun, car-

riage and projectile. The discussion of the formulas deduced will bring out many important questions, such as the proper relation of weight of charge to weight of projectile and length of bore, the best size and shape of powder grains for different guns and their effect upon the maximum muzzle pressures, the velocity of recoil etc." James M. Ingalls, *Interior Ballistics*, 3rd ed. (New York: John Wiley, 1912), 1.

7. S. H. Mauskopf, "Chemistry and Cannon: J.-L. Proust and Gunpowder Analysis," *Technology and Culture* 31(1990): 398–426.

8. Benjamin Thompson, Count Rumford, "Experiments to Determine the Force of Fired Gunpowder," (1797) in *The Collected Works of Count Rumford*, vol. 4, ed. Sanborn C. Brown (Cambridge: Harvard University Press, 1968–1970), 397. The entire paper, originally published in the *Philosophical Transactions of the Royal Society of London*, is contained on pages 395–471. I have detailed Rumford's research in "From Rumford to Rodman: The Scientific Study of the Physical Characteristics of Gunpowder in the First Part of the Nineteenth Century," in *Gunpowder: The History of an International Technology*, ed. Brenda J. Buchanan (Bath: Bath University Press, 1996), 277–293.

9. Brett D. Steele, "Muskets and Pendulums: Benjamin Robins, Leonhard Euler, and the Ballistic Revolution," *Technology and Culture* 35(1994): 348–382.

10. Pressure in multiples of atmospheric pressure was determined by the weight that had been lifted off divided by 0.73631 pounds (normal atmospheric pressure on the cross-section of the bore). Rumford, "Experiments," 438.

11. Rumford graphed the results from which he computed a density-pressure relationship of

$$y\ (\text{pressure}) = x\ (\text{density})^{1+0004x}$$

from which he extrapolated a pressure of 29,177.9 atmospheres for the explosion of full bore of 25.642 grains. Ibid., 447.

12. Ibid., 454.

13. He suggested substituting hemp charcoal (used in Spain) for the traditional black alder. The ballistic tests of powder made with hemp by the mortar-eprouvette were ambiguous.

14. Mauskopf, "Chemistry and Cannon," 417–420.

15. G. Piobert, *Traité d'Artillerie Théorique et Pratique: Partie Théorique et Expérimentale* (Paris: Bachelier, 1847), 289 ff, esp. 292. The figure of 29,178 atmospheres was "definitive." Ibid., 326. However, Piobert did modify Rumford's density-pressure function.

16. Andrew Noble, "On the Tension of Fired Gunpowder" (1871), in *Artillery and Explosives*, ed. Sir Andrew Noble (London: John Murray, 1906), 54 ff. See also Ingalls, op. cit., 4–5.

17. I concentrate on France for my analysis.

18. Caused by English innovations in powder-making and fully mechanized powder production.

19. See Mauskopf, "From Rumford to Rodman;" Patrice Bret, "Le Dernier des Procédés Révolutionnnaires: La Fabrication et l'Expertise de la Poudre Ronde (1795–1830)," *Annals of Science* 50(1993): 325–347.

20. Piobert, op. cit.

21. A.-J. Morin, "Notice sur le Général Piobert," *Mém. Acad. Sci. de l'Inst. de France* 38(1873): 102 ff.

22. Mauskopf, "From Rumford to Rodman."

23. In fact, Piobert cited Proust too, but Rumford received far greater space and approbation.

24. For details, see G. S. Louis, *The Origins and Discovery of the First Nitrated Organic Explosives* (doctoral Dissertation, University of Wisconsin, 1977); S. L. Norman, *Guncotton to Smokeless Powder: The Development of Nitrocellulose as a Military Explosive, 1845–1929* (doctoral Dissertation, Brown University, 1988).

25. See Maurice Crosland, *Gay-Lussac, Scientist and Bourgeois* (Cambridge, UK: Cambridge University Press, 1978), 181–188, for background. It was a gravimetric and volumetric analysis of gunpowder explosion, carried out by dropping small quantities of gunpowder into a tube heated to redness and set up to receive the gases. Nothing was particularly novel in his results except for his eccentrically large volumetric result of 450 volumes of the original powder, as compared with previous standard estimates of approximately 250 volumes. Reported in Andrew Noble and Frederick Abel, "Researches on Explosives, part I, (1876) in Noble, op. cit., 103–104. The authors stated that they had "been unable to obtain the original of this report" and gave reference to Piobert's *Traité d'Artillerie Théorique et Pratique*.

26. I am using the English translation: "On the Chemical Theory of Gunpowder," translated by E. Atkinson, *London, Edinburgh and Dublin Philosophical Magazine and Journal of Science* (4th series) 15(1858): 489.

27. Ibid., 490.

28. J. Uppmann and E. von Meyer, *Traité sur la Poudre, les Corps Explosifs et la Pyrotechnie*, translated by E. Désortiaux (Paris: Dunod, 1878), 461–462, where they move directly from the analyses of Gay-Lussac and Chevreul of the 1820s to that of Bunsen and Schischkoff.

29. Leon Nikolaevich Shishkov (Russian spelling) (1830–1908) was a Russian artillerist who specialized in the chemistry of munitions. Graduating from the Mikailov Artillery School in St. Petersburg in 1851, he was later a professor there (1860–1865), where he organized a model chemical laboratory. In the middle and late 1850s, he worked on the chemistry of munitions: first on fulminating and fulminic acids and their salts, then in J.-B. Dumas's laboratory in Paris on nitro compounds (1856–1857), and ultimately with Bunsen in Heidelberg (1857) on

black powder. Source for this information: *Biograficheskii Slovar' Deyatelei Estest-voznaniya i Tekhniki* (Biographical Dictionary of Individuals in Science and Technology), (Moscow: Bolshaya Sovetskaya Entsiklopediya, 1959), vol. 2, 381. I thank my colleague, Richard Rice, for this reference.

30. Georg Lockemann, *Robert Wilhelm Bunsen* (Stuttgart: Wissenschaftliche Verlagsgesellschaft M.B.H., 1949), 141. See the discussion of the paper later.

31. John W. Servos, *Physical Chemistry From Ostwald to Pauling* (Princeton: Princeton University Press, 1990), 11–12. Bunsen was particularly important in developing physical instruments, such as the spectoscope (in collaboration with Kirchhoff). Servos wrote that "Bunsen's labors on instruments ... gave him a broad knowledge of contemporary physics, and he is said to have stressed the value of physical study in his lectures." However, Servos went on to stress his eschewal of interest in physical theory.

32. Pierre Antoine Favre (1813–1880) and Johann Theobald Silbermann (1806–1865). Significantly, Silbermann was an engineer and physicist. They employed a mercury calorimeter of their own devising and, although "the results were not very accurate," they were frequently cited in the decades after their publications. The evaluation of their results is by J. R. Partington, *A History of Chemistry*, vol. 4 (London: Macmillan, 1962–1964), 611. See R. G. A. Dolby, "Thermochemistry Versus Thermodynamics: The Nineteenth century Controversy," *History of Science* 22(1984): 375–400. See also, Helga Kragh, "Julius Thomsen and classical Thermochemistry," *British Journal for the History of Science* 17(1984): 255–272, who terms their measurements "accurate" (256).

33. Products: nitrogen and carbon dioxide as gases; potassium sulfide as the sole solid residue.

34. Bunsen and Schischkoff, "Chemical Theory of Gunpowder," 489–490. At saturation reaction, one gram of powder (containing 0.7484 grams of saltpeter, 0.1184 of sulfur, and 0.1332 of charcoal) is predicted to produce 330.92 cc of gases at (presumably) 0°C and 760 mm pressure, consisting of 82.52 cc of nitrogen and 248.40 cc of carbon dioxide. This was much too high.

35. Composition by weight: saltpeter, 78.99%, sulfur, 9.84%; charcoal, 11.17; composed of carbon, 7.69%; hydrogen, 0.41%; oxygen, 3.07%. Ibid., 491.

36. Ibid., 498–499.

37. Ibid.,489 and 504. For the explosion of 1 gm of powder, the standard model predicted 0.4078 gm of potassium sulfide; in fact, only 0. 0213 gm were found out of a solid residue of 0.6806 gm. The bulk of the solid residue obtained was potassium sulfate (0.4227 gm) and carbonate (0.1264 gm). They obtained more potassium hyposulfite (0.0327 gm) than potassium sulfide. Concerning this product, see further (note 70).

38. Less than 1.5 : 1 rather than 3 : 1. Ibid., 503. This was determined from the observed weight ratios of 52.67–41.12 for carbonic acid gas–nitrogen. Although these were the main gaseous products, Bunsen and Schischkoff were struck by the

appearance of a small but measureable amount of free oxygen in their gaseous products.

39. Ibid., 505.

40. Favre offered "several interesting reflections on the theory of the deflagration of gunpowder" although he regretted that a complete thermochemical analysis of nitric acid formation was not yet feasible. Assuming the simple stoichiometry of gunpowder explosion standard in his time (products were nitrogen, carbon dioxide and potassium sulphide), Favre computed, from the heats of reactions that he had, that 530 "units" of heat" were produced when one gram of "ordinary" powder exploded. P.-A. Favre, "Recherches thermo-chemiques sur les combinaisons formées en proportions multiples," *Journal de pharmacie et de chimie*, [3] 24(1853): 338–339.

41. Bunsen and Schischkoff, "Chemical Theory of Gunpowder," 506–507 and Plate III, figure 4.

42. Ibid., 505–509. Their own experimental data gave a value of 643.9 calories (modern terminology). However, Bunsen and Schischkoff suggested that secondary oxidation of some of the minor products (hydrogen, hydrogen sulfide, carbon monoxide) took place in the calorimeter; using the heats of reaction data for the oxidation of these products, Bunsen and Schischkoff suggested that 24.4 calories arose from these secondary reactions and thus had to be subtracted from the experimental calorimetric results.

They further noted that the data of Favre and Silvermann for heats of reactions for all the oxidations that produced the primary products equalled 1,039.1 rather than 619.5 calories. They attributed the differences to the heat absorbed during the explosion by the vaporization of the nitrogen fixed in the saltpeter. Ibid., 508.

43. Ibid., 508–510. The solid products were considered to have remained solid throughout the explosion process. They considered these figures only as approximate upper limits, because the specific heats of solid bodies increase with temperature.

44. Ibid., 505.

45. Benjamin Robins, *New Principles of Gunnery* (1742; reprinted in 1972 by Richmond, Surrey: The Richmond Publishing Co., Ltd., 14.

46. Piobert, *Traité d'Artillerie*, 280–281.

47. The formula they used was

$$p_0 = \frac{V(1 + 0.00366t)}{\dfrac{G_p}{S_p} - \dfrac{G_r}{S_r}}$$

where G_p is the weight of the powder (1 gm); S_p is its specific weight (0.964); G_r is the weight of the solid residue (0.6806 gm); S_r is its density at temperature of 3,340°C (1.50); V is the volume of the gaseous products at 0°C and atmospheric pressure (they say "1 m"?) (193.1 cc); and $t = 3340°C$. The density of the solid residue at 3,340°C was determined "by a method not yet published, which one of us has used to determine the volatilization and expansion of rocks melted at high temperatures."

If the density of the solid residue were taken to be 1, the computed pressure would be 3414.6. The additional 1000 atmospheres of pressure was thus ascribed to the expansion of the powder residue by heating. However, from the results of their own heating of such residue in a hydrogen flame, they asserted that the vapor pressure from (solid) powder residue would be negligible. See Bunsen and Schichkoff, "Chemical Theory of Gunpowder," 509–511.

48. Ibid., 511.

49. Specifically, to 322 of that *Traité*.

50. Their figure was "67,410 metre kilogrammes." Ibid., 511–512.

51. Andrew Noble and Frederick Abel, "Researches on Explosives, part I (1875), in *Artillery and Explosives*, ed. A. Noble, 109–113. Upmann and von Meyer termed it "une série d'éxperiences relatives aux produits de la combustion de la poudre, qui ont servi de type pour toutes les recherches ultérieures sur ce sujet" (*Traité sur la Poudre,* 462). See 466–473 for brief accounts of these researches. Some attempted to move beyond Bunsen and Schischkoff to determine the chemical reaction under the conditions operative in large guns.

52. O. F. G. Hogg, *Artillery: Its Origins, Heyday and Decline* (Hamden, CT: Archon, 1970), 82 ff; J. F. C. Fuller, *The Conduct of War, 1789–1961: A Study of the Impact of the French, Industrial, and Russian Revolutions on War and Its Conduct* (New Brunswick, NJ: Rutgers University Press, 1961).

53. Noble, "Tension of Fired Gunpowder," 67.

54. Hogg, *Artillery*, 136, gives the date of the first of these as 1869, founded at the order of the secretary of state for war as a continuation of an earlier Sub-Committee of the Ordnance Select Committee (defunct). Finding out very much about this committee (officially, the Committee on Gunpowder and Explosive Substance) is exceedingly difficult. However, the minutes of the Ordnance Select Committee (1860–1869), the reports of this committee (1863–1869), and the reports of the Committee on Gunpowder and Explosive Substances are to be found at the Royal Artillery Library, Old Royal Military Academy, Woolwich Arsenal. The first "Preliminary Report" of the Committee on Gunpowder gives a brief history; annual "progress reports" appear throughout the 1870s, with a "final report" in 1881. Both Noble and Abel served on the committee throughout its existence. Younghusband, president of the committee until 1880, was super-intendant of the Royal Gunpowder Factory.

55. Noble, "Tension of Fired Gunpowder," 65. In fact, the pressure estimate of Bunsen and Shischkoff was found to be somewhat lower than actual measurement, which yielded 40 tons per square inch (6100 atmospheres) and higher pressure under oscillation waves in large guns. Ibid., 86.

56. Regarding Noble, see *Dictionary of National Biography*, 24, *Twentieth Century, 1912–1921*, 412–413. Noble made use of the resources of his company to carry out his research.

57. For biography details on Abel, see the obituary by J. Spiller, *Transactions of the Journal of the Chemical Society*, 87(1905): 565–570; also, *Dictionary of National Biography*, 23, *Twentieth Century, 1901–1911*, 5–6. When Abel took over as ordnance chemist in 1854, "there was some uncertainty as to his duties" because the position had lapsed for many years prior. O. F. G. Hogg, *The Royal Arsenal: Its Background, Origin, and Subsequent History*, vol. 2, (London: Oxford University Press, 1963), 749.

58. Noble and Abel, "Researches on Explosives, Part I," 100.

59. Ibid. Even the moderate range of pressure estimates (between 2,200 and 29,000 atmospheres) is described as "sufficiently startling as regards a physical fact of so much importance." And "the views regarding the decomposition of gunpowder are nearly as various."

60. The first and fifth objects. Ibid., 113.

61. The second, seventh, and eighth objects. Ibid.

62. The third and fourth objects. Ibid.

63. Ibid.

64. The first object. Ibid.

65. This was the most ambitious attempt to fire an explosive in a closed vessel to date. Arthur Marshall, *Explosives*, 2nd ed., vol. 2 (London: J. & A. Churchill, 1917–1932), 440–441.

66. Description, Noble and Abel, "Researches on Explosives, Part I," on 114; illustration on Plate X, following 230.

67. "In one or two cases, it was found impossible to open the cylinder until melted iron had been run round it, so as to cause it to expand." Ibid., 115.

68. Ibid., 118.

69. Oleaginous, smelling of hydrogen sulfide and ammonia, exceedingly deliquescent, blackening in air. Ibid.

70. In reviewing Noble and Abel's work, Marcellin Berthelot argued against the formation of one of their solid products, potassium hyposulfite, during the explosion reaction on thermochemical grounds and suggested that it might be an artifact of postexplosion oxygenation through contact of the residue with the atmosphere. "Sur l'hyposulfite de potasse," *Comptes rendus hebdomadaires des Séances de l'Académie des Sciences* 82(1876): 400–403. Noble and Abel conceded Berthelot's point for at least a portion of the observed potassium hyposulfite in the second installment of their paper, explaining that the exposure to the air probably took place during the

difficult procedure of cutting out the solid residue from the explosion vessel. However, they defended their view that some potassium hyposulfite was produced as a primary product of gunpowder explosion. "Researches on Explosives, Part II," 239 ff.

Subsequently, Heinrich Debus challenged the presence of potassium hyposulfite among the reaction products from the viewpoint of analytical procedure. He claimed that the hyposulfite was an artifact of the use of copper oxide to test for potassium sulfide in the residue (the method of Bunsen and Schischkoff). When zinc chloride was substituted as the testing agent, as recommended by Debus, very little hyposulfite was manifest. Andrew Noble and Frederick Abel, "Fired Gunpowder. Note on the Existence of Potassium Hyposulphite in the Solid Residue of Fired Gunpowder," *Proceedings of the Royal Society of London*, 30(1879–1880), 198–208, reprinted in "Researches on Explosives, Part II," 309–323. Debus had been an assistant to Bunsen before coming to England.

Noble and Abel thereupon conceded fully to Berthelot. Ibid., 319.

71. From the Waltham-Abbey Gunpowder Works. The English powders were: pebble-powder, rifle large-grain, rifle fine-grain, and fine-grain. These powders differed from the sporting powder used by Bunsen and Schischkoff in the ratio of the components, principally in a smaller proportion of saltpeter (roughly 74%) and a larger proportion of charcoal (roughly 14%). Ibid., 128. As their names indicate, they also differed from sporting powder (and from each other) in the size of the powder grain. For an account of the nature and manufacture of these powders, see Oscar Guttman, *The Manufacture of Explosives*, vol. 1 (London: Wittaker & Co., 1895), 243.

72. Noble and Abel, "Researches on Explosives, Part I," Table 2 (128); Table 3 (titled "Showing the analytical results obtained from an examination of the Solid and Gaseous Products," 130–131); and Table 4 (titled "Showing the composition by weight of the products of combustion of 1 gramme of Fired Gunpowder," 132–133).

73. Noble and Abel secured a greater proportion of CO (carbonic oxide) than did Bunsen and Schischkoff, but the former team recognized that this could have been the effect of the difference in powder composition. Ibid., 134.

74. Ibid., 148.

75. Ibid., 210 (under "Summary of Results"). They went on to note that the proportions of the residue compounds were "quite as much affected by slight accidental variations in the conditions which attend the explosion of one and the same powder in different experiments as by decided differences in the composition as well as in the size of grain of different powders."

This position was soon challenged by Marcellin Berthelot and Heinrich Debus. Berthelot thought that the production of the principal products of gunpowder explosion (potassium sulfide, sulfate and carbonate, as solids and oxide of carbon [carbon monoxide], carbonic acid [carbon dioxide] and azote [nitrogen]) could be quantitatively represented by five simultaneous equations, each of which represents what transpires in a determinate proportion of the exploded powder ... "Sur l'Explosion de la Poudre," *Comptes Rendus*, 82(1876): 469–475. Debus argued that the variability of Noble and Abel's results stemmed from the actual variability in the composition of each of the powders they used and in the chemical inter-action between one of the products, potassium disulfide, and the iron walls of the explosion vessel. Debus proposed a two-stage reaction for gunpowder explosion, which he collapsed into one grand equation. "The Bakerian Lecture on the Chemical Theory of Gunpowder" (delivered February 8, 1882), *Proceedings of the Royal Society of London*, 33(1881–1882): 361–370.

76. Termed *mechanical and physical properties*: density, hardness, grain size, and form. Ibid., 146.

77. Ibid., 150. Already in 1864, this had led the Government Committee on Explosives (on which Abel served) to limit the alterations in gunpowder for heavy guns to physical characteristics; the chemical composition was left unchanged. Ibid. 146, footnote.

78. Ibid., 161. This was for rifle large-grain powder; for fine-grain, the corre-sponding pressure was 5,870 atmospheres. By *density*, I mean (again) the volumet-ric ratio of powder to container. They also determined a theoretical pressure curve for various densities of gunpowder based on the assumption that pressure would vary as the density and that the volumetric ratio of solid powder to liquid products was a constant (0.65 or 0.6). Taking this constant as 0.65, the highest theoretical pressure was the same as that measured (6,350 atmospheres); taking it to be 0.6, the highest theoretical pressure was 6316 atmospheres. Ibid., 168–169.

79. Which took into account the work performed by the gunpowder explosion.

80. Ibid., 177. Entire section: 173–211.

81. The pressure was measured directly by Noble crusher-gauges (a variant of the Rodman gauge) inserted in the walls of the gun barrel and indirectly by an elec-troballistic chronoscope. Noble and Abel did notice that the crusher-gauge some-times gave higher pressure readings than those computed from the chronoscope data; they attributed this to the local generation of pressure waves along the wall of the gun barrel, especially for faster-burning powders. Ibid., 175–176.

82. Explosion temperature calculated by dividing the heat produced in the reac-tion by the (weighted) specific heats of the products.

83. Their evidence came from inspection of the contents of some of their explo-sion vessels after these had been tilted within the first moments of the explosion. One example: "Accordingly, in Experiment 40, the cylinder being about two-thirds full with F.G. (fine grained power), 30 seconds after the explosion the vessel was tilted so as to make an angle of 45 degrees. Two minutes later it was restored to its first position.

   On subsequent examination, the deposit was found to be lying at the angle of 45 degrees, and the edges of the deposit were perfectly sharp and well defined." Noble and Abel, "Researches on Explosives, Part I," 156.

84. Ibid., 165. What should be pointed out is that what actually took place, physically and chemically, in explosion under high pressure remained a very con-

tentious issue. The French, for example, maintained that all products were gassified at the moment of explosion and that dissociation occurred, something that Noble and Abel denied.

85. From their density-pressure data, Noble and Abel were able to compute a number that expressed the ratio of the volume of liquid products to original powder volume (they adopted a value of 0.6). Assuming an explosion in a container of 1 cc. completely filled with 1 gm of powder (at gravimetric density of 1), the resultant liquid products would occupy 0.6 cc.; the gaseous products 0.4 cc. However, other data of theirs indicated that the gaseous products of 1 gm of powder would occupy 280 cc under atmospheric pressure; hence, the pressure of this gas would be 700 atmospheres. From this data, they could calculate the gas constant, R, and thus the temperature of the explosion. Ibid., 167–171.

86. Ibid., 171–172. The then-current estimate of temperature of fusion was 2000°C; of a volatilization, 3200°C.

87. Ibid., 197ff. See especially 197–204 and the graphs (XIX, XX, following 230). The formulas for calculating temperature and pressure are on 199; that for work on 203. These values of many of the data they used (e.g., the specific heats of the permanent gases at constant pressure and constant temperature) were corrected, after criticism, in the subsequent memoir.

88. Noble and Abel, "Researches on Explosives, Part I," ibid., 204.

89. Ibid., 205. Table 21, 206–207.

90. By the formula

$$v = \frac{\sqrt{2gW}}{w}$$

where W was the determined work value for any given volumetric expansion, multiplied by the "factor of effect," and w was the weight of the projectile.

91. Upmann and von Meyer, *Traité sur la Poudre*, 473.

92. For example, Tenney L. Davis, *The Chemistry of Powder and Explosives* (New York: John Wiley & Sons, 1943), 42–43. A detailed account of their work was also provided as the starting point for the analysis of interior ballistics in the textbook on ordnance and gunnery used at the US Military Academy, West Point, at the turn of the century. The account is not without major emendations; for example, the physical hypothesis behind the work function of Noble and Abel (nongaseous products serving as a heat reservoir for the gases) is criticized, and the function itself is amended. Lawrence L. Bruff, *A Text-Book of Ordnance and Gunnery*, 2nd ed. (New York: John Wiley, 1903), 19–41.

93. Guttmann, *Manufacture of Explosives*, 22–23.

94. See Jean Jacques, *Berthelot, 1827–1907, Autopsie d'un Mythe* (Paris: Belin, 1987), 141 ff, for an account of the background of the Franco-Prussian War to Bertholet's taking up research on munitions and publishing his first essay on *La*

Force de la Poudre et des Matières Explosives. In addition to objecting to Nobel and Abel's analysis of the solid products of explosion (note 70) and to their assertion about the inherent variability in the explosion reaction (note 75), Berthelot also criticized Noble and Abel's calorimetrical data. Jules-Arthur Morin and Marcellin Berthelot, "Rapport sur le Mémoire publié par M. le capitaine Noble . . . et par M. Abel . . . sous le titre de 'Researches on explosives fired gun powder [*sic*]'," *Comptes Rendus*, 82(1876): 487–493, especially 490. This was summarized in Berthelot's review of the second part of Noble and Abel's paper: "Observations sur le mémoire de MM. Noble et Abel relatif aux matières explosives," *Mémorial des Poudres et Salpêtres*, 1(1882–83): 252–261.

95. The first successful smokeless powder was developed by Vieille.

96. My translation. Paris: Gauthier-Villars, 1883, *x*. Similar credit is given to the work of Bunsen and Schischkoff by Émile Sarrau concerning the experimental determination of the heat of combustion of explosives substances. *Recherches Théoriques sur les Effets de la Poudre* (Paris: Tanera, 1874), 38.

14

LABORATORY PRACTICE AND THE PHYSICAL CHEMISTRY
OF MICHAEL POLANYI
Mary Jo Nye

ROOTS OF THE TACIT DIMENSION OF SCIENCE

It is well-known in contemporary science studies that the ideas of
Michael Polanyi (1891–1976) about the practice of science have informed the work of historians and sociologists of science from Thomas
S. Kuhn to Steve Fuller.[1] In his book *Personal Knowledge*, first published
in 1958, Polanyi attacked all varieties of positivist philosophy:[2]

> I start by rejecting the ideal of scientific detachment [although in]
> the exact sciences, this false ideal is perhaps harmless, for it is in fact
> disregarded there by scientists.... I regard knowing as an active
> comprehension of the things known, an action that requires skill.

As in his earlier essays of the 1940s, Polanyi argued in 1958 that science
is not a prescriptive "method" of reasoning but a social community of
investigators working within a set of institutions that includes journals,
textbooks, laboratories, and lecture rooms.

A master-pupil apprenticeship is foremost among the social practices that fosters mutual reliance and mutual discipline among members
of the community. Scientific knowledge is personal knowledge with
tacit and consensual dimensions that lead to verification and validation
of facts and theories. Scientific controversies are decided not by objective logic but by persuasion and conversion to a different language and
a different world.[3] Only those people who have worked their way
through the apprenticeship system of science have the expertise that
qualifies them to exercise the authority of natural science.

At the time that he began writing his essays on this subject,
Polanyi was professor of physical chemistry at the Victoria University
of Manchester, a post he exchanged in 1948 for a chair of "social
studies" at Manchester before retiring in 1959 to Merton College at
Oxford as a senior research fellow. His brother, Karl Polanyi (1886–
1964), was a notable economist, and his son, John Polanyi (b. 1929), is

a distinguished physical chemist who shared the Nobel Prize for Chemistry in 1986 with Dudley Herschbach and Yuan T. Lee. The views of Polanyi, originally a Hungarian Jew, should be rooted in political and economic events of Central Europe and the Soviet Union during the 1920s and 1930s, in his differences of opinion with his brother, Karl, and in his private religious experience with Christianity.[4] Clearly, however, a crucial foundation for Polanyi's epistemology must also lie in his scientific work or, more exactly, in his scientific career that spanned a period of active scientific research from 1910 to 1949.[5] His investigative experiences, like his epistemology, are exemplary of the everyday disciplinary practice of laboratory science and specifically of physical chemistry.

Keeping in mind Polanyi's emphasis on the importance of scientific apprenticeship in forging the values of science among its practitioners, I begin by examining aspects of his education and his early researches from 1910 to 1930. In this period, Polanyi moved through the ranks from scientific novice to director of the department for reaction kinetics in physical chemistry under Fritz Haber at the Kaiser Wilhelm Institut in Berlin-Dahlem, laboring to win a respected position within the German research community.

I then analyze in some detail, in the central section of this chapter, Polanyi's research during the 1920s in x-ray crystallography, examining a striking case of the gradual development and adaptation of new instrumentation from one scientific field (physics) to another (chemistry). Here we clearly see some of the connections between the material and conceptual dimensions of scientific practice in the laboratory.

Polanyi's application of x-ray techniques for studying the structure of material substances met initial resistance from organic chemists, although it was to transform the practice of theoretical and industrial chemistry in the long run. The work also failed initially to gain respect among theoretical physicists whose opinions Polanyi valued, even though Polanyi and his collaborators were in fact helping to lay foundations for what was to become solid-state physics and materials science.

That Polanyi's researches of the 1920s took place within the fiercely competitive and industrially oriented German scientific community of the time was absolutely constitutive of the postcritical philosophy of *The Tacit Dimension* and *Personal Knowledge*. That Polanyi felt his early work failed to make revolutionary breakthroughs and only embodied everyday scientific practice played a role in his formulation of his later powerful analysis of the scientific enterprise. How this is the case is the subject of the last section of this chapter.

EARLY APPRENTICESHIP IN PHYSICAL CHEMISTRY

While undertaking medical studies in Budapest during the years 1908 to 1913, Polanyi carried out experimental work in biochemistry[6] and then entered the Technische Hochschule in Karlsruhe to study chemistry with Georg Bredig and Kasimir Fajans in 1913, the year in which he completed his medical degree. In August 1914, Polanyi entered the Austrian army as a military surgeon, but spent much of the war period on leave or on light duty, largely for reasons of ill health. In 1917 he completed a doctoral thesis in physical chemistry, which he defended at the University of Budapest.

The years 1918 and 1919 were ones of turmoil in the Budapest of the defeated Austrian-Hungararian monarchy. Both Michael and Karl Polanyi, as well as other members of their family, were active in political organizations and discussions during these difficult years. Michael Polanyi briefly worked as an assistant at the University of Budapest for Georg de Hevesy, and he spent time as well in Karlsruhe, where he met a chemistry student, Magda Kemeny, whom he married in 1921.[7]

In September of 1920, Polanyi took a position in Berlin under Reginald Herzog at the Kaiser Wilhelm Institut für Faserstoffchemie (Institute for Fiber Chemistry), which was housed in the buildings of the Kaiser Wilhelm Institute for Physical Chemistry and Electrochemistry.[8] By this time, Polanyi had published papers in several areas of thermodynamics and made efforts to bring his work to the attention of well-placed German scientists in physical chemistry and theoretical physics. Polanyi's efforts at networking in his early career are worth some consideration in light of his later writings about the social practices of the scientific community.

First, after publishing a paper in 1913 with J. Baron on the application of the second law of thermodynamics to animal processes, Polanyi wrote a theoretical paper extending Nernst's heat theorem and Einstein's quantum theory for specific heats to the domain of very high pressures at ordinary temperatures. Polanyi predicted that at infinitely high pressure the entropy would go to zero, because the decreasing volume has an effect on energy frequencies similar to that of decreasing temperature.[9]

Writing from Budapest, Polanyi asked his friend, Alfred Reis, who was at the Karlruhe Technische Hochschule, to transmit this manuscript to Georg Bredig with the request that the work be sent to Albert Einstein in Zurich. Bredig advised Polanyi that Einstein's opinion was favorable, while himself offering some stylistic suggestions and advising Polanyi to contact Karl Scheel in Berlin about publishing the work in the German Physical Society's *Verhandlungen*.[10]

Walther Nernst was less encouraging about a paper that Polanyi sent him about the derivation of Nernst's theorem. He especially objected to Polanyi's historical introduction to the paper, in which Polanyi attributed to Einstein alone the prediction that specific heats will tend toward zero as temperature declines toward absolute zero. Nernst wrote Polanyi that he had pointed out this possibility in the heat theorem 2 years before Einstein and that this behavior of specific heats was self-evident, given the decrease of entropy to zero at absolute zero. Nernst also questioned the clarity of Polanyi's discussion of cyclical processes in his manuscript, and then broke off correspondence in the fall of 1913 with the comment that the only way to solve their points of disagreement was by speaking personally.[11] Polanyi learned at least two behavioral norms from this correspondence: the scientist's preoccupation with priority and the significance of personal contact in resolving controversy.

For his doctoral thesis, Polanyi presented work from the thermodynamic study of adsorption of gases on the surface of colloidal droplets and solids, using data in the published literature on adsorption of CO_2 by charcoal. Extending the work of Arnold Eucken, Polanyi argued that an adsorbed gas behaves in accordance with its normal equation of state as attractive forces are exerted between the adsorbing solid and the atoms or molecules that locate themselves in several layers at the surface. His theory used a function representing the volume available at a given energy of the adsorbed molecules, and he derived the way the adsoption isotherm (the relation of gas pressure to the volume of gas adsorbed at a given temperature) must vary with temperature.[12]

At approximately the same time (1916–1918), Irving Langmuir at the General Electric Laboratory in Schenectady, New York, published experimental results on the adsorption of gases on mica surfaces and on water, arguing that the surface layer is monomolecular in constitution with a structure determined by electrostatic forces. Langmuir, who had taken his doctoral degree with Nernst in 1906, was to receive the Nobel Prize in 1932 for his work in surface chemistry.

Polanyi found himself in the next few years arguing defensively about his own thermodynamic theory of adsorption, while also beginning to extend his interests into the related fields of heterogeneous catalysis and reaction rates. By the early 1920s, as G. N. Lewis's and Langmuir's chemical theories of the electron-pair bond were becoming more widely known, as were quantum mechanical models of the atom, support for Polanyi's thermodynamic adsorption theory languished.

Haber's invitation to Polanyi to give a full account of the adsorption theory at an Institute colloquium resulted in considerable criticism from both Haber and Einstein. Polanyi later said, "Professionally, I survived the occasion only by the skin of my teeth."[13] Given the then-current preoccupation among physical chemists and theoretical physicists with atomic theories and quantum mechanics, Polanyi began to doubt that he could even have published his more classical 1916 paper in a German scientific journal if he had written it 5 years later.[14] Polanyi also resented the fact that Eucken criticized his adsorption theory while adopting some of Polanyi's assumptions.[15] Not until 1930 did Polanyi find in Fritz London a colleague who was able to help him reconcile his adsorption theory with quantum mechanics in an approach that was to be adopted later as an alternative and more general theory than Langmuir's.[16]

What did Polanyi learn from this experience with the adsorption theory? He later wrote,[17]

My verification could make no impression on minds convinced that it was bound to be specious.... There must be at all times a predominantly accepted scientific view of the nature of things, in the light of which research is jointly conducted by members of the community of scientists.

Polanyi chose not to introduce students to his theory because they could not pass the required examinations. "I could not undertake to force on them views totally opposed to generally accepted opinion."[18]

At the same time during which he was pursuing the problem of adsorption, Polanyi also initiated work on chemical reaction velocities and kinetics, beginning with papers published in 1920. He showed that the rate at which Br_2 molecules dissociate is 300,000 times too slow for explaining the observed rate for reaction of a bromine atom with a hydrogen molecule in the formation of hydrogen bromide. By way of explaining energy of activation, Polanyi proposed an external energy source, perhaps, he said, in the ether.[19]

This notion was not as far-fetched as it now sounds. French physical chemist Jean Perrin, for example, broached a radiation theory of chemical activation in a long article in *Annales de Physique* in 1919. The notion of etherial forces was still fairly common in chemistry; a letter from Reginald Herzog to Polanyi in October of 1920 mentions the notion of "a kind of ether" (*ätherartig*) bonding for secondary valences in some molecules.[20]

Radiation theory was to be the subject of discussions at a Faraday Society symposium in 1921 and at the first and second international Solvay chemistry conferences in 1922 and 1925.[21] A correspondent encouraged Polanyi in his line of argument and wrote him from Würzburg that he should look at an earlier paper by René Marcellin regarding photochemistry energy values.[22] Max Born wrote Polanyi, however, of his skepticism about this theory of reaction velocity and that James Franck shared Born's misgivings.[23]

When Born wrote Einstein asking his opinion, Einstein, who already had told Perrin that he thought the radiation theory was doubtful, replied "Polanyi's ideas make me shudder. But he has discovered difficulties for which I know no remedies."[24] Invited to a special session of the American Chemical Society in Minneapolis in 1926, Polanyi found opinion hardening in favor of Berlin colleague Max Bodenstein's theory of chain reactions and Oxford physical chemist Cyril Hinshelwood's kinetic theory of collisions.

Other approaches by Polanyi to reaction mechanisms and reaction velocities were to prove more successful in the late 1920s and 1930s. Among these was his development of the experimental method, called highly dilute flames, for studying the course of reaction of two reactants (e.g., sodium vapor and chlorine gas) in an evacuated vessel. This work was closely related to Haber's interest in chemiluminescence in the early 1920s.[25] Henry Erying, who arrived in Berlin in 1929 as a National Research Council Fellow, began experimenting with these alkali-halogen flames and then turned with Polanyi to theoretical calculations.

Using semiempirical methods based on quantum mechanical principles on the one hand and experimental data for vibrational frequencies and dissociation energies on the other, Eyring and Polanyi constructed a three-dimensional diagram or surface for following potential energy in the $H + H_2$ reaction. Their paper, published in 1931, led to further work by Eyring at Princeton, H. Pelzer and Eugene Wigner at Göttingen, and Polanyi and Meredith Gywnne

Evans at Manchester in formulations of transition-state theory for understanding all types of chemical and physical processes. This was the work that really cemented Michael Polanyi's reputation, and it was an area of research later taken up by his son, John.[26]

This striking success came in the early 1930s, along with the very handsome offer of a professorship and a new laboratory at the University of Manchester. However, another line of research, with results judged ambiguous at the time, predated the work in reaction kinetics. This work—its instruments, experiments, and theories—warrants examination in some detail.

X-RAY DIFFRACTION AND CRYSTAL STRUCTURE

In October of 1920, Polanyi joined the staff of the Kaiser Wilhelm Institute (KWI) for Physical Chemistry and Electrochemistry in Berlin-Dahlem. The Physical Chemistry Institute was one of the two original chemistry institutes established in 1912; by the early 1930s, the Kaiser Wilhelm Gesellschaft was operating some 38 research institutes in Germany, employing more than 1200 professional scientists, including eleven Nobel Prize winners.[27] Workers in Berlin-Dahlem met at Harnack House for dining, colloquia, and socializing. Polanyi joined Otto Hahn and other colleagues, too, for gymnastics exercises there.[28]

The Institute for Fiber Chemistry owed its origins within Haber's Institute to the fear after the First World War that foreign competition and lack of access to essential materials and equipment were harming the German textile industry.[29] Polanyi was first assigned here.

Polanyi was no stranger to industry and engineering. His father had been trained in civil engineering at the Zurich Polytechnic Institute before becoming a railroad entrepreneur. Eugene Wigner, who collaborated with Polanyi in the 1920s, received his introduction to Polanyi through chemical engineer Paul Beer, who worked in a chemical factory with Wigner's father.[30] Polanyi exerted considerable effort to obtain patents on some of his work in chemical processes, and he served as a consultant in industry. His correspondence from 1919 on includes exchanges with industrial firms about the manufacture of hydrogen and nitrogen and the fabrication of lanolin[31] and electric lamps.[32] By 1933, Polanyi had a regular income from the Tungsram Factories (later General Electric) in Budapest for his work as consultant.[33]

By Polanyi's own account, when he joined the KWI and called on Herr Geheimrat Fritz Haber, Haber immediately mentioned Polanyi's recent speculative papers on radiation and reaction kinetics with the admonishment, "Reaction velocity is a world problem. You should cook a piece of meat." Polanyi took this to mean that he must prove his capacity as a "craftsman" before daring to have the ambition to make a world-class discovery.[34]

Reginald Herzog had just the problem for Polanyi. The cotton fiber is almost pure cellulose. Chemist Heinrich Ost had shown in 1910 that cellulose undergoes conversion into dextroglucose on acid hydrolysis. Herzog and his collaborator, Willi Jancke, had just made x-ray diffraction photographs of cellulose fiber, using a technique that Jancke had learned from Paul Scherrer in Göttingen.[35] Polanyi's assignment was to study the cellulose x-ray diffraction photographs and to interpret them.

Polanyi had no experience with x-ray crystallography, although his good friend, Alfred Reis, was now working in this field at Karlsruhe.[36] The theory and practice of x-ray crystallography had just been highlighted during the summer of 1920, when Max von Laue gave his long-delayed Nobel Lecture in Stockholm (for the physics award of 1914), discussing how his postulate of the interference of x-ray waves had been confirmed by Walter Friedrich and Paul Knipping in Arnold Sommerfeld's Munich laboratory in 1912.[37] Laue's theoretical analysis had convinced his colleages that the photographic spots produced by polychromatic or "white" x-irradiation falling on a crystal of copper sulfate were the result of diffraction.

A crystal is an array of atoms arranged in a pattern that repeats periodically in three dimensions. Metals are crystalline, but the individual crystals are small, and faces are not apparent. If a point is chosen at random within the crystalline pattern, all points identical with this point constitute the set of lattice points, each with exactly the same surroundings. If lattice points are connected by straight lines, the two-dimensional space is made up of parallelograms. In three dimensions, the space is divided into parallelepipeds, and the parallelepiped that may be used to generate the entire lattice by repetition is the unit cell. Six numbers specify the size and shape of a unit cell, and a point within the unit cell is located by triplicates of numbers where, for example 0, 1, 2, 3, and so on refer to the corners of unit cells of length 0 to 1, 1 to 2, 2 to 3, and the like (figure 14.1). In 1848, Bravais had shown the existence of only 14 different ways of arranging points in a space lattice, of which 3 are cubic space lattices: simple (primitive), face-centered, and body-centered.

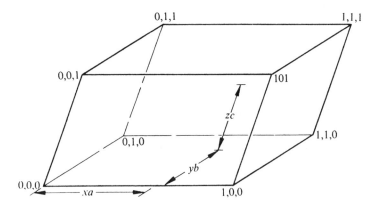

Figure 14.1
Location of a point with coordinates x, y, z. Numbers indicate coordinates of unit cell corners. [From Donald E. Sands, *Introduction to Crystallography* (New York: Dover, 1975), 8, figure 1–6.]

In x-ray waves, which have wavelength comparable to the distance between atoms in the crystal lattice, the radiation's waves are scattered by the lattice atoms (actually, their electrons) and emerge in phase at some points. The angles at which these diffracted rays or points are observed are a function of the crystal's periodic structure (unit cell) and the wavelength of the radiation. For example, in the Laue formulation, in one dimension only,

$$a \left(\cos \alpha_0 - \cos \alpha \right) = h\lambda$$

where a is the repeat distance (unit cell) in one dimension, α_0 is the angle between the incident radiation and the unit cell axis A, α is the corresponding angle for the diffracted radiation, λ is the wavelength, and h is an integer.[38]

In contrast to Laue, William H. Bragg preferred to think of x-rays as reflected from planes in the crystal, with the reflected beam flashing up only at a particular angle of incidence and this angle becoming greater the closer together the planes are. Thus, each spot in the x-ray diffraction photograph corresponds to a particular set of planes, and the spots farthest out in the photographic pattern correspond to the most closely spaced sets of planes. The angle θ for a spot is determined by the wavelength λ of the x-rays and by the interplanar spacing d in the crystal. The x-ray spectrometer that Bragg and his son, W. Lawrence Bragg, had built by their mechanic at Leeds relied on ionization for registering the intensity of reflection of the x-rays, rather than the blackening of photographic film. Their instrument opened up the exploration of x-ray spectra emitted by different elements and opened up the analysis of crystal structure. Their mathematical formula became widely known as the *Bragg equation*:[39]

$$n\lambda = 2d \sin \theta$$

While trying to obtain evidence of the regular spacing of electrons on atomic orbits in Max Born's laboratory in Göttingen, Peter Debye obtained with Paul Scherrer some characteristic diffraction lines from a fine-grained powder of lithium fluoride. They used monochromatic x-irradiation and a cylindrical diffraction camera. The powder sample, in rotation at the center of the camera, was bathed in radiation, and a cylindrical film around the circumference of the camera recorded the diffracted images.[40] Albert Hull in Schenectady, New York, independently published a theory of powder crystal analysis a year later, in 1917, as part of an investigation of the crystal structure of iron, from which single crystals had not been produced.[41]

It was the Debye-Scherrer method that Jancke employed with Herzog at Berlin to investigate cellulose in fibers of cotton, ramie, and twine and to study silk. They used monochromatic radiation from the so-called Coolidge tube (developed by William D. Coolidge, also at the General Electric Laboratory in Schenectady), which produced strong monochromatic x-rays at constant conditions over matters of hours.[42]

At first, Jancke irradiated a pulverized salt of cellulose, but then he switched to using a compressed bundle of cellulose fibers. Herzog and Jancke obtained a table of values for Bragg's parameter of sin $(\theta/2)$. Their photograph of a bundle of ramie fibers showed a pattern of spots symmetrical to two mirror-planes, one plane passing through the primary beam and the axis of the fibres and the other oriented to the normal.[43] Polanyi concluded that the four-point diagram was due to a group of parallel crystals arranged around the axis of the fiber and that the fiber as a whole has rotational symmetry around its axis. His results appeared in a joint publication with Herzog and Jancke (figure 14.2).[44]

Herzog put Polanyi in charge of a research group that was to develop x-ray diffraction studies of both natural fibers and metals, with Polanyi foreseeing the possibility of exploiting the analogy between the diffraction patterns for extended fibers and for metals under stress. Polanyi's group, first housed in the Fiber Institute and, from 1923, in the Physical Chemistry Institute, was one of three teams in Berlin-

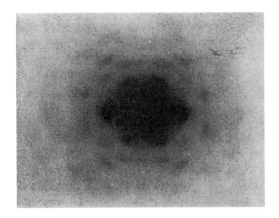

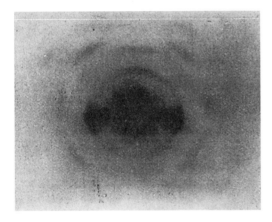

Figure 14.2
X-ray diffraction photographs of a bundle of flax fiber illuminated by a beam of
x-rays at 90 degrees and 45 degrees, respectively, at distances 3.7 cm and 4.5 cm
from the fiber to the photographic plate. (From R. O. Herzog, Willi Jancke, and
Michael Polanyi, "Röntgenspektrographische Beobachtungen an Zellulose: II,"
Zeitschrift für Physik, 3(1921): 347.)

Dahlem in the x-ray field, including Paul Becker and Jancke working
with Herzog, J. Böhm and H. Zocher working with Herbert Freund-
lich, and Hermann F. Mark, Erich Schmid, Ernst von Gomperz, and
Karl Weissenberg working with Polanyi.[45] Eugene Wigner joined
Polanyi's x-ray group at the Physical Chemistry Institute but switched
in 1925 to a focus on chemical reaction rates for his dissertation.[46]

Polanyi's work in x-ray crystallography came to have three
aspects: the development of improved instruments for x-ray diffraction,
the investigation of the strength and properties of materials, and the
architecture of large molecules.

The Instruments

First, the instruments: Mark later recollected, at the beginning of the
1920s, the absence of x-ray tubes in Berlin that could be operated at
high intensities over long periods without careful supervision. Additionally, he lacked precision cameras in which crystals or crystalline
objects could be conveniently mounted for irradiation in all possible
orientations and lacked instruments for working at very low or very
high temperatures or in other extreme conditions. During the next
decade a well-equipped x-ray crystallography laboratory was created
in Berlin-Dahlem, which became a world center for this kind of
research.[47]

Using the Coolidge tube and working with Schmid and Weissenberg, Polanyi began developing improved methods for taking x-ray
photographs, including refinements of the rotating crystal method that
were applied during 1922 to 1923 to study the plastic flow of zinc
crystals and the unknown crystal structure of white tin (figure 14.3).[48]

In the rotating crystal method, a crystal (or fiber) is placed at C
(see figure 14.3A and B), with one of its principal axes (a unit cell axis)
parallel to the vertical arrow in the figure, about which it is rotated. A
monochromatic beam and a small single crystal (0.1 mm) are used. The
crystal is placed in an incident beam of x-rays and the cross section of
the diffracted beam is determined by the dimensions of the crystal. If
the axis of rotation is a fairly prominent zone axis, the reflection spots
are found drawn out on long curves known as *layer lines*. All diffracted
beams for which the path difference is wavelength 1 will make the same
angle with the C axis and so appear on layer line 1.

The crystal is rotated in a predetermined manner so that one after
another of the sets of planes comes into the correct reflecting position.
The diffracted beams are recorded on a flat plate (giving layer lines that

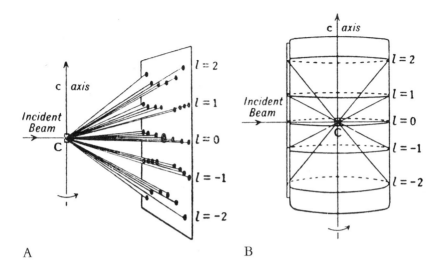

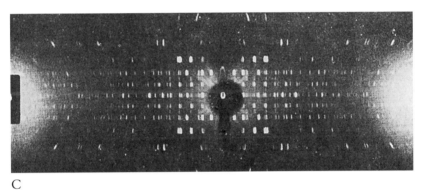

Figure 14.3
(A) Layer lines of a rotation photograph on a flat plate. (B) Layer lines of a rotation photograph on a cylindrical film. (C) Rotation photograph. [From Sir Lawrence Bragg, *The Development of X-ray Analysis* (New York: Hafner Press, 1975), 139, figures 1 and 2.]

are hyperbolas) or on a cylindrical film (resulting in straight parallel lines).[49]

Polanyi later recollected that he had the idea of using an elongated Debye camera to include higher layer lines. He also recollected that Weissenberg derived a mathematical treatment of the layer-line relationship, in which the distances between the layer lines provided direct information about the identity period of the lattice under investigation, so that the size of the repeating unit could be calculated. As early as 1921, Polanyi worked out a simple geometrical formula for determining the identity period (i.e., the distance between the identical diffracting centers) of a lattice from the layer lines before he and Weissenberg worked out a more general equation in 1922.[50] Mark was in control of the first experiments in the improved procedure, and Polanyi published with Weissenberg and E. Schiebold a detailed account of the method in 1924 as improvements continued to be made.[51]

The choice among the ionization, powder, and rotating crystal methods was made on the basis of aims and materials, and the x-ray method quickly became a standard analytical tool for checking step by step the progress of a wide array of investigations of the solid state. It revolutionized both research and analytical practice in the academic and industrial chemical laboratory.

From the start, Polanyi and his collaborators were keen to stake out shares of credit for the new theories and instruments of structure studies by x-ray diffraction. Thus, at the conclusion of a coauthored article in the *Zeitschrift für Physik* in 1924, credit was partitioned in the following way:[52]

> The rotating crystal method therefore has first been used by E. Shiebold in the above-mentioned dissertation on structure determination. At the same time he already was using the constant of the layer-line index, which was communicated independently by Polanyi. From the latter also originated the calculation of the identity intervals from the layer-line distances, which was worked out in general by him in collaboration with Weissenberg.

Polanyi's understandable ambition to be recognized for his role in the development of this new instrumentation seems to have been somewhat thwarted, however. Sometime after the Second World War, Polanyi assigned one of his assistants (or perhaps a secretary) to peruse x-ray crystallography textbooks and handbooks published during the period of the late 1920s to the early 1940s, to see what was said about the rotation method. Most often, the books failed to mention Polanyi's name. Bragg, Bernal, Hull, Shiebold, and others were cited, but Polanyi was not.[53]

Study of Solid Materials

Using both the powder and the rotating crystal methods, with improvements made by Weissenberg, Polanyi and his coworkers undertook a program of research in solid-state analysis. They constructed a variety of machines for treating fibers, particularly cellulose and silk, by swelling, stretching, relaxing, and drying them. Their studies of rigidity, extensibility, elasticity, melting, softening, and swelling were applied to both metal wires and fibers. They drew out molten metals into single-crystal wires and studied stress-strain properties using a stretching apparatus (*Dehnungsapparat*) devised in 1925.[54]

Polanyi and Mark discovered the slip properties of single-crystal tin, and Schmid worked out a law for the shear stress component along the slip direction in a slip plane[55] (figure 14.4). These studies clearly had both industrial and theoretical interest, and Polanyi continued them throughout the 1920s and 1930s, developing in 1932 the concept of dislocation (*Versetzung*) for describing the strength of crystals.[56] Polanyi found that every process that destroyed the ideal structure of crystals increased material strength: in dislocation, "[t]en atoms on one side are opposed by eleven atoms on the other side of the line."[57]

Polanyi presented his theory of dislocation in April 1932 to members of A. F. Joffe's research institute in Leningrad. The theory took notice of the fact that diffraction lines from powdered samples of cold-worked metals are diffuse and that diffraction spots in single-crystal photographs are elongated. Though a material such as diamond is a reasonably perfect crystal, metallic crystals were now proposed to consist of small units slightly out of alignment with one another. Thus, many properties of metals depend on imperfection of the crystal structures (e.g., reducing the tensile strength of rocksalt 200-fold from its theoretical values).[58]

On returning to Berlin from Leningrad, Polanyi learned that a similar idea about dislocation was to appear in a thesis, under the direction of Paul Becker, by Hungarian engineer Egon Orowan. Orowan, who was aware of Polanyi's work on the subject, suggested that they write a joint paper, but Polanyi preferred that they publish side-by-side papers in the *Zeitschrift für Physik*. In 1934, G. I. Taylor, a Royal Society professor at Cambridge, published another version of the theory of dislocation.[59]

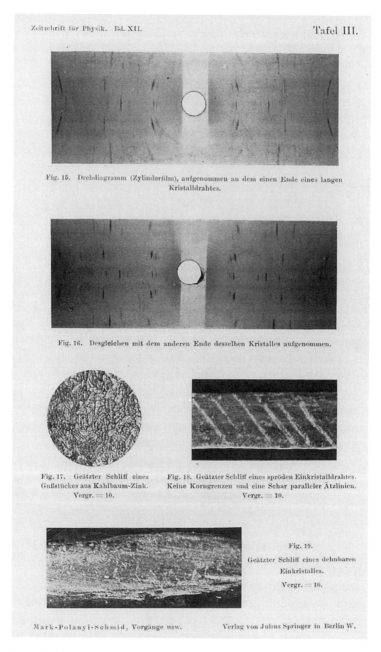

Figure 14.4
Illustrations from H. Mark, M. Polanyi, and E. Schmid, "Vorgänge bei der Dehnung von Zinkkristallen: II. Mitteilung. Quantitative Festlegung des Dehnungs-
mechanismus," *Zeitschrift für Physik*, 12(1923): Tafel III.

Polymeric Structure of Cellulose

What can be said about the original project on cellulose? In further evaluating the photographs by Herzog and Jancke, Polanyi found that all but two dots forming the fiber diagram lay on a series of hyperbolas, with each hyperbola composed of dots reflected by planes having identical indices with respect to the crystal axis parallel to the fiber. He concluded that the formula for the series of these hyperbolas was a function of the identity period parallel to the axis of the fiber.[60]

At this point, Polanyi found himself in a discussion with Herzog about matters of priority in valence theory, with Herzog writing Polanyi to clarify the history of ideas about subsidiary valences within molecules. Colloid chemist Kurt Hess should not be given credit for the hypothesis of subsidiary or secondary valences (*Nebenvalenzen*) within large molecules of cellulose, proteins, and the like, wrote Herzog, because this idea has been long established within Johannes Thiele's school. Further, Herzog himself should receive just as much acknowledgement as Hess with respect to the possibilities of the linking of atoms by secondary valences. Herzog also drew Polanyi's attention to the recent résumé on the subject in Hermann Staudinger's new paper about polymerization and long-chain molecules.[61]

What precipitated this letter appears to have been the fact that Polanyi was moving toward postulating a mechanism for the linking of glucose groupings within cellulose into larger aggregates that would constitute the "molecule" of cellulose. According to Hermann Mark, on 7 March 1921, Polanyi gave a lecture in the Haber colloquium at Harneck House, evaluating x-ray data currently at hand within a theoretical framework of hypotheses about the structure of colloidal substances of high molecular weight.

Some recent work regarding both cellulose and proteins favored the assumption that the molecular units of glucose or amino acids, respectively, are polymerized into very long-chain molecules adding up to molecular weights for cellulose or protein in the tens of thousands. As Herzog had mentioned to Polanyi, Hermann Staudinger (at the Zurich Polytechnic Institute) had just published a controversial paper to this effect, proposing long-chain structures not for cellulose but for polystyrene, polyoxymethylene, and rubber. Also in 1920, Karl Freudenberg at Heidelberg reported that the yield of cellobiose during cellulose degradation favored long-chain structure.[62]

The notion of long-chain structures of very high molecular weight was at odds, however, with the point of view held by most

German organic chemists, rooted in the protein work of Emil Fischer, who had thought impossible the notion that natural proteins have molecular weights higher than some 4000. If evidence, for example from freezing point depression, seemed to point to higher weights, most chemists favored explanation of what they called "pseudo-high molecular weight" by an "aggregate theory" of colloids. The idea, as held by both Herzog and Kurt Hess, was that secondary valences physically hold together aggregates of molecules into a substance of apparently high molecular weight that is broken down easily by heating.[63]

Heinrich Wieland was at this time a prominent chemist who had worked at the KWI in Berlin during the war. He was to be awarded the Nobel Prize in Chemistry in 1927 for his studies of the structure of bile acids. He told Staudinger, who later succeeded Wieland in the chair of organic chemistry at Freiberg, ". . . [D]rop the idea of large molecules; organic molecules with a molecular weight higher than 5000 do not exist. Purify your products, such as rubber, then they will crystallize and prove to be low molecular compounds."[64]

In his talk at the 7 March colloquium, Polanyi reported that his analysis of the cellulose diffraction patterns suggested that the molecular structure was either one giant molecule composed of a single file of linked hexoses or it was an aggregate of hexobiose-anhydrides. Both structures, he said, were compatible with the symmetry and size of the elementary cell, the repeat unit within the three-dimensional space lattice of the crystal.[65] After his remarks ensued what Polanyi later called "a storm of protest from all sides."[66] However, most chemists were to draw from Polanyi's work the conclusion that the small size of the unit cell within crystalline cellulose proved that cellulose is a chemical substance of low molecular weight on the grounds that the molecule cannot be larger than the unit cell.[67] For Polanyi, this was a missed opportunity: "I failed to see the importance of the problem," he later reflected.[68]

Technically, Polanyi and his Kaiser Wilhelm Institute colleagues who were applying x-ray diffraction to structural chemistry were in the vanguard in the 1920s. Structure determination of large organic molecules was to bring scientific fame to some other x-ray crystallographers, among them James Bernal, W. T. Astbury, Max Perutz, and Dorothy Hodgkin. In the 1920s, however, some continued to disparage the diffraction method for structural chemistry, among them Henry Armstrong. The irascible Armstrong railed against Lawrence Bragg's x-ray proof that each atom of sodium in a crystal of salt is surrounded symmetrically by six chloride atoms and vice versa:

This statement is more than "repugnant to common sense." It is absurd to the nth degree, not chemical cricket. Chemistry is neither chess nor geometry, whatever X-ray physics may be. Such unjustified aspersion of the molecular character of our most necessary condiment must not be allowed any longer to pass unchallenged.[69]

For whatever reason, Polanyi seems not to have risen to the challenge by his critics at the Haber colloquium, retreating from the disciplinary authority of the Berlin organic chemists and colloid chemists.

When in 1926 Mark became director of the I. G. Farbenindustrie research laboratory in Ludwigshafen, he continued work with Kurt Meyer on the crystal structure of cellulose. Two Americans—botanist L. O. Sponsler and x-ray crystallographer W. H. Dore—published a detailed structure for a long-chain cellulose molecule in 1926. They proposed an alternating 1-1, 4-4 bonding of the glucose units in the chain, rejecting a 1-4 bonding despite irrefutable chemical evidence of a 1-4 glucosidic bond in cellobiose, the cellulose degradation product that is two linked glucose units. Not surprisingly, many organic chemists renewed their objections to the claims of x-ray crystallographers for solving important structural problems.[70]

When Polanyi's attention was drawn to Sponsler and Dore's interpretation, he responded that their structure was impossible, given his own early diffraction data and the geometry of the four-point system.[71] Shortly afterward, in 1928, Meyer and Mark proposed a compromise theory in which long-chain molecules are held together by special Van der Waals—type forces in "micelles" or "banks."[72] By the mid-1930s, Staudinger's hypothesis of macromolecules, backed by new molecular weight determinations from Thé Svedberg's ultracentrifuge in Uppsala, began to prevail.[73]

REFLECTIONS OF A NORMAL SCIENTIST

In 1929, Polanyi was head of the kinetics department of Fritz Haber's Institute, where he had moved in 1923. As we have seen, Polanyi and his coworkers were investigating an array of closely related problems, many of them directly tied to industrial needs, with many of his assistants gaining the experience by now considered essential for employment in German industrial laboratories. In early 1929, however, as can be seen clearly in the diary he kept during the year, Polanyi was often depressed. To be sure, economic and political reasons justified his depression, as (according to the diary) did family reasons. None of them were out of the ordinary, however, just as Polanyi's worry about his family's income was not unusual among his peers. What is of significance to us are the reasons why Polanyi might have been glum in 1929 about the course of his laboratory work during the last decade.

On the one hand, not only did organic chemists seem skeptical about Polanyi's application of x-ray diffraction to molecular structure, but they paid scant attention to his investigations of the mechanical behavior of organic solids. As Mark wrote,[74]

> We, in our fiber research, were interested in ... strength, elasticity, water absorption, and abrasion resistance. Until this point, for solid organic compounds, interest was focused on melting point, solubility, color, surface activity, ... but never on mechanical behavior. We soon recognized that the solution viscosity of polymeric systems is not their most important property, but the enormous influence of polymer chain length on all physical and mechanical characteristics is ... [important, as can be seen in] ... the hardness of ivory and ebony, the elasticity of kangaroo tails, the toughness of alligator skins, and the softness of cashmere and vicuna wool.

Even more daunting to Polanyi, on the other hand, was the reaction by his colleagues in the physics community to his work in physical chemistry. Again, relying on Mark,[75]

> Truthfully, the results of our studies failed to impress the leading members of the scientific community in the Kaiser Wilhelm Institute, including Max von Laue, Fritz Haber, O. Hahn, Lise Meitner, James Franck, K. F. Bonhoeffer, and others who were preoccupied with radioactivity, atomic and molecular quantum phenomena, and catalysis.

Einstein and other Berlin physicists sometimes visited Polanyi's laboratory in the 1920s, but one of their main purposes was to ask whether Polanyi's x-ray research group could find the increase in wavelength of x-rays scattered by electrons, as predicted in 1923 by Arthur H. Compton, an effect that W. Duane had not been able to duplicate at Harvard University. Mark and H. Kallman did this successfully in 1924.[76]

Accordingly, in correspondence, Polanyi wrote despondently that the scientific problems of the strength of materials (*Festigkeit*), magnetism, electrical conduction, viscosity, and the like were areas in which no physicist wanted to study unless they had some relation to atomic physics.[77] "Dirt physics" was the term of opprobrium used by Wolfgang Pauli for the study of processes in real solids.[78] Egon Orowan, by training an electrical engineer, later recalled, "[P]lasticity was a prosaic and even humiliating proposition in the age of De Broglie, Heisenberg and Schrödinger, but it was better than computing my sixtieth transformer."[79]

In short, Polanyi found himself unable to launch an experimental research program or, in the 1920s, to make a theoretical breakthrough in the 1920s that counted as a real discovery, as a fundamental natural law, or as an achievement for which he, like Laue or Haber or Nernst or Einstein or Franck (or indeed Langmuir in 1932 and Debye in 1936) could enjoy a worldwide reputation and even the award of a Nobel Prize. Polanyi worried, too, that he was not able to keep up with such colleagues as Wigner in quantum mechanics.[80] He wrote to himself that the things that interested him he could not do, whereas the things that he could do did not interest him.[81]

He might still be able to create success, he wrote; it was partly a matter of technical questions. In 40 years, everything would be decided.[82] If only he could make a solid discovery!, he mused.[83] A Sunday afternoon reading of Paul De Kruif's chapter in *The Microbe Hunters* on Robert Koch's discoveries failed to lift Polanyi's spirits and only made him feel worse.[84] His birthday came and went; he was 38 years old.[85] The next weeks brought only more work.[86]

In the previous year (1928), Polanyi had published some 18 scientific papers, coauthored with eight different coworkers or assistants; in 1929, the published papers fell to three, rising again to eight in 1930 with six coauthors, including Schmid, Eyring, and London. In 1932, his *Atomic Reactions* appeared in English and, in 1933, during his last year in Berlin, 14 papers appeared from his laboratory, with nine different coauthors, including Japanese physical chemist J. Horiuti, who was to remain a longtime friend. His collaborators extolled him as research leader and colleague, Erich Schmidt typically saying 60 years later that "Polanyi was perhaps the most inspiring man I ever met."[87]

The year 1933 was decisive for Polanyi, as for so many Central Europeans. He left Berlin permanently in October of 1933, but what is particularly of interest for our purposes is that he chose to stay as late as mid-January of 1933.

On 13 January, 1933, Michael Polanyi ended 10 months of discussions with English scientists and administrators by declining a chair in physical chemistry at the University of Manchester.[88] He gave as his reason a bout of rheumatism triggered by his last visit to damp and smoggy Manchester.[89] Only 3 weeks earlier, Polanyi had received a letter from a business manager at the University of Manchester informing him that tenants on the site on which Polanyi's new laboratory was to be built had received 3 months notice to leave.[90] The vice-chancellor of the university had sent him a document attesting that Polanyi might resign his Manchester post with no dishonor if he were unable to adapt himself happily to English life.[91] He was to receive an annual salary of £1,500 and contributions toward a retirement fund,[92] a newly constructed laboratory of physical chemistry estimated to cost £40,000, and apparatus and funds for some 20 research assistants and coworkers. He was to have light teaching duties and absolute independence in running his laboratory.[93]

Polanyi had informed Haber of negotiations with Manchester from the very beginning of the process. Rather, a draft of a letter to this effect asked Haber for reassurances about Polanyi's future in Germany given the fact that he, Polanyi, was a foreigner and a Jew.[94] After Polanyi's first visit to Manchester, Haber, who like Polanyi was a Jew converted to Christianity, encouraged Polanyi to accept the position in Manchester, given the uncertainty of the future in Germany.[95]

Despite the political events in Germany, Polanyi's research program had been going well in the last few years. Meredith Evans was planning to come to Polanyi's Berlin laboratory from Manchester on a Rockefeller Fellowship,[96] Richard Ogg would be arriving from Harvard on a National Research Council Fellowship,[97] and D. E. Moelwyn-Hughes would be visiting from Cambridge and Trinity College at Oxford.[98] Polanyi's Berlin laboratory was well-organized and busy and, when he finally left Berlin in the fall of 1933, he shipped to Manchester much of his prized Berlin equipment, including high-vacuum mercury pumps and a precision machine lathe.[99] Polanyi in 1933 was a craftsman who had cooked more than one piece of meat and still was on the watch for his world-class discovery. He found unlikely the possibility that it would come in Manchester, so far outside the Berlin scientific community that had nourished and disciplined him.

Fifteen years later, Polanyi retired from an active career in physical chemistry. He spoke and wrote about the need to break with the rationalist account of science that had characterized the epistemological writings of Henri Poincaré, Ernst Mach, Pierre Duhem, Karl Pearson, Bertrand Russell, and Karl Popper. An important origin of *The Tacit Dimension* and of *Personal Knowledge* lies in Polanyi's Berlin career in what Thomas Kuhn has called the practice of "normal science."

"The example of great scientists is the light which guides all workers in science," wrote Polanyi in 1962,[100]

> but we must guard against being blinded by it. There has been too much talk about the flash of discovery and this has tended to obscure the fact that discoveries, however great, can only give effect to some intrinsic potentiality of the intellectual situation in which scientists find themselves. It is easier to see this for the kind of work that I have done than it is for major discoveries.

Polanyi's experiences during his early career took place in the extremely competitive, indeed aggressive, scientific community of Weimar Germany in one of its greatest research institutions. From the beginning of his career, Polanyi was aware of the difficulties of attaining a secure and prominent position in Germany. Asking advice from Hungarian compatriot Theodor von Kármán in the spring of 1920, Polanyi was warned that the mood in the Technische Hochschulen was unfavorable to foreigners and that the *Habilitation* was no guarantee of becoming a professor.[101]

Polanyi's early correspondence and contacts with such distinguished scientists as Nernst and Herzog demonstrated to him an obsession with discovery, originality, and priority. The reception of Polanyi's first publications, which spanned several areas of physical chemistry, impressed on him how scientific work must fit within a community of opinion about what is acceptable and what is significant. His first assignments at the Kaiser Wilhelm Institute served as an apprenticeship in laboratory practice and laboratory culture. His participation in colloquia, his correspondence and conversations, and his trips abroad all integrated him into the mores and customs of the scientific community, including the partitioning of credit and value.[102]

The quest for novelty and recognition was not an easy one. "The scientist's surmises and hunches," he wrote, "involve high stakes, as hazardous as their prospects are fascinating. The time and money, the prestige and self-confidence gambled away in disappointing guesses will soon exhaust a scientist's courage and standing. His gropings are weighty decisions."[103]

Polanyi's experiences taught him that the scientist is not "indifferent to the outcome of his surmises," that the scientist "risks defeat but never *seeks* it," and that the working scientist's quest for discovery is passionate, rooted both in a craving for success and in beliefs about a real world. This postcritical philosophy of science is historically embedded in the culture of Weimar science, specifically in Fritz Haber's Institute of

Physical Chemistry, for all the fact that it is generalizable in time and place.

Finally, much of Polanyi's scientific work in the 1920s may be said to have been a practice suspended between the construction of instruments and the exercise of their use in research programs. Polanyi navigated the use of instrumentation across the boundaries of disciplinary fields and communities—between x-ray physics and structural chemistry, between atomic physics and materials science, and between fundamental research and applied science. His work in the development and adaptation of x-ray diffraction techniques to studies of molecular structure and mechanical behavior demonstrates the creative interplay between the material and conceptual dimensions of science. It also exhibits both the lag in time between instrumental and theoretical innovation and the scientific community's ranking of experimental investigations within a hierarchy of disciplinary values, dominated in the 1920s by the interpretive framework of atomic physics and quantum theory.

ACKNOWLEDGMENTS

This chapter makes considerable use of the Michael Polanyi Papers, which are held at the Special Collections of the Regenstein Library at the University of Chicago. I am grateful for permission to have consulted these papers. I thank Frederic L. Holmes, Trevor Levere, and especially Erwin N. Hiebert for suggestions about this chapter and Gabor Pallo and Endre J. Nagy for helpful conversations about Michael and Karl Polanyi. My research for this project was supported by the National Science Foundation grant no. SBR-9321305 and by the Thomas Hart and Mary Jones Horning Endowment.

NOTES

1. Thomas S. Kuhn, *The Structure of Scientific Revolutions*, 2nd ed. (Chicago: The University of Chicago Press, 1970), 44, n. 1; Steve Fuller, *Social Epistemology* (Bloomington: Indiana University Press, 1988); "Social Epistemology and the Research Agenda of Science Studies" in *Science as Practice and Culture*, ed. Andrew Pickering, (Chicago: University of Chicago Press, 1992): 390–428, on 393–394; and Sheila Jasanoff et al., eds., *Handbook of Science and Technology Studies* (London: Sage, 1995), especially 7.

2. Michael Polanyi, *Personal Knowledge: Towards a Post-Critical Philosophy* (Chicago: University of Chicago Press, 1958).

3. See Michael Polanyi, *Science, Faith and Society* (Chicago: University of Chicago Press, 1946), especially 47, 54; *The Tacit Dimension* (New York: Doubleday, 1966), especially 24–25; and *Personal Knowledge*, especially 150–151, 202.

4. See the journal *Tradition and Discovery: The Polanyi Society Periodical* (St. Joseph, MO: Missouri Western State College, 1966). See also Kari Polanyi-Levitt, *The Life and Work of Karl Polanyi* (Montréal: Black Rose Books, 1990); Endre J. Nagy, "After Brotherhood's Golden Age: Karl and Michael Polanyi," in *Humanity, Society and Commitment: On Karl Polanyi*, ed. Kenneth McRobbie (Montréal: Black Rose Books, 1993), 81–112.

5. See bibliographies of publications in E. P. Wigner and R. A. Hodgkin, "Michael Polanyi, 12 March 1891–22 February 1976," *Biographical Memoirs of Fellows of the Royal Society* 23(1977): 413–448; Marjorie Grene, ed. *The Logic of Personal Knowledge, Essays Presented to Michael Polanyi on his Seventieth Birthday 11th March 1961* (London: Routledge and Kegan Paul, 1961); and Harry Prosch, *Michael Polanyi: A Critical Exposition* (Albany: State University of New York Press, 1986).

6. About chemistry of the hydrocephalic fluid and about the blood serum during starvation: publications of 1910 and 1911: see bibliography (cited in n. 5).

7. Magda K. Polanyi did not forget chemistry. See Magda Polanyi, *Technical and Trade Dictionary of Textile Terms*, 2nd, enlarged ed. (Oxford: Pergamon Press, 1967).

8. Michael Polanyi, "Curriculum Vitae," composed June 1933, in Michael Polanyi Papers (MPP), University of Chicago Regenstein Library Special Collections, Box 2, Folder 12 (2 : 12). Of great utility is John M. Cash, *Guide to the Papers of Michael Polanyi* (Chicago: The Joseph Regenstein Library, September 1977). Also, Wigner and Hodgkin (1977), 413–415.

9. Michael Polanyi, "Eine neue thermodynamische Folgerung aus der Quanten-hypothese," *Verhandlungen der deutschen physikalischen Gesellschaft* 15(1913): 156–161, esp. 157; and "Neue thermodynamische Folgerungen aus der Quanten-hypothese," *Zeitschrift für physikalische Chemie* 83(1913): 339–369, discussed in William T. Scott, "Michael Polanyi's Creativity in Chemistry," in *Springs of Scientific Creativity*, eds. Rutherford Aris et al. (Minneapolis: University of Minnesota Press, 1983), 279–307, on 282–283. See also Wigner and Hodgkin (1977), 416.

10. Letters from Polanyi to Alfred Reis, 11 December 1912; Georg Bredig to Polanyi, 1 February 1913; Georg Bredig to Polanyi 12 February 1913 (MPP, 1 : 2). More than 30 years later, Percy Bridgmann wrote to Polanyi that he was surprised to have just learned of this work and thought it unfortunate that it had escaped general notice: letter from Percy Bridgman to Polanyi, 19 December 1946 (MPP, 5 : 2).

11. Letter from Walter Nernst to Polanyi, 22 August 1913; Nernst to Polanyi, 30 August 1913; draft of letter from Polanyi to Nernst, 3 September 1913; and Nernst to Polanyi, 15 October 1913 (MPP, 1 : 2). See Polanyi, "Neue thermodynamische

Folgerungen," 340–341. Regarding Nernst, the heat theorem, and Einstein, see Diana Barkan, *Walther Nernst and the Transition to Modern Physical Science* (Cambridge: Cambridge University Press, 1999).

12. Michael Polanyi, "Adsorption von Gasen (Dämpfen) durch ein festes nicht-flüchtiges Adsorbens," *Verhandlungen der deutschen physikalischen Gesellschaft* 18(1916): 55–80. See Scott, "Michael Polanyi's Creativity," 283.

13. Regarding the adsorption theory, see Michael Polanyi, "The Potential Theory of Adsorption," *Science* 141(1963): 1010–1013, reprinted in Michael Polanyi, *Knowing and Being*, ed. Marjorie Grene (London: Routledge and Kegan Paul, 1969), 87–96; quotation on 89.

14. Ibid., 93.

15. See Polanyi, "The Potential Theory of Adsorption," in *Knowing and Being*, ed. M. Grene, 95, n. 2; and letter from A. Eucken to Polanyi, 31 March 1922 (MPP, 1 : 17) and carbon copy of letter from Polanyi to Eucken, 4 April 1922 (MPP, 1 : 19).

16. Fritz London and Michael Polanyi, "Über die atomtheoretische Deutung der Adsorptionskräfte," *Die Naturwissenschaften* 18(1930): 1099–1100. Fritz London was in Berlin for most of the period from 1927 to 1933. He developed a theory of chemical reactions as activation processes and investigated the characteristics of molecular forces using the newer quantum mechanics. See Kostas Gavroglu, *Fritz London: A Scientific Biography* (Cambridge: Cambridge University Press, 1995).

17. Polanyi, "The Potential Theory of Adsorption," in *Knowing and Being*, ed. M. Grene, 91–92.

18. Ibid., 94.

19. Michael Polanyi, "Über die nichtmechanische Natur der chemischen Vorgänge," *Zeitschrift für Physik* 1(1920): 337–344; "Zur Theorie der Reaktionsgeschwindigkeit," *Zeitschrift für Physik* 2(1920): 90–110; and "Zum Ursprung der chemischen Energie," *Zeitschrift für Physik* 3(1920): on 31–32: "We want to show that a still unknown kind of energy source coming from empty space is possible . . ."

20. Letter from Herzog to Polanyi, 8 October 1920 (MPP, 1 : 11). See Helge Kragh, "The Aether in Late 19th-Century Chemistry," *Ambix* 36(1989): 49–65.

21. See my discussion in Mary Jo Nye, *From Chemical Philosophy to Theoretical Chemistry: Dynamics of Matter and Dynamics of Disciplines, 1800–1950* (Berkeley: University of California Press, 1993), 121–129, and Keith J. Laidler, "Chemical Kinetics and the Origins of Physical Chemistry," *Archives for History of Exact Sciences* 32(1985): 43–75, and *The World of Physical Chemistry* (Oxford: Oxford University Press, 1993), 263–265.

22. Letters from Hans Halban to Michael Polanyi, 20 June 1920 and 28 July 1920 (MPP, 1 : 9 and 1 : 10, respectively). The Austrian (and later French) physicist, Hans Halban, was born in 1908, so it is unlikely that this correspondent is he.

23. Letter from Max Born to Polanyi, 13 June 1921 (MPP, 1 : 14).

24. Quoted in Scott, "Michael Polanyi's Creativity," n. 10, 305–306 from Einstein to Born, 30 December 1921, in *Born-Einstein Letters* (New York: Walker, 1971).

25. Dietrich Stoltzenberg, *Fritz Haber: Chemiker, Nobelpreisträger, Deutscher, Jude* (Weinheim: VCH, 1994), 475.

26. See Henry Eyring, "Physical Chemistry: the Past 100 Years," *Chemical and Engineering News* 54(1976): 88–104, on 90–93; and Jeffry Ramsey, "Between the Fundamental and the Phenomenological: The Challenge of 'Semi-Empirical' Methods," *Philosophy of Science*, 64(1997): 627–653. Henry Eyring and Michael Polanyi, "On the Calculation of the Energy of Activation," *Naturwissenschaften* 18(1930): 914–915; Henry Eyring and Michael Polanyi, "Über einfache Gasreaktionen," *Zeitschrift für Physikalische Chemie* B12(1931): 279–311; M. G. Evans and Michael Polanyi, "Some Applications of the Transition State Method to the Calculation of Reaction Velocities, Especially in Solution," *Transactions of the Faraday Society* 31(1935): 875–894. For a contemporary discussion of transition-state theory, with mention of Michael and John Polanyi, see A. Maureen Rouhi, "Chemists Hone New Techniques for Probing Transition-State Events," *Chemical and Engineering News* 22(April 1996): 36–41.

27. About the early years, see Jeffrey Allan Johnson, *The Kaiser's Chemists: Science and Modernization in Imperial Germany* (Chapel Hill: University of North Carolina Press, 1990). Morris Goran's *The Story of Fritz Haber* (Norman: University of Oklahoma Press, 1967), 99–106, is now superseded by Stoltzenberg's *Fritz Haber*; regarding Haber's Institute, see chapter 11, 439–526.

28. Otto Hahn, *My Life: The Autobiography of a Scientist*, translated by Ernst Kaiser and Eithne Wilkins (New York: Herder and Herder, 1970), 94–95.

29. Herman (or Hermann, in earlier publications) F. Mark, *From Small Organic Molecules to Large: A Century of Progress* (Washington, DC: American Chemical Society, 1993), 20.

30. Interview of Eugene P. Wigner with Thomas S. Kuhn, 21 November– 4 December 1963, Session II, 3 December 1963, 5 in transcripts in Sources for History of Quantum Physics, AIP Niels Bohr Library.

31. Letter from Aktien Gesellschaft für Betriebsökonomie (signed Arthur), dated 22 April 1920, 18 May 1920 (MPP, 1 : 8).

32. Letter to Polanyi from patent division of Badische-Anilin and Soda Fabrik (BASF), 20 July 1920; letter to Polanyi from Philips Glühlampenfabriken Aktien Gesellschaft, Eindhoven, Holland, 30 July 1921 (MPP, 1 : 15); and letter to Polanyi from Vereinigte Glühlampen und Electricitäts-Actien Gesellschaft, Budapest, 15 May 1923 (MPP, 1 : 19).

33. Michael Polanyi's Affidavit, signed 22 December, 1937 before Norris Haselton, Vice Consul of the United States of America in Manchester (MPP, 3 : 10).

34. Michael Polanyi, "My Time with X-Rays and Crystals" in *Knowing and Being*, ed. M. Grene, 97.

35. Regarding Jancke, see E. E. Hellner and P. P. Ewald, "Schools and Regional Development: Germany" in *Fifty Years of X-Ray Diffraction* ed. P. P. Ewald (Utrecht: Oosthoek, 1962), 456–468, 461–462; and R. O. Herzog and W. Jancke, "Roentgenspektrographische Beobactungen an Zellulose," *Zeitschrift für Physik* 3(1920): 196–198.

36. See Alfred Reis, "Zur Kenntnis der Kristallgitter," *Zeitschrift für Physik* 1(1920): 204–220; see comment about Reis by Herman Mark, *From Small Organic Molecules*, 80.

37. Regarding the discovery and confirmation of x-ray diffraction, see Paul Forman, "The Discovery of the Diffraction of X-Rays by Crystals: A Critique of the Myths," *Archives for History of Exact Sciences* 6(1969): 38–71.

38. Sir Lawrence Bragg, *The Development of X-Ray Analysis*, eds. D. C. Phillips and H. F. Lipson (New York: Hafner Press, 1975), 20; also Donald E. Sands, *Introduction to Crystallography* (New York: Dover, 1975), 89.

39. Lawrence Bragg, *Development of X-Ray Analysis*, 31–32. See design of the spectrometer, figure 2, p. 31. Also John C. Kendrew, "The Three-Dimensional Structure of a Protein Molecule," *Scientific American* (reprint), 1961.

40. Paul Scherrer, "Reminiscences," in Ewald, *Fifty Years of X-Ray Diffraction*, 643.

41. Albert W. Hull, "Autobiography," in Ewald, *Fifty Years of X-Ray Diffraction*, 584.

42. P. Ewald, "The Immediate Sequels to Laue's Discovery" in Ewald, *Fifty Years of X-Ray Diffraction*, 80.

43. O. Herzog and Willi Jancke, "Röntgenspektrographische Beobachtungen an Zellulose," *Zeitschrift für Physik* 3(1920): 196–198.

44. O. Herzog, Willi Jancke, and M. Polanyi, "Roentgenspektrographische Beobactungen an Zellulose II," *Zeitschrift für Physik* 3(1920): 343–348.

45. Scott, "Michael Polanyi's Creativity," 288; Hermann Mark, "Recollections of Dahlem and Ludwigshafen," in Ewald, *Fifty Years of X-Ray Diffraction*, 603.

46. Eugene Wigner, *The Recollections of Eugene P. Wigner as Told to Andrew Szanton* (New York: Plenum, 1992), 78–80. The thesis title (1925) was "Bildung und Zerfall von Molekülen" or "Formation and Decay of Molecules."

47. Mark in Ewald, *Fifty Years of X-Ray Diffraction*, 603.

48. Hermann Mark, Michael Polanyi, and E. Schmid, "Vorgänge bei der Dehnung von Zinkkristallen," parts I, II, and III, *Zeitschrift für Physik* 12(1922): 58–77; 78–110; 111–116; and H. Mark and M. Polanyi, "Die Gitterstruktur,

Gleitrichtungen und Gleitebenen des Weissen Zinns," *Zeitschrift für Physik* 18(1923): 75–96. See Polanyi, "My Time with Crystals," in *Knowing and Being*, ed. M. Grene 100.

49. Lawrence Bragg, *Development of X-Ray Analysis*, 139. Also Kathleen Lonsdale, *Crystals and X-Rays* (New York: D. Van Nostrand, 1949), 79; and Sands, *Introduction to Crystallography*, 90–92, 95–97.

50. See Mark, *From Small Organic Molecules to Large*, 24; Michael Polanyi, "Faserstruktur im Röntgenlichte," *Naturwissenschaft* 18(1921): 337–340, Eq. 2 and figures 1 and 3; M. Polanyi, E. Schiebold, and K. Weissenberg, "Über die Entwicklung des Drehkristallverfahrens." *Zeitschrift für Physik* 24(1924): 337–340, on 339.

51. Polanyi, "The Potential Theory of Adsorption," in *Knowing and Being*, ed. M. Grene, 100; and Mark, *From Small Organic Molecules*, 24. See Polanyi, Schiebold, and Weissenberg, "Über die Entwicklung des Drehkristallverfahrens."

52. Ibid., 340.

53. Handwritten notes (4 pages) about x-ray crystallography (MPP, 43 : 4). Sir Lawrence Bragg's 1975 *Development of X-Ray Analysis* refers, however, to the "classical rotation photograph method" "developed by Shiebold and Polanyi" (138–139).

54. See Scott, "Michael Polanyi's Creativity," 291.

55. Mark (1993), 25; and E. E. Hellner and P. P. Ewald, "Schools and Regional Development," in *Fifty Years*, ed. P. P. Ewald, 462.

56. Scott, "Michael Polanyi's Creativity," 292. Michael Polanyi, "Über eine Art von Gitterstörung, die einem Kristal plastisch machen könnte," *Zeitschrift für Physik* 89(1934): 660–664.

57. Polanyi, "Über eine Art non Gitterstörung," (1934), cited earlier. See Ernest Braun, "Mechanical Properties of Solids," in *Out of the Crystal Maze: Chapters from the History of Solid-State Physics* eds. Lillian Hoddeson et al. (New York: Oxford University Press, 1992), 317–358, 327–331.

58. See W. Hume-Rothery, "Applications of X-Ray Diffraction to Metallurgical Science," in Ewald, *Fifty Years of X-Ray Diffraction*, 190–211, 198–199; and Kathleen Lonsdale, "X-Ray Diffraction and Its Impact on Physics," in Ewald, *Fifty Years of X-Ray Diffraction*, 221–247, on 232–233.

59. Polanyi, in Ewald, *Fifty Years of X-Ray Diffraction*, 636. G. I. Taylor, "The Mechanism of Plastic Deformation of Crystals," *Proceedings of the Royal Society* A145(1934): 362–415; and E. Orowan, "Zur Kristallplastizität," *Zeitschrift für Physik* 89(1934): 605–659. Regarding recent studies of dislocation, see Robert F. Service, "Materials Scientists View Hot Wires and Bends by the Bay," *Science* 272(26 April 1996): 484–485.

60. Michael Polanyi, "Faserstruktur im Röntgenlichte." *Naturwissenschaft* 9(1921): 337–340. See Polanyi, "My Time with X-Rays," in *Knowing and Being*, ed. M. Grene (1969), 99.

61. Letter from Reginald Herzog to Polanyi, 8 October 1920 (MPP, 1 : 11).

62. Herman F. Mark, "Polymer Chemistry: The Past 100 Years," *Chemical and Engineering News* 54(1976): 179.

63. See Yasu Furukawa, "Hermann Staudinger and the Emergence of the Macromolecular Concept," *Historia Scientiarum* 22(1982): 7–9.

64. Quoted from Staudinger's *Arbeitserrinerungen*, 54, in Yasu Furukawa, *Inventing Polymer Science: Staudinger, Carothers, and the Emergence of Macromolecular Chemistry* (Philadelphia: University of Pennsylvania Press 1998), p. 67.

65. Polanyi, "My Time with X-Rays," in *Knowing and Being*, ed. M. Grene 99.

66. Ibid.

67. Mark, "Polymer Chemistry," 179.

68. Ibid.

69. Quoted from *Nature*, 1927, in L. Bragg, *Development of X-Ray Analysis*, 149.

70. Olenus Lee Sponsler and Walter Harrington Dore, "The Structure of Ramie Cellulose as Derived from X-ray Data," *Colloid Symposium Monographs* 4(1926): 174–202, translated in *Cellulosechemie* 11(1930): 186. See Mark, "Polymer Chemistry," 179.

71. Polanyi, reply to "Irrtümliche: Bestimmung des Zellulose–Raumgitters," *Naturwissenschaft* 15(1928): 263–264.

72. Kurt Meyer and Herman Mark, "Über den Bau des kristallisierten Anteils der Cellulose," *Berichte der deutschen chemische Gesellschaft* 61(1928): 593–614.

73. Cellulose now is estimated to have an average molecular weight of 400,000, corresponding to 2500 glucose units. As Meyer and Mark proposed in 1928, the cellulose fiber is said to be made up of micelles, or bundles of parallel chains held together by hydrogen bonds. The diameter of a micellar unit corresponds to 100 to 200 cellulose chains, rather than the 40 to 60 that Meyer and Mark proposed, and the length of the chains is approximately 200 glucose units, rather than 30 to 50. Both the mechanical strength of cellulose and its chemical stability (in contrast to protein) is considered to result from the micellar structure. See Louis F. Fieser and Mary Fieser, *Introduction to Organic Chemistry* (Boston: D. C. Heath, 1957), 280. Also see Furukawa, "Staudinger, Carothers, and . . . Macromolecular Chemistry," 60.

74. Mark, *From Small Organic Molecules*, 25.

75. Ibid., 29.

76. Ibid., 31.

77. Carbon copy of letter to Wichard von Moellendorff [W], 27 March 1929 (MPP, 2:5).

78. Michael Eckert and Helmut Schubert, *Crystals, Electrons, Transistors: From Scholar's Study to Industrial Research*, translated by Thomas Hughes (New York: American Institute of Physics, 1990), 184–185.

79. Egon Orowan, "Dislocations in Plasticity," in *The Sorby Centennial Symposium on the History of Metallurgy*, ed. C. S. Smith (New York: Gordon and Breach, 1965): 359–376, quoted in Braun, "Mechanical Properties," in *Out of the Crystal Maze*, ed. L. Hoddeson et al., 331.

80. Diary, 9 January (MPP, 44:4).

81. Diary, 10 January (MPP, 44:4).

82. Diary, 21 January (MPP, 44:4).

83. Diary, 26 February (MPP, 44:4).

84. Diary, 3 March (MPP, 44:4).

85. Diary, 11 March (MPP, 44:4).

86. Diary, 8 April (MPP, 44:4).

87. Erich Schmid, interview with E. Braun, summer 1982, quoted in Braun, "Mechanical Properties," in *Out of the Crystal Maze*, ed. L. Hoddeson et al., 330.

88. Carbon copy of letter from Polanyi to Arthur Lapworth, 13 January 1933 (MPP, 44:4). On 30 January 1933, Adolf Hitler was appointed chancellor of Germany.

89. Carbon copy of letter from Polanyi to Arthur Lapworth, 13 January 1933, and carbon copy of letter from Polanyi to F. G. Donnan, 17 January 1933 (MPP, 2:11).

90. Letter to Polanyi from E. D. Simon, 22 December 1932 (MPP, 2:10).

91. Letter from Walter H. Moberly, vice chancellor, University of Manchester, to Polanyi, 15 December 1932 (MPP, 2:10).

92. Handwritten letter from Arthur Lapworth to Polanyi, 1 March 1932 (MPP, 2:8).

93. Letter from A. J. Allemand to Polanyi, 15 May 1932 (MPP, 2:8) and 29 November 1932 (MPP, 2:10).

94. Handwritten draft of letter from Polanyi to Haber, 16 April 1932 (MPP, 2:8).

95. Letter from Fritz Haber to Polanyi, 27 June 1932, in response to a note from Polanyi of 26 June 1932 (MPP, 2:9).

96. Letter from M. G. Evans to Polanyi, 20 February 1933 (MPP, 2 : 11).

97. Letter from Richard A. Ogg, Jr., to Michael Polanyi, 5 May 1933 (MPP, 2 : 12).

98. Letter from D. E. M. Hughes to Polanyi, 7 May 1933 (MPP, 2 : 12).

99. Scott, "Michael Polanyi's Creativity," 299.

100. Polanyi, "My Time with X-Rays," in *Knowing and Being*, ed. M. Grene, 97.

101. Letter from von Kármán to Polanyi, 17 March 1920 (MPP, 1 : 7).

102. See Polanyi, *The Tacit Dimension*, 63–75.

103. Ibid., 76.

BIBLIOGRAPHY

Bragg, [Sir] Lawrence. *The Development of X-Ray Analysis*, eds. D. C. Phillips and H. F. Lipson. (New York: Hafner Press, 1975).

Eckert, Michael, and Helmut Schubert. *Crystals, Electrons, Transistors: From Scholar's Study to Industrial Research*, translated by Thomas Hughes (New York: American Institute of Physics, 1990).

Ewald, P. P., ed. *Fifty Years of X-Ray Diffraction* (Utrecht: Oosthoek, 1962).

Eyring, Henry. "Physical Chemistry: the Past 100 Years." *Chemical and Engineering News* 54(1976): 88–104.

Forman, Paul. "The Discovery of the Diffraction of X-Rays by Crystals: A Critique of the Myths." *Archives for History of Exact Sciences* 6(1969): 38–71.

Furukawa, Yasu. "Hermann Staudinger and the Emergence of the Macro-molecular Concept." *Historia Scientiarum* 22(1982): 1–18.

Gavroglu, Kostas. *Fritz London: A Scientific Biography* (Cambridge: Cambridge University Press, 1995).

Hahn, Otto. *My Life: The Autobiography of a Scientist*, translated by Ernst Kaiser and Eithne Wilkins (New York: Herder and Herder, 1970).

Hoddeson, Lillian, et al. *Out of the Crystal Maze: Chapters from the History of Solid-State Physics* (New York: Oxford University Press, 1992).

Kragh, Helge. "The Aether in Late 19th-Century Chemistry." *Ambix* 36(1989): 49–75.

Laidler, Keith J. "Chemical Kinetics and the Origins of Physical Chemistry." *Archives for History of Exact Sciences* 32(1985): 43–75.

Laidler, Keith J. *The World of Physical Chemistry* (Oxford: Oxford University Press, 1993).

Lonsdale, Kathleen. *Crystals and X-Rays* (New York: Van Nostrand, 1949).

Mark, Herman F. "Polymer Chemistry: the Past 100 Years." *Chemical and Engineering News* 54(1976): 176–189.

Mark, Herman F. *From Small Organic Molecules to Large: A Century of Progress* (Washington, DC: American Chemical Society, 1993).

Nagy, Endre J. "After Brotherhood's Golden Age: Karl and Michael Polanyi," in *Humanity, Society and Commitment: On Karl Polanyi*, ed. Kenneth McRobbie (Montréal: Black Rose Books, 1993), 81–112.

Nye, Mary Jo. *From Chemical Philosophy to Theoretical Chemistry: Dynamics of Matter and Dynamics of Disciplines, 1800–1950* (Berkeley: University of California Press, 1993).

Polanyi, Michael. "Neue thermodynamische Folgerungen aus der Quantenhypothese." *Zeitschrift für physikalische Chemie* 83(1913): 339–369.

Polanyi, Michael. "Zum Ursprung der chemischen Energie." *Zeitschrift für Physik* 3(1920): 31–35.

Polanyi, Michael, with R. O. Herzog and Willi Jancke. "Roentgenspektrographische Beobachtungen an Zellulose II." *Zeitschrift für Physik* 3(1920): 343–348.

Polanyi, Michael. "Faserstruktur im Röntgenlichte." *Die Naturwissenschaften* 18(1921): 337–340.

Polanyi, Michael, with Hermann Mark and E. Schmid. "Vorgänge bei der Dehnung von Zinkkristallen: I, II, and III," *Zeitschrift für Physik* 12(1922): 58–77, 78–110, 111–116.

Polanyi, Michael, with E. Schiebold and K. Weissenberg. "Ueber die Entwicklung des Drehkristallverfahrens." *Zeitschrift für Physik* 24(1924): 337–340.

Polanyi, Michael, with Henry Eyring. "Ueber einfache Gasreaktionen." *Zeitschrift für physikalische Chemie* B12(1931): 279–311.

Polanyi, Michael. "Ueber eine Art von Gitterstörung, die einem Kristal plastisch machen könnte." *Zeitschrift für Physik* 89(1934): 660–664.

Polanyi, Michael. *Science, Faith and Society* (Chicago: University of Chicago Press, 1946).

Polanyi, Michael. *Personal Knowledge: Towards a Post-Critical Philosophy* (Chicago: University of Chicago Press, 1958).

Polanyi, Michael. "The Potential Theory of Adsorption." *Science* 141(1963): 1010–1013.

Polanyi, Michael. *The Tacit Dimension* (New York: Doubleday, 1966).

Polanyi, Michael. *Knowing and Being*, ed. Marjorie Grene (London: Routledge and Kegan Paul, 1969).

Polanyi-Levitt, Kari. *The Life and Work of Karl Polanyi* (Montréal: Black Rose Books, 1990).

Sands, Donald E. *Introduction to Crystallography* (New York: Dover, 1975).

Scott, William T. "Michael Polanyi's Creativity in Chemistry," in *Springs of Scientific Creativity*, ed. Rutherford Aris et al. (Minneapolis: University of Minnesota Press, 1983), 297–307.

Stoltzenberg, Dietrich. *Fritz Haber: Chemiker, Nobelpreisträger, Deutscher, Jude* (Weinheim: VCH, 1994).

Wigner, Eugene. *The Recollections of Eugene P. Wigner as told to Andrew Szanton* (New York: Plenum, 1992).

Wigner, E. P., and R. A. Hodgkin. "Michael Polanyi, 12 March 1891–22 February 1976." *Biographical Memoirs of Fellows of the Royal Society* 23(1977): 423–448.

Index

Note: Page numbers followed by an italic *f* indicate figures; page numbers in italics indicate the beginning of an article by the author.

Abel, Frederic, xv, 335. *See also* Noble, Andrew
"Research on Explosives" (with Andrew Noble), 346, 351
Academy of Sciences (Paris), 103n.56, 140, 159, 160, 176, 211, 218, 298
Achard, F. K., 216
Acids
oxalic acid study, 256–258, 260–261
oxygen theory of, 145–146, 170, 225–227
Adsorption theory, 369–371, 392n.13
Affinity theory, 246
Agricola, Gnaeus Julius, ix, xi
Air. *See also* Atmospheric air analysis; Gases
atmospheric, 83, 85, 86–91, 107, 122–123, 126, 128, 130, 131n.14
early understandings of, 76, 79, 81*f*, 82–86
measuring the "goodness of," xiii, 105, 107, 110–112, 116–117
in mineral water, 90–91
Alchemy, xi, 1–2, 73n.32
apparatus design in, 59–64
chrysopoeian alchemy, 45, 58, 60–62, 65, 70–71, 73n.22
historiography and image of, 2, 35, 39–40, 49–50n.1, 55–57, 56, 58, 71n.1
the mercurialists, 65–71
"otherness" from chemistry (discontinuous view), 55–57, 64, 70, 71, 75

the Philosophers' Stone, 58, 65
as a precursor of chemistry (continuity view), 35, 49, 55, 71
search for gold, 45, 58, 60–62, 65, 68, 70–71, 73n.22
texts, 11–12, 14–18, 28, 30–31n.22, 56, 58
Alchemical imagery
decodable into experimental activities, 60–62, 61*f*, 64–65, 72n.5, 74n.42
Jungian interpretation of, 55, 56, 64–65, 70, 71n.1
Alchemical operations, 2, 10, 26, 28, 65–68, 66*f*, 67*f*, 69*f*
assaying metals, 30*f*, 35, 44–49, 60–62, 71, 73n.22
furnaces used in, 25–26, 32n.45, 59, 63
replicability of, 2, 55, 57, 58–59, 63, 68–70 (*see also* Hermes Tree)
testing for salts and alums, 37–39, 49, 51n.13, 52n.17
Alcograde, 173–174
Alcohol thermometer, 192
Alcohol titration, 161–165, 175
Alembic
artifacts, 5–7, 10, 11*f*, 12*f*, 14, 19*f*, 19, 22(1.9)*f*
historiography of, 5–6, 28
manuscript illustrations of, 16*f*, 23*f*, 24*f*
Rosenhut alembic, 25, 26*f*
Alkaloids, discovery of, 280
Alkylation research, 311–313, 318
Allegorical imagery. *See* Alchemy

American Chemical Society, 273
Ammonia process of water analysis, 322–324
Analytical balance. *See* Balance
Anderson, Robert G. W., ix, xi, xix, 1–2, *5*
Annales de Chimie, 285–286
Apothecaries, 87, 160
Apparatus, viii–ix, 91. *See also* Alembic; Blowpipe; Ceramic apparatus; Distillation apparatus; Furnace; Glass apparatus; Instruments; Metal Apparatus
animal bladders, 94–95, 95*f*, 219
continuity of chemical -, xi–xii, xiv, 14–15, 28, 75, 150
crucible, 20, 40, 43*f*, 45
cucurbit, 10, 11, 12*f*, 14, 20, 23
fuses, 328
household objects as, 91–92
rubber apparatus, 231, 328
vessels, viii, 9, 13, 244, 314–318, 315*f*, 317*f*
Archimedes law, 154, 176
Areometer, 165–166, 167, 171. *See also* Hydrometer
Fahrenheit's, 157, 159, 163
Lavoisier's, 159–160, 161*f*
whether it is different from the hydrometer, 156–160
Argand, Aimé, 225
Argand lamp, 225, 235–236n.55
Argon, 121, 133n.34
Aristotelian elements, 75, 86, 87
Armstrong, H. E., 321, 325, 328–329, 384–385
Artifacts of chemical apparatus
cucurbit, 7*f*, 14, 20
kiln sites, 18, 23
Mesopotamian, 8*f*, 9–11, 11*f*, 12*f*
Ashmolean Museum, 25
Ashmole, Elias, *Theatrum Chemicum Britannicum*, 40, 42*f*, 56
Assaying, 1, 20, 40, 53n.27. *See also* Mineralogy
alchemical, 30*f*, 35, 45–49, 49, 58, 60–62, 65, 70–71, 73n.22
precision of early, 40, 44, 53n.27

Atmospheric air analysis, 87–91, 107, 122–123, 126, 128, 130, 131n.14. *See also* Gasometric analysis
Boyle's studies, 83, 85, 86–87
Atomic theory, 240, 259, 262–263, 266, 275, 301n.6
Dalton's work leading to, 239, 245, 250–251, 253, 255, 257, 262–264, 266
Atomic weights, 240, 298–299
Avicenna (pseudo-), 44–45

Bachelard, Gaston, 176, 200
Balance
analytical balance, 243–244
compared with the hydrometer, 169, 175, 177
historiography of, 35–39, 49, 131n.6
importance of, 243–244, 245
Lavoisier's, ix, xi, 137, 202–204
Norton's, 40, 41*f*, 43*f*, 44, 49
precision balance, ix, xi, xii–xiii, 44, 107–108
use in alchemy, xi, 35, 39–40, 49
"Balance sheet method," 144, 147, 170, 205, 239, 244
Baretta, Marco, *Enlightenment of Matter*, 56–57
Basil Valentine (attr.) alchemical corpus, 59–63, 61*f*, 72n.13
Baumé, Antoine
laboratory of, 25, 87
use of hydrometer, 157, 160–165, 164, 165, 172
Becher, Johan Joachim, x, 12, 17
Becher-Shaw portable furnace, 17–18, 25
Beddoes, Thomas, *Elective Attraction*, 203
Bensaude-Vincent, Bernadette, xi, xiv–xv, xix, 77, *153*
Bérard, Jacques Étienne, salts analysis, 257–258, 263–265, 265*f*, 270n.57
Beretta, Marco, 169, 200
Bergman, Torbern, 201, 211–212
table of affinities, 160, 170
Berthelot, Marcellin, 351, 354, 362–363n.75
Berthollet, C. L., 257, 263, 265, 266

Berzelius, Jacob, x, xi, 36, 273, 281–282, 302n.15
  atomic weights focus, 240, 298–299
  organic theory and research, 274–276, 279–280, 284, 287–288, 291–292, 294, 314
  spark eudiometer, 119f, 119–120
Berzelius Museum, 127
Biringuccio, Vannoccio, Pirotechnia, 48–49
Black, Joseph
  heat transfer theory, 193–199, 200, 205
  isolation of "fixed air" (carbon dioxide), 86–87, 90, 100, 107–108, 141
  pneumatic apparatus of, 82–83
  portable furnace of, 25
  theory of acids, 170
  on uses of thermometer, 186–187
Blades, John, 220
Blagden, Charles, 172
Blowpipe, x, 50n.5, 211
  platinum, 212, 213–214
  use in alchemy, 2, 35, 39
Boerhaave, Herman
  theory of corpuscular heat, 190–193, 196–197, 198
  on uses of thermometer, 186–187, 190, 191, 205
Bories, Pierre, 157, 159, 164–165, 169
Boyle, Robert, 56, 68, 70, 72n.5, 74n.41, 79, 80, 101nn.13–14
  Medicina Hydrostatica, 166–167
  Sceptical Chymist, 189
  studies of atmospheric air, 83, 85, 86–87
  Touching Cold, 188–189
  use of hydrometer, 153, 156, 157
  "Boyle's Law," 83
Bragg, William H., 375–376, 384–385
Bredig, Georg, 369, 370
Bret, Patrice, 171
Brisson, Mathurin, 160
Brock, William, 137
Brownrigg, William, 80, 90, 91
  beehive shelf, 91, 92f

Brunschwig, Hieronymus, Liber de arte Distallandi de Compostis, 7f, 7, 11, 14, 15f, 15, 24f, 28
Bucquet, Jean-Baptiste, 141, 143
Buffon, Comte de, 211
Bulb-apparatus. See Kaliapparat
Bunsen, Robert, xv, 135n.57, 289, 335–336, 358n.31
  caloric measurment of fired gunpowder, 343–344, 350, 355, 364nn.85 and 86
  "A Chemical Theory of Gunpowder" (with Leon Schischkoff), 336, 341
  Gasometry, 130
  measuring explosive force of fired gunpowder, 344–345
  thermodynamic studies of gunpowder, 337, 341–342, 343f, 349, 359n.42, 359–360n.47
Burning coal tests, 38–39, 52n.21
Butane ("ethyl"), 314, 318

Caloric measurmement of fired gunpowder, 343–344, 350, 355, 364nn.85 and 86
Caloric theory of heat, 200–204, 205
Calorimeter, 358n.32
  of Lavoisier and Laplace, 137, 148–149, 201–204, 205
Carius method, 318
Castles as artifact sites, 18–21, 28, 33n.47
Cavallo, Tiberius, ix
Cavendish, Henry, 79, 80, 95, 96f, 97, 110, 195, 201
  eudiometer of, 117–119, 118f, 119f, 126–128
  gasometric analysis of, 116–117, 120–121, 126, 132n.26
  isolation of "inflammable air," 87, 120, 121
  Philosophical Transactions, 126
  pneumatic apparatus, xiv, 95, 96f
  precision balance, 108
  studies of atmospheric air, 120–121, 126, 128, 133n.32

Ceramic apparatus, 9, 13, 18, 19, 21, 22(1.9)*f*, 26, 93
Chaulnes, Louis d'Albert de, 221
Chemical apparatus. *See* Apparatus; Artifacts; Instruments
Chemical nomenclature. *See* Nomenclature
Chemical revolution, 75, 137, 150, 169, 197, 239
Chemistry. *See also* Experimental operations; *and by field*
defined, 75, 190–191, 197
as a mixed science, 335, 355n.3
origins and history of, vii–viii, x–xiii, 102, 153, 185, 239–241, 273 (*see also* Innovation; Precision; Social Context)
archaeological evidence, 5, 9–11, 28
chemical revolution, 75, 137, 150, 169, 197, 239
continuity of early apparatus, xi–xii, xiv, 1, 2, 14–15, 28, 35, 75, 150
continuity view of, 35, 49, 55, 71
controversies, 285–287, 294–296
discontinuous view of, 55–57, 64, 70, 71, 75
the Enlightenment, xiii, 55, 156, 275–76
the pre-electronic era, 311, 324
pre-Lavoisier era, 79, 91
Chevreul, M.-F., 340
Chrysopoeia, 58, 65, 68, 73n.22
Clayfield, William, 125
Collesson, Jean, 65–66, 68, 70
Combustion analysis, 81*f*, 138, 145, 146, 201–204, 326–327. *See also* Kaliapparat
Compton, Arthur H., 386
Conservation of mass, 244
Conservatoire des Arts et Métiers, ix, 122, 137
Continuity
of alchemy and chemistry, 35, 49, 55, 71
of chemical apparatus, xi–xii, xiv, 1, 2, 14–15, 28, 35, 75, 150
in theories of heat, 194, 200, 204

Copper digester, 316–318, 317*f*, 318*f*, 328
tap, 219
Crawford, Adair, 194, 200
Cronstedt, A. F., 214
Crosland, Maurice, xix, 76, 79, 298–299
Crucible, 20, 40, 43*f*, 45
Crusher gauge, 347, 363n.83
Crystalline structure, 374–376, 384–385. *See also* X-ray crystallography
Cucurbit artifacts, 10, 11, 12*f*, 14, 20, 23
Culture of precision, 169–170, 176–177, 182–183n.58, 201, 248, 276, 302–303n.15
Cupel/cupellation, 20, 36, 40, 45

Dalton, John, xii, 128
atomic theory of, 239, 245, 250–251, 253, 255, 257, 262–264, 266
multiple combining proportions research, 243–247, 250, 252*f*, 253–254, 257, 268n.21
nitrogen-oxygen reactions study, 246–248, 251, 253, 255–256
studies of atmospheric air, 248–249
Daumas, Maurice, 138, 149
Scientific Instruments, 153
Davy, Humphrey, 251, 252, 253, 259, 296
Debye, Peter, 376, 380
de Chaulnes, Louis d'Albert, 221
de la Fond, J. A. Sigaud, 221–223
della Porta, G. B., 7
De Luc, Jean André, 165–166
studies of heat expansibility, 194–195, 197
De Perfecto Magisterio, 45
"Dephlogisticated air," 97. *See also* Oxygen
Diamond heating, 140, 216
Dickinson, Edmund, 56
Diderot, Denis and Jean D'Alembert, *Encyclopédie*, ix, xi, 82
Digester, copper, 316–318, 317*f*, 318*f*, 328

Dijon Academy, 211, 212, 216, 217, 218–219
"Discontinuous" view of alchemy, 55–57, 64, 70, 71, 75
Disease and air quality, 105, 107
Disinfection apparatus, 29f, 225–231, 227f, 230f, 236nn.58 and 67
Distillation, 1, 14, 163
  in inert atmospheres (Frankland), 319, 320f
  process and methods of, 6, 15, 17, 24, 28, 80
Distillation apparatus, 5–11, 28, 29nn.2,3, and 6, 61–62, 62. See also Alembic; Pelican; Retort and receiver
  bain-marie, 14–15f, 16(1.6)f
  ceramic vessels, 9, 13
  cucurbit artifacts, 10, 11, 12f, 14, 20, 23
  distillation bases, 23f, 23, 24f
  Moor's head still, 15–16(1.7)f, 27f
Distilled water, 224–225
Dobbs, B. J. T., Janus Faces Genius, 56
Donovan, Arthur, 137, 187, 199
Dore, W. H., 385
Duclo, Gaston (Claveus), 68
Dumas, J.-B., 276, 278–279, 285–287, 288
  organic analysis of, 276, 277–279, 280, 287, 290–291, 305n.45, 308n.76
Dumotiez's gasometer, 125, 126

École Polytechnique, 340
Electrical eudiometer, 130, 133n.31
Electrochemistry, 326, 371
Elements, 121, 244–245, 259. See also by element
Elliot, John, 220, 221
Enlightenment, xiii, 55, 70, 156, 275–276
Equipment. See Apparatus; Instruments
Erker, Lazarus, 43f
Ether concept, 89, 371–372
"Ethyl" (butane), 314, 318
Eucken, Arnold, 370
Eudiometer, xi, xiii, xv, xvi, 77, 105, 110f, 116–118, 117f, 131–132n.17

electrical, 130, 133n.31
  Fontana's, 112–116, 113f, 114f, 115f, 126
  Frankland's use of, 314–318, 315f
  iron, 217
  Pepys' eudiometer, 128–130, 129f
  spark eudiometer of Berzelius, 119f, 119–120
  Volta eudiometer, 217, 233n.20
Experimental operations, x–xi, 95, 111, 149, 205, 326
  evolution of the experimental method, 240–241
  laboratory practice as a craft, 240, 367, 388–390
  laboratory vs. field research, 336, 338, 341, 343, 346, 354
  selection of experimental method, 249–250, 264, 348
  theory interacting with, vii–viii, xv–xvi, 243–244
Explosives. See Gunpowder
Eyring, Henry, 372

Fahrenheit, Daniel Gabriel, 157, 159, 163, 191–192, 195, 204
Faraday, Michael, 108, 126, 130, 132n.26, 212, 231
Fermentation, 147–148
Fiber Institute, 369, 373, 376, 389
Field vs. laboratory research, 336, 338, 341, 343, 346, 354
Find, J. A. Sigaud de la, 221–223
"Fire air" (oxygen), 94, 95, 97
Fischer, Emil, 384
"Fixed air" (carbon dioxide), 82, 168
  isolation of, 86–87, 90, 100, 107–108, 141
  in the reduction of metals, 97–100, 139–145, 143f
Flame tests, 2, 38–39, 52n.21
  highly dilute flames, 373–373
Fondazione Scienza e Tecnica, 127
Fontana, Felice, 111
  eudiometers of, 112–116, 113f, 114f, 115f, 126
Forbes, R. J., 6
Fourcroy, A. F., 99

Fox, Robert, 187
Frankland, Edward, xii, xiii, xv, 240, 332
  chemical laboratory of, 311, 327–328
  condensers, 312–314, 313f, 331
  copper digester, 316–318, 317f, 318f, 328
  reaction vessels, 314–318, 315f, 316f
  use of Bohemian glass, 315, 328, 331
  water analysis apparatus, 320f, 322f, 323f, 328–329, 330
  combustion analysis, 326–327
  distillation in inert atmospheres, 319, 320f
  "ethyl" (alkylation) research, 311–313, 318
  gas collection methods, 318–320, 319f
  How to Teach Chemistry, 331
  water analysis for organic compounds, 321–326, 320f, 322f, 323f
French, John, The Art of Distillation, 8f, 12, 14–17, 16f, 23f, 22(1.10f, 23, 24, 30–31n.22
French Gunpowder Association, 340
Furnace
  Agricola's furnaces, ix, xi
  ceramic furnace, 26
  early or alchemical, 8–9, 13–14, 15, 17, 25–26, 32n.45, 45, 62–64
  portable, 17–18, 25, 25–26
  reverbatory, 139, 144–145, 145f
  used by alchemists, 25–26, 32n.45, 59, 63

Gases, 80, 82, 90. See also Air; Atmospheric air analysis; Gasometric analysis
  as "air," 76, 79, 81f, 82–86
  large-scale production of (Guyton), 225–231
  understanding and defining, 82–88, 97–100, 100n.4, 108, 110, 244
Gasometer, xi, 105, 123–128, 125f, 127f, 130
  hydrostatic (Martinus), 123–125, 124f
  Lavoisier's, ix, 122–123, 123f, 126, 137

Gasometric analysis, xvi, 105, 219. See also Atmospheric air analysis
  gas collection methods and apparatus, 317f, 318f, 318–320, 319f, 328
  gravimetric measurement of gases, 107–108
  of gunpowder (see also Gunpowder), 342–343
  nitric oxide measurement, xv, 110–112, 116–117
  volumetric measurement, 81f, 83–84, 84f, 107
Gay-Lussac, Jules L., 248, 277, 298, 341
  organic analysis, 274–276, 284
  testing of fulminates, 278, 341, 356n.25
Geber (pseudo-), Summa Perfectionis, 45–49, 50n.2
General Electric Laboratory, 370
Glass apparatus, viii, 11f, 13, 14, 18, 21, 94, 283. See also Pelican
  Bohemian glass, 315, 328, 331
  glass blowing skills, 329
  glass tap or cock, 219, 220–221
  glass tubes and cylinders, xvin.4, 119f, 119–120
  ground glass apparatus, ix, 77–78, 93, 211, 213, 221–223
  Nooth's apparatus, 219– 221, 220f
  Tschirnhausen lens (Lavoisier), 140
Glass houses and kilns, 18, 2&
Glowing coal tests, 38–39, 52n.21
Gold. See also Alchemy
  alchemical search for, 45, 58, 60–62, 65, 68, 70–71, 73n.22
  assaying for, 44–46, 58
Golinski, Jan, xv–xvi, 137, 185
Göttling, J. F. A., 215–16
Gravimetric techniques, xiv. See also Atomic theory
  Berzelius' atomic weights, 240, 298–299
  measurement of gases, 107–108
  specific gravity measurement, 163–164, 166–170, 172
Grosses Distillierbuch, 7f, 7, 11, 14, 15f, 15, 24f, 28

Guerlac, Henry, 139, 187
Gunpowder, invention of smokeless, 365
Gunpowder, 162, 164, 170–171, 327
analysis of residues, 348–349, 350–351, 361–362n.70, 362–363n.75
caloric measurmement of fired, 343–344, 350, 355, 364nn.85 and 86
gasometric analysis, 342–343
measuring explosive force of fired, 335, 336, 339–340, 357n.25, 360n.55
Bunsen and Schischkoff studies, 344–345
Noble and Abel studies, 345–351, 352f, 353f, 363n.78, 364n.90
Rumford's apparatus for, 337–340, 339f
Proust's research on, 337–341
testing of fulminates (Gay-Lussac), 278, 341, 356n.25
thermodynamic approach to studying (Bunsen and Schischkoff), 335, 337, 341–342, 343f, 349, 359n.42, 359–360n.47
Guyton de Morveau, Louis Bernard, 126, 211
disinfection apparatus, 29f, 225–231, 227f, 230f, 236nn.58 and 67
eudiometers, 217–218, 219, 222, 223f, 224
ground glass apparatus, ix, 77–78, 213, 222
platinum apparatus, 77, 211, 212, 213–214
studies of gases, 219, 225–231
portable laboratory of (nécessaire chemique), 212–214, 213f, 215f, 224f, 224–226, 235n.54
use of distilled water, 224–225

Haber, G. Fritz, 373–374, 388
Haber Institute, 385, 389–390
Hales, Stephen, 75, 90
on fixed air, 168
pneumatic apparatus, 76, 77, 83–84, 84f, 85f, 105, 106f, 108, 109f, 110, 133–134n.35, 138

pneumatic apparatus modified by Lavoisier, 138–140, 139f, 141, 142f
pneumatic studies of, 79, 90, 97, 108
Halleux, Robert, 10, 36
Halogen, 318
Hancock, Thomas, 231
Hassenfranz, Jean, 172, 173
Hauch, Wilhelm, 125f, 125, 126, 132n.21
Heat. See also Calorimeter; Furnace; Thermometer
caloric theory of, 200–201, 203–204
chemical theory of, 196–197, 198, 199, 203
corpuscular theory of, 190–193, 196–197
debate over the nature of, 185, 189
as "elemental fire," 190–193
expansibility of heated materials, 194–195, 197
latent transfer theory of, 193–195, 196–200
thermal equilibrium theory, 194, 198–199
Helmont, J. B. van, 79, 82, 99
Henry, E.O., 219
Hermes' Tree, 65–69, 66f, 67f, 73n.29
reproduced in modern laboratory, 69–71, 70f
Herzog, Reginald, 374, 376, 383, 389
History of chemistry. See Chemistry, origins and history of
Hobbe, Thomas, 189
Hoffmann, August Wilhelm, 277, 316
Holmes, Frederic L., vii, xii, xiii, xx, 76, 137, 300
Homberg's hydrometer, 158–159, 163
Hope, Thomas Charles, 128
Housz, Ingen, 89, 107
Hull, Albert, 376
Hydrogen ("inflammable air"), 87, 89, 98, 120, 121
Hydrometer, xi, 77, 158–159, 163, 174. See also Areometer
balance compared with, 169, 175, 177
comparisons of rival instruments, 163, 182n.48

Hydrometer (cont.)
early uses of, 153, 154–56, 155*f*
efforts to standardize, 171, 172–177, 201
heterogeneity and problems of replicability, 171–174, 176–177
scientific vs. commercial uses of, 158–160, 162, 182n.50
use in titrating alcohol, 161–165, 175
Hydrostatic balance, 154, 163
Hydrostatic gasometer, 123–125, 124*f*
Hypathia, 153

Iatrochemistry, 44
Ice clorinometer, ix, 77
"Inflammable air" (hydrogen), 87, 89, 98, 120, 121
Innovation
adoption of *Kaliapparat*, 287–291, 297–298
adoption of novel technologies, xii–xiv, 287–291
"black boxing" of thermometer, 185–186, 201–202, 204–206
particular vs. universal adaptations, 177, 186
technology transfer, 328–329, 368
Instrument-makers, 18, 21, 93, 108, 133–134n.35, 187
M. Boulley, 228, 230*f*, 230
Dumotiez brothers, 125, 226, 229*f*, 235n.50
Daniel Gabriel Fahrenheit, 157, 159, 163, 191–192, 195, 204
M. Fortin, 108, 138
Pierre Bernard Mégnié, 138, 218–219
James Nasmyth, 316
Hasledine Pepys, 126, 127*f*, 127, 128–130, 129*f*
Josiah Wedgewood, 93, 202–203
Instruments, x, 29n.4, 91, 273. *See also*
Balance; Eudiometer; Gasometer; Hydrometer; *Kaliapparat*; Thermometer
evolution of, 55, 137, 211, 311
precision of, xii–xiii, xiv, xv, xvi, 76, 154, 160, 291
sensory information and, 187–89, 193,
205, 206n.6, 243
theory and experimentation interacting with, vii–viii, x–xi, xvi, 75–76, 99–100, 176, 243, 299
Irvine, William, and heat capacities of bodies, 199–200
Isfahan apparatus, 11, 12 *f*

Jancke, Willi, 374, 376
Janety, Marc Étienne, 216
Jungian alchemical interpretation, 55, 56, 64–65, 70, 71n.1. *See also* Alchemy

Kaiser Wilhelm Institute, 369, 373, 376, 389
Kaliapparat, xii, xiv, 291. *See also* Liebig, Justin
adoption of, 287–291, 297–301
benefits of, 281–285, 298
controversy over, 285–287
Liebig's design and method, xiii–xiv, 240, 270, 273, 288, 292–296, 299, 307n.72
Kallman, Mark and H., 386
Kekulé Laboratory, xi, 127, 285
Kerr, Robert, 204
Key (alchemical), 68
Kiln sites, 18, 23
Kirwan, Richard, 170–172, 203
Knightons glasshouse, 18
Kopp, Hermann, *Geschichte der Chemie*, 36–37
Kuhn, Thomas S., 367, 388
Kunckel, Johann, *Ars Vitraria*, 36, 39
Kunth, Carl S., 276

Laboratories, 78, 127, 305n.53. *See also*
Apparatus; Instruments
Frankland's chemical laboratory, 311, 327–328
Lavoisier's simple research apparatus, x, 76, 121, 137–138, 140, 144–145, 145*f*, 146, 147–148 (*see also* Lavoisier, chemical laboratory)
portable laboratory (Guyton), 212–214, 213*f*, 215*f*, 224*f*, 224–225, 226, 235n.54

Priestley's laboratory and chemical apparatus, ix, 29n.4, 90–93, 93f, 94, 95, 110f, 131–132n.17, 137, 138, 221

Laboratory artifact sites, 18–21, 25, 28, 29nn.4 and 5, 33n.47. *See also* Artifacts

Laboratory practice, 59, 64
as a craft, 240, 367, 388–390
vs. field research, 336, 338, 341, 343, 346, 354

Laboratory vs. field research: *See also* Experimental operations

Landriani, Marsilio, 112, 117f

Langmuir, Irving, 370

Laplace-Lavoisier calorimeter, 137, 148–149, 201–204, 205

Laplace-Lavoisier ice clorinometer, ix, 77

Latour, Bruno, 185, 186

Lavoisier, Antoine Laurent, xiv, 49, 77, 107, 149, 269n.36, 337
analysis and synthesis of air, 145, 146–148
"balance sheet method," 144, 147, 170, 205, 239, 244
calibrating thermometric measurement, 200–201
caloric theory of heat, 200–204, 205
on "chemical geometry," 168
chemical laboratory of, ix, xi, 121–122, 133–134n.35, 137, 140, 143f
areometer, 159–160, 161f
calorimeter (Lavoisier-Laplace), 137, 148–149, 201–204, 205
demonstration apparatus, 123, 137–138
fermentation experiments, 147–148
gasometer, ix, 122–123, 123f, 126, 137
marble basin, 147, 148f
metal apparatus, 141, 142f
modifications to Hales' pneumatic apparatus, 138–140, 139f, 141, 142f
pneumatic apparatus, 147–148, 148f
precision balance, ix, xi, 137, 202–204

research apparatus, x, 76, 121, 137–138, 140, 144–145, 145f, 146, 147–148
water synthesis apparatus, 137
combustion studies and theory, 138, 145, 146, 201–204
conservation of mass law, 244
gasometric analysis of, 122f, 122–123
oxygen theory of acids, 145–146, 170, 225–227
precision emphasis of, 146, 168, 173, 200–201
proposes new nomenclature, 98–100, 150
reduction of metals, 97–100, 139–145, 143f
simplifying research apparatus, x, 76, 121, 137–138, 140, 144–145, 145f, 146, 147–148
"table of simple substances," 244-245
on uses of the hydrometer, 154, 159, 164, 168–169
on uses of thermometer, 186–187
works by
Calorique, 89
General Considerations on the Nature of Acids, 146
Method of Chemical Nomenclature, 100
Opuscules Physiques et Chymiques, 142f, 144, 170
Traité élémentaire, x–xi, 98–100, 106f, 122, 137, 171, 203–204

Lavoisier-Laplace calorimeter, 137, 148–149, 201–204, 205

Lavoisier-Laplace ice clorinometer, ix, 77

LeFebvre, Nicaise, 189

Lemery, Nicolas, *Elements of Chemistry*, 75

Lenoir, Paul Étienne, 217

Levere, Trevor H., vii, xi, xiii, xv, xvi, xx, 76–77, *105*

Levey, Martin, 9

Libavius, Andreas, *Alchemia*, xi, 153
Liber de arte Distallandi de Compostis, 7f, 7, 11, 14, 15f, 15, 24f, 28

Liebig, Justin, 273–274, 276–279, 281, 282, 303n.20. *See also Kaliapparat*

Liebig, Justin (cont.)
Liebig condenser, 313*f*, 313
organic analysis critique and research, 278, 280–281, 285–289, 292–296, 301n.3, 304–305n.40, 305n.45, 307n.72
Louvre pelican, 24
Love, Rosaleen, 190
Löwig, Carl, 298
Luc, Jean-André de. *See* de Luc, Jean-André
Lundgren, Anders, 39

Macquer, Pierre-Joseph
Dictionaire de Chymie, 88–91, 99, 154
Elements of Theoretical Chemistry, viii, ix
Maets, Carolus de, 63
Magellan, Jean H. de, 219–220
Maier, Michael, *Atalanta Fugiens*, 65, 67*f*, 70
Mallet, J. W., 324
Martine, George, 195–196
Mass, conservation of (Lavoisier), 244
Mauskopf, Seymour H., xiv, xx, 240–241, 254, *335*
A System of Chemistry, 254–255, 257
Mayow, John, 79, 80, 81*f*, 90
McKie, Douglas, 187
Medicine, 105, 107, 225–226
Mercurialists, 65–71
Mesopotamian artifacts, 8*f*, 9–11, 11*f*, 12*f*
Metal apparatus, 25, 26*f*, 26, 27*f*, 34n.59. *See also* Furnace; Platinum apparatus
brasssware, viii, ix
copper digester, 316–318, 317*f*, 318*f*, 328
copper tap, 219
iron eudiometer, 217
ironware, ix
Lavoisier's, 141, 142*f*
pewter alembic, 25, 26*f*
platinum blowpipe, 212, 213–214
Metallurgy, 20, 35, 39, 240. *See also* Assaying
Metric system, 168, 173–177. *See also* Quantifying spirit

Metzger, Hélène, 190
Meyer's theory of acids, 170
Microbiology, 326
Middleton, W. E., 185
Mineralogy, 43*f*, 155, 187–188, 211–213, 214. *See also* Assaying
Mineral water analysis, 90–91, 158–160, 175, 219–220
Mitscherlich, Eilhard, 292, 294, 295
Mohr, Fredrich, 219, 313
Monasteries as artifact sites, 18–21, 28
Montigny, 157
Moorhouse, Stephen, 21, 29n.3, 32n.40
Mortimer, Cromwell, 193
Morveau. *See* Guyton de Morveau, Louis Bernard
Multiple combining proportions research, xii, 239, 245, 266
Dalton's work, 243, 244, 245–247, 250, 252*f*, 253–254, 257, 268n.21
oxalic acid study of Thomson, 256–258, 260–261
salts analysis of Bérard, 257–258, 263–265, 265*f*
salts analysis of Wollaston, 259–263, 260*f*, 261*f*
Wollaston, 258–260, 262–263
Musée des Techniques, ix, 122, 137
Museum of the History of Science (Oxford), 25, 27*f*

Nernst, Walther, 369, 370, 389
Newman, William R., xi, 2, 3, *35*, 62–63, 74n.42
Newton, Isaac, 26, 56, 80
Nitric oxide
"goodness of air" measurement, xiii, xv, 105, 107, 110–112, 116–117
isolation of, 110–111
Nitrocellulose powder, 351
Nitrogen, 89, 97
Nitrogen-oxygen reactions study (Dalton), 246–248, 251, 253, 255–256
Nitrous air. *See* Nitric oxide
Noble, Andrew, 335

analysis of gunpowder residues, 348–349, 350–351, 361–362n.70, 362–363n.75
measuring force of fired gunpowder, 345–351, 352*f*, 353*f*, 363n.78, 364n.90
"Research on Explosives" (with Frederic Abel), 346, 351
Nomenclature
conservativism of early chemists re:, 87, 97–98, 103n.56, 263
Lavoisier proposes new, 98–100, 150
Nooth, J.M., 219–221, 220*f*
Norton, Thomas, 59, 64
the Norton balance, 40, 41*f*, 43*f*, 44, 49
Ordinal of Alchemy, 8–9, 11, 12–13, 29n.23, 40, 52n.23
Novel apparatus. *See* Innovation
Nye, Mary Jo, xxi, 241, *367*

Odling, W., 324
Offenbahrung (*Basil Valentine* corpus), 59–63
Operations. *See* Experimental operations
Organic chemistry, 240, 327
Berzelius' theory and research, 274–276, 279–280, 284, 287–288, 291–291, 296–297, 314
Dumas, 276, 277–279, 280, 287, 290–291
Liebig, 278, 280–281, 285–289
national schools of, 296–297, 299–301, 384, 389–390
organic analysis, xii, xiv, 273, 290, 301n.6
water analysis for organic compounds (Frankland), 320*f*, 321–324, 322*f*, 323*f*
Oxalic acid study, 256–258, 260–261
Oxygen, 97, 123, 222
as "fire air," 94, 95, 97
nitrogen-oxygen reactions study, 246–248, 251, 253, 255–256
oxygen theory of acids, 145–146, 170, 225–227
preparation of, 94, 95*f*

Paracelsian of Hessen, *Stein der Uhralten*, 59–60, 62, 63
Paris Academy of Sciences, 103n.56, 140, 159, 160, 176, 211, 218, 298
Parker, William, 219
Partington, J. R., *History of Chemistry*, 81*f*, 97
Pauli, Wolfgang, 387
Pelican, 20, 24, 28, 29n.6
the double pelican, 6–7, 7*f*, 14
Pelletier, P. J., 277, 287
Pelouze, J., 297
Pepys, Hasledine, 126, 127*f*, 127, 128–130, 129*f*
Pharmacy/pharmacists, 228, 277
apothecaries, 87, 160
Philalethes, Eirenaeus, 62–64, 67, 68, 70. *See also* Starkey, George
Ripley Reviv'd, 63–64, 74n.42
Philosophers' Stone, 58, 65. *See also* Alchemy
Philosophical Transactions, 126, 129*f*, 202, 256, 257
Phlogiston theory, x, 88–91, 99, 120–121, 244, 273
Physical chemistry, 241, 335, 337, 367. *See also* Gunpowder; X-ray crystallography
Piobert, Guillaume, 340–341, 344
Platinum apparatus, 211
crucibles, viii
Guyton's, 77, 211, 212, 213–214
introduction of, 211, 216, 259
Plot, Robert, 105
Pneumatic apparatus, xii, xiv, 96*f*, 218–219, 221–223
animal bladders used in, 94–95, 95*f*, 219
Black's apparatus, 82–83
Cavendish's apparatus, xiv, 95, 96*f*
Hales' apparatus, 76, 77, 83–84, 84*f*, 85*f*, 105, 106*f*, 108, 109*f*, 110, 133–134n.35, 138
the pneumatic trough, 105, 221
Pneumatic chemistry, 97, 99, 128, 130, 219. *See also* Black, Joseph; Cavendish, Henry; Hales, Stephen
respiration studies, 81*f*, 107

Poisonous gas, 90
Polanyi, Michael, 241, 367
critique of positivism, 367, 388
dislocation/diffraction theory, 381,
383f, 383, 384–385, 395n.50
Personal Knowledge, 368, 388
polymeric strucutre of cellulose studies,
383–387, 396n.73
reaction velocities and kinetics work,
371–373
scientific apprenticeship views and
experiences, 367–368, 369–373
social context of chemistry and, 367,
368–369, 385–390
The Tacit Dimension, 368, 388
on thermodynamic theory of
adsorption, 369–371, 392n.13
x-ray crystallography research, 369,
374–381, 377f, 379f, 382f
Portable laboratory (nécessaire chemique),
212–214, 213f, 215f, 224f, 224–225,
226, 235n.54
Pottery apparatus, See Ceramic apparatus
Precision. See also Quantifying spirit;
Replicability; Thermometric
calibration
culture of, 169–170, 176–177, 182–
183n.58, 201, 248, 276, 302–
303n.15
efforts to standardize hydrometer, 171,
172–177, 201
of instruments, xii–xiii, xiv, xv, xvi,
76, 154, 160, 291
Lavoisier's emphasis on, 146, 168, 173,
200–201
the metric system, 168, 173–177
Precision balance, xii–xiii, 44, 107–108,
175, 177
Lavoisier's, ix, xi, 137, 202–204
Priestley, Joseph, 79, 80, 84, 95, 107,
128
chemical apparatus and laboratory, ix,
29n.4, 90–93, 93f, 94, 95, 110f,
131–132n.17, 137, 138, 221
conservatism in nomenclature, 97–98,
103n.56
Different Kinds of Air, 94, 221

Directions for Impregnating Water, 91
isolation of nitrous oxide, 110–111
mineral water analysis, 219–220
pneumatic/gasometric analysis of, 88–
89, 91–92, 94, 97–100, 187–188,
218
Principe, Lawrence M., xxi, 2–3, 55
Pringle, John, Observations on the Diseases
of the Army, 105, 107
Proust, Joseph-Louis, 336
gunpowder research, 337–341, 341
Purity/purification, 15, 95, 97
Pyrometry, 165–166

Quantifying spirit, 76–77, 90, 107,
155–156, 168, 174. See also
Precision; Tables
the "balance sheet method," 144, 147,
170, 205, 239, 244
the metric system, 168, 173–177

Ramsay, William, 121
Ramsden, Jesse, xii
Raspail, F. V., 297
Rayleigh, Lord, 121
Rāzi, Muḥammad ibn Zakariyya al-, 37,
45
Reaction vessels, 244, 314–318, 315f,
317f
Reagents, 95, 97
Réaumur thermometer scale, 200–201,
204
Receiver. See Retort and receiver
Reflux condenser, 312–314, 313f,
331
Regnault, V., 297
Replicability
of experimental operations, xii, xiv, 1,
58–59, 243, 269n.43, 270n.48, 290,
296, 304n.32
issues raised by differing results, 171–
177, 247–250, 255, 257–258
of some alchemical operations, 2, 55,
57, 58–59, 63, 65–69, 70
Research, laboratory vs. field, 336, 338,
341, 343, 346, 354. See also
Experimental operations

Respiration studies, 81*f*, 107
Retort and receiver, 9–10, 22(1.10)*f*, 80,
    93, 141, 142*f*, 144, 145*f*, 146
Guyton's portable retort, 214, 215*f*
Ripley, George, *Compound of Alchemie*,
    63–64, 65, 67–68
Roberts, Lissa, 188
Robins, Benjamin, 338, 344, 345
Robison, John, 196–198
Rocke, Alan J., xii, xxi, 240, *273*
Rose, Heinrich, 279
Rosenhut alembic, 25, 26*f*
Rouelle, G. F., x, 86, 138, 139*f*, 201
Royal College of Chemistry, 330, 345
Royal Committee on Gunpowder and
    Explosive Substances, 345–346,
    360n.54
Royal Society (London), 80, 90, 195,
    259
Rubber apparatus, 231, 328
Rumford, Count, 337–340, 339*f*
Russell, Colin A., xii, xxi, 240, *311*

Sagredo, Francesco, 188
Salts analysis
    Bérard studies, 257–258, 263–265,
    265*f*
    Wollaston studies, 259–263, 260*f*, 261*f*
Salts-testing, alchemical, 37–39, 49,
    51n.13, 52n.17
Sarrau, Émile, 351, 354
Schaffer, Simon, 56, 155–156
Scheele, C. W., 80, 95, 256
    On Fire and Air, 94, 95*f*
    "preparation of oxygen ("fire air"), 94,
    95*f*
Scherrer, Paul, 376
Schischkoff, Leon, 335, 357–358n.29.
    See also Bunsen, Robert
    "A Chemical Theory of Gunpowder"
    (with Robert Bunsen), 336, 341
Schloss Oberstockstall castle, 20–21
Science Museum (London), 10, 25, 26,
    230
    artifacts featured in, 11*f*, 12*f*
Scientific controversy, 285–287, 294–
    296

Scotus, Michael, *Ars Alchemy*, 37–39,
    49
Sensory information and instruments,
    187–189, 193, 205, 206n.6, 243
Shapin, Stephen, 56
Shaw, Peter, 17–18
Silberer, Herbert, 55
Sirkap archaeological site, 9–11
"Slippery substances," 75. *See also* Gases
Smeaton, William A., 77, *211*
Smithsonian Institution, ix
Social context, 239, 243, 329–330, 340
Polanyi's views and experiences re:,
    367, 368–369, 385–390
    World War, II, 387–388
Société Royale des Sciences, 162
Sponsler, L. O., 385
Stability. *See* Continuity
Stahl, Georg Ernst, x, 86, 169
Starkey, George, 62, 68, 70. *See also*
    Philalethes, Eirenaeus
    Marrow of Alchemy, 69–70
Starkey-Philalethes tract, 62–64
Staudinger, Hermann, 383, 385
Stein der Uhralten (*Basil Valentine* corpus),
    59–60, 61*f*, 62, 63
Stoichiometry, 275, 342
Strength of materials studies (Polanyi),
    383–387
Suda (10th c. encyclopedia), 36
Summa Perfectionis (Psuedo-Geber), 45–
    49, 50n.2
Swedish Enlightenment, 275–276
Szabadváry, Ferenc, 35, 44, 45–46,
    50n.5, 53n.27

Tables
    Bergman's table of affinities, 160, 170
    Lavoisier's "table of simple substances,"
    244–245
    Martine's thermometric conversion
    table, 195–196
    Réaumur's conversion table, 160, 170
    Tabula Smaragdina, 7–8
Taranto, Paul of, *Theoretica et Practica*,
    36, 45–46
Taylor, Brook, 194, 197, 231

Technology transfer, xiii–xiv, 328–329, 368, 390. *See also* Innovation

Terminology. *See* Nomenclature

Thackray, Arnold, 253

Thenard, L. J., 274, 276

Thermodynamic theory of adsorption, 369–371, 392n.13

Thermometer, xiii, xv–xvi, 164–165, 185. *See also* Calorimeter
  air thermometer, 188
  alcohol thermometer, 192
  "black boxing" (ensemble use) of, 185–186, 201–202, 204–206
  mercury thermometer, 191, 192, 201
  uses of, 186–187, 189–190

Thermometric calibration, 185–186, 191–192, 204–205
  Lavoisier's efforts to attain, 200–201
  Martine's conversion table, 195–196

Thoelde, Johann (publ.)
  Offenbahrung, 59–63
  Stein der Uhralten, 59–60, 62, 63

Thompson, Benjamin, 337–340, 339*f*

Thomson, Thomas, xii, 248, 254–256
  A System of Chemistry, 257, 259
  Organic Bodies, 297
  oxalic acid study, 256–258, 260–261, 269nn.36 and 43

Tschirnhausen lens (Lavoisier), 140

Turner, Edward, 296

Twelve Keys (Basil Valentine corpus), 59–60, 61*f*, 62, 63

Twentieth century chemistry, 239

"Universal language," 195

Usselman, Melvyn C., xii, xxi, 239–240, *243*, 270n.48

Valentinian alchemical corpus, 59–63, 61*f*, 72n.13

van Helmont, J. B., 79, 82, 99

Van Marum, Friedrich, 124*f*, 124

Van Marum, Martinus, 123–125, 124*f*

Vegetable Staticks, 84*f*, 85*f*

Vessels, chemical, viii, 9, 13
  reaction vessels, 244, 314–318, 315*f*, 317*f*

Vickers, Brian, 56

Vielle, Paul, 351, 354, 365n.95

Vocabulary. *See* Nomenclature

Volta, Alessandro, eudiometer, 217, 233n.20

Volumetric techniques, xiv

von Engeström, Gustav, 214

von Hessen-Kassel, Moritz, 25–26

von Humboldt, Alexander, 274, 277, 303n.20

von Kalbe, Ruelein von, *Nutzlich Bergbuchley(n)*, 49

von Laue, Max, 374–375

von Suchten, Alexander, 68

von Todtenfeldt, Hertodt, 63

Vulcanization, 231

Wanklyn, J. A., 322, 324, 325

Water analysis
  ammonia process of, 322, 322–323, 324, 325
  Frankland's apparatus for, 320*f*, 322*f*, 323*f*, 328–329, 330
  for organic compounds, 321–326
  water synthesis apparatus, 137

Watt, James, 125

Wedgewood, Josiah, 93, 202–203

Weimar science, 389

Wieland, Heinrich, 384

Wine trade and alcohol titration, 161–165, 175

Wise, M. Norton, xiii

Wöhler, Friedrich, 279, 280, 281, 285, 288–289

Wollaston, William Hyde, xii, 247–248, 258–59, 296
  on integral multiple proportions, 258–260, 262–263
  salts analysis, 259–263, 260*f*, 261*f*, 270n.48

X-ray crystallography, 241, 262–263, 373
  crystalline structure, 374–376, 384–385
  diffraction/dislocation theory of, 381, 383*f*, 383

instrumentation, 378–381
Polanyi's research in, 369, 374–381,
    377f, 379f, 382f
rotating crystal method, 378, 378–381,
    379f, 380

Young, Arthur, 211, 218

Zeitschrift für Physik, 380, 381, 382f
Zwölff Schlüssel (*Basil Valentine corpus*),
    59–60, 62, 63

QD 53 .I57 2000
Instruments and
experimentation in the

DATE DUE
